# Social and Conceptual Issues in Astrobiology

*Edited by*

KELLY C. SMITH AND CARLOS MARISCAL

OXFORD

UNIVERSITY PRESS

# OXFORD
## UNIVERSITY PRESS

Oxford University Press is a department of the University of Oxford. It furthers
the University's objective of excellence in research, scholarship, and education
by publishing worldwide. Oxford is a registered trade mark of Oxford University
Press in the UK and certain other countries.

Published in the United States of America by Oxford University Press
198 Madison Avenue, New York, NY 10016, United States of America.

CIP data is on file at the Library of Congress
ISBN 978-0-19-091565-0

1 3 5 7 9 8 6 4 2

Printed by Integrated Books International, United States of America

# Contents

# Social and Conceptual Issues
in Astrobiology

# INTRODUCTION

# 1

# To the Humanities and Beyond

## Exploring the Broader Questions in Astrobiology

*Kelly Smith and Carlos Mariscal*

Astrobiology is a scientific discipline emerging before our eyes. Scientists (and, increasingly, humanists and other experts as well) from a variety of fields are just beginning to address the many raised by the real possibility of life on other planets. While new, astrobiology's recent success has been nothing short of amazing—consider that, in just the past 25 years, we have learned:

- The building blocks of life are found basically everywhere in our universe.
- Getting these building blocks to engage in the kinds of complex chemistry we associate with life is far easier than we used to think.
- Planets where life could potentially evolve are extremely common—we have confirmed nearly 4,000 "exoplanets" since 1988, and that number will increase exponentially as new telescopes come online in the next few years.

In short, this is not your grandmother's exobiology, nor is it the kind of purely speculative enterprise we should leave to science fiction. We have progressed to the point where it makes perfect sense for NASA's chief scientist to declare, "I think we're going to have strong indications of life beyond Earth within a decade, and I think we're going to have definitive evidence within 20 to 30 years" (Stofan, 2015).

To be sure, there are myriad *scientific* questions that astrobiologists have only begun to address. But make no mistake: this is *not* a purely scientific enterprise. The discovery of life elsewhere would surely rank as of the greatest discoveries of all time and would disrupt our sense of who we are and our place in the universe. It would be a gestalt shift every bit as drastic as the Copernican or Darwinian revolutions. Given that clear fact, it is curious that relatively little research on the broader social and conceptual aspects of astrobiology has been undertaken by scholars outside the small community of space scientists. But a fertile field awaits early adopters from other disciplines, with many profound and largely un-explored questions waiting to be addressed by relevant experts. Some of these research questions fall squarely within traditional humanities, while others

Kelly Smith and Carlos Mariscal, *To the Humanities and Beyond*. In: *Social and Conceptual Issues in Astrobiology*. Edited by: Kelly C. Smith and Carlos Mariscal, Oxford University Press (2020).
© Oxford University Press.
DOI: 10.1093/oso/9780190915650.001.0001

span the boundary between empirical science and other fields. For example, the current volume includes discussion of this sampling: Just what are our ethical obligations toward different sorts of alien life? What is "life" in the most general sense? How would the existence of extraterrestrial life impact religion? What does the public think about astrobiology, and how should this impact our decision-making? What can we say about the nature of an alien "intelligence," given our limited experience on Earth? What sorts of cultural ideas inform (or misinform) our attitudes toward alien worlds? How can legal frameworks adapt to the new challenges of exploration and colonization of space? In what sense does astrobiology challenge standard models of science, given its unique nature and limitations?

We need scientific information and analysis if we are to make progress answering these questions. But we must also encourage other experts, with other perspectives, to weigh in: philosophers, historians, theologians, social scientists, legal scholars, and so on. The present volume grew out of work by just such a diverse group of scholars, many of whom initially connected at the 2016 and 2018 meetings of the newly formed Society for Social and Conceptual Issues in Astrobiology. These meetings were vibrant, friendly events in which scholars from dozens of disciplines actively engaged with each other's ideas and approaches. From the initial 30 founding members, the society's ranks have swelled to over 150 in just three years. This collection is designed to trace the outlines of some of the debates that are just beginning to emerge among these early adopters. They will surely evolve and change as the field matures, but we hope to capture here the exciting, formative stages of these conceptual battles.

We start off in Part I with two historical surveys to ground our discussions. First, the eminent historian of astronomy, **Steve Dick**, offers a whirlwind tour of the search for life from the ancient Greek atomists to the emergence of modern astrobiology. He pays special attention to recent attempts to pursue the kinds of broader questions this volume addresses, concluding that these kinds of studies have a bright future indeed. **Derek Malone-France**, a philosopher of religion, then adds his own sweeping survey of the history of questions concerning life on other worlds within theology. He traces the development of critical concepts from a diverse array of philosophical and religious thinkers from Aristotle to Emerson, ending with an analysis of contemporary religious movements that explicitly incorporate alien life (e.g., Raelieans, Solarians, and Scientologists).

In Part II, we gather several perspectives on the knotty conceptual issues surrounding the subject matter of astrobiology: life. **Lucas Mix**—a biologist, philosopher, and theologian—starts off the section with an exploration of the concept of life across traditions from science to philosophy to theology. He identifies three distinct "hard problems" of life: moving from non-life to life, from life to sentience, and from sentience to rationality. Philosopher **Emily Parke** takes a more pragmatic stance. She points out that the concept of "life" is used in several rather distinct ways: sometimes as an all-or-nothing phenomenon and other times as a matter of degree; sometimes referring to individual organisms and other times to communities; sometimes based on specific chemistries and other times on functions. She notes how confusion could be limited by more clearly specifying the question to which life is the answer. **Cole Mathis**, a physicist, draws inspiration from quantum physics to describe a statistical category he terms the "living state." This can be rigorously defined independent of the unique details of terrestrial environments and may allow new scientific hypotheses regarding the nature of life on Earth as well as its origins. Finally, historian **Luis Campos** summarizes some of the research he undertook as the Blumberg Chair in Astrobiology at the Library of Congress. Musing on the historical development of two related fields, synthetic biology and astrobiology, he suggests that their long period of coevolution makes it impossible to fully understood either in isolation.

Part III ventures into some philosophical issues in astrobiology, beginning with a piece by astrobiologist **Sean McMahon** that examines Carl Sagan's famous dictum, "extraordinary claims require extraordinary evidence," in the context of some specific problems in astrobiology. He argues that, while Sagan's dictum is a justified skeptical response to claims that are known to be highly improbable or contrary to well-substantiated science, it is irrational and contrary to scientific objectivity to demand extraordinary evidence for those that are merely amazing or bizarre and thus Sagan's dictum must be handled with caution in astrobiology. The philosopher **Jason Howard** tackles a different sort of problem: What can we know about "intelligence" in a very general sense, and how does this inform debates concerning extraterrestrial life? He concludes that our concept of "conceptual intelligence" is rooted in basic principles of logic that go far beyond the details of our biology. The idea that science is our best means to uncover the truth about the universe relies on this belief, and this would be equally true of other intelligences as well. Then another of our editors, **Carlos Mariscal**, joins forces with **Tyler D.P. Brunet** (both philosophers) to examine the concept of an *extremophile*. They delineate five different ways to think about extremophiles, concluding that the concept is especially prone to the vagueness and arbitrariness that plague other biological categories (e.g., life, species, genes), since it unavoidably involves debatable assumptions about life's nature and limits.

Part IV centers on ethical questions raised by astrobiology. **Brian Green**, an ethicist and philosopher, begins with an outline of the ethical frameworks provided by some of the leading figures in the contemporary debate, identifying resonances between their views and traditional concepts in ethical philosophy such as natural law and virtue theory. Despite their differences, he argues that they converge on a broadly applicable ethical framework: protect alien life in proportion to its capacity for excellence. Next, the philosopher **Adam Potthast** examines how classical ethical theory might apply to the kind of alien life we are most likely to encounter: nonrational, nonsentient life. He concludes that we would not have ethical obligations to such life per se, but the obligations we clearly do have to our fellow humans entails that we cherish, promote, and protect extraterrestrial life—perhaps even more so than similarly situated terrestrial life. One of us, the philosopher and sometime biologist **Kelly Smith**, ends the section by asking about the ethical justification for attempting to message extraterrestrial intelligence. This has become an incredibly pressing question as multiple large-scale efforts to do this are being planned, despite the fact that most of the debate to date has conflated discussions of risk with proper ethical analysis. He concludes that, until appropriate efforts have been made to seek consensus among those affected (the global public), it is immoral to proceed with these projects.

We wrap things up in Part V with an examination of some of the many social, cultural, and legal issues in the field. **Linda Billings**, an expert in communications with extensive experience working with NASA, explores how the scientific search for evidence of extraterrestrial life has affected our conception of the *terrestrial* biosphere. She argues that we face a choice of perspectives to guide our actions in space: the prevailing "manifest destiny" attitude of exploitation for human ends or her preferred alternative of *astroenvironmentalism* that calls for us to preserve pristine extraterrestrial environments for their own sake. Next, the philosopher **Jim Schwartz** asks the sociological question that has largely been ignored: How much does the public care about life in space? He summarizes the findings of a wide range of opinion surveys and argues that the only conclusion supported by the data is that the public likely has lukewarm feelings toward the importance of extraterrestrial life, and that feeling may be even cooler toward astrobiology. Finally, lawyer **Christopher Newman** provides a detailed account of the current state of space law and, in particular, the mechanisms that might be used to deal with the discovery of extraterrestrial life. He concludes that existing planetary protection and contamination rules, with their clearly anthropocentric bias (preserve the science), are outmoded and the resulting regulatory gap should be filled with nonbinding *soft laws*.

The topics presented in this volume are the vanguard of a new scholarly frontier. They will morph and change as more scholars join the fray and as new questions take shape from our experiences with new satellites, landers, and commercial space ventures. The current situation is thus pregnant with exciting possibilities, and there is much low-hanging fruit to be plucked. We hope some of our readers find the ideas in this collection sufficiently interesting to help pluck them.

# Reference

Stofan, E. [NASA] (2015, April 7). Water in the Universe [Video File]. Retrieved from: https://youtu.be/eiAT41aHaH4 on November 29, 2019, timestamp: 55:30-55:36.

# PART I
# HISTORICAL BACKGROUND

# 2

# Astrobiology and Society

## An Overview

*Steven J. Dick*

The histories of astrobiology, exobiology, and the plurality of worlds tradition are now well known, at least in broad outline for Western civilization. Fifty years ago, this was not the case; a graduate student wishing to write a dissertation on the plurality of worlds tradition in a History of Science Department would be told (actually was told!) that two barriers precluded writing on that topic: it wasn't science, and it had no history worth writing. As will be evident in the next section, that was far from the truth and was based on a parochial view of both science and history. Today, life on other worlds is considered one of the signal historical themes in Western intellectual history, and the question can profitably be asked whether this idea is a trope only of the West or if a substantial tradition remains to be uncovered in other cultures and, if not, why not.

In the last three decades a new question has arisen: What are the implications of astrobiology for society? When one considers that astrobiology encompasses research on the origin and evolution of life, the existence of life beyond Earth, and the future of life on Earth and beyond, the scope of that deceptively simple question becomes clear. It embraces not only the religious, ethical, legal, and cultural concerns inherent in those subjects, but also the meaning of life and even human destiny in a universe where humans are unique—or not. Particularly in the area of extraterrestrial life—which has been a focus for astrobiology and society concerns in terms of implications—the issues have been global and contentious. The consequences have long been vividly played out in science fiction by classic authors such as Arthur C. Clarke in *Childhood's End* or *2001: A Space Odyssey*, and by more recent writers like Ted Chiang in "Story of Your Life" and its film adaption *Arrival*.

How can we even approach such questions as the impact of discovering life beyond Earth, whether microbial or intelligent? How can we transcend anthropocentrism when we address concepts such as life and intelligence, culture and civilization, technology and communication? And in what areas is humanity most likely to be transformed by such a discovery? We cannot answer these questions in this chapter, but there is now a surprisingly substantial literature

Steven J. Dick, *Astrobiology and Society*. In: *Social and Conceptual Issues in Astrobiology*. Edited by: Kelly C. Smith and Carlos Mariscal, Oxford University Press (2020). © Oxford University Press. DOI: 10.1093/oso/9780190915650.001.0002 that does address them. As with astrobiology, it is prudent for current researchers in the subject to be aware of this much shorter history, whether to contest or expand it. After a brief history of astrobiology and its predecessors (covering over 2,500 years), we provide an overview of this literature on astrobiology and society. Substantial as it may seem, it is only the leading edge of what is sure to become an entire discipline of its own, especially if life is actually discovered out there among the stars.

## A Brief History of Astrobiology

The idea of inhabited worlds dates back at least to the ancient Greeks and was rationally discussed as a part of natural philosophy, mainly in the context of cosmological worldviews (Dick, 1982). The ancient Greek atomist and Aristotelian world views came to opposite conclusions about many worlds, with the atomists championing an infinite number and Aristotle arguing for a single world, in the sense of *kosmos*, an ordered system that included the entire visible universe. In the 16th century, the Copernican heliocentric cosmology made the Earth a planet and the planets potential worlds, changing the very definition of world from *kosmos* to planet. The idea of world as planet was elaborated in Descartes's vortex cosmology filled with planetary systems and in Newton's gravitationally ruled universe, which he filled with inhabitants, albeit mainly for reasons of natural theology.

The extraterrestrial life debate in the 18th and 19th centuries was waged not so much on a cosmological scale as on a scale of world views a level or more below the cosmological. Although sometimes discussed by the elaboration of Newtonian science such as the Laplacian nebular hypothesis, more often it fell in the domain of philosophical explorations, both secular and religious. If cosmological world views gave birth to the idea of extraterrestrial life, then philosophy and literature, in their traditional role of examining the human condition, explored the ramifications of the idea borne of that cosmological context. These debates have now been discussed in great detail (Crowe, 1986), and a universe full of life as expressed in literature and the arts has also been well explored (Guthke, 1990).

This long history demonstrates the sustained human interest in the subject, both from the point of view of the natural philosophy and science of the times, as well as its cultural aspects. Such was the allure of the subject that philosophers and scientists alike tried to tackle it employing a variety of general arguments, including plenitude, analogy, and the principle of mediocrity (Dick, 2013a, 2013b). Only in the 20th century were techniques applied that were equal to the problem (Dick, 1996, 1998). Building on the ideas of the Soviet biochemist Alexander

Oparin and J. B. S. Haldane, in the 1950s origins of life studies took a great leap forward with the Miller–Urey experiment. In the 1960s, the first search for extraterrestrial intelligence (SETI) searches were made with radio telescopes. In the 1970s, life on Earth was found in extreme environments such as hydrothermal vents deep in the ocean, demonstrating an astonishing flexibility and tenacity. By the 1980s complex organic molecules were found in molecular clouds—the birthplaces of stars and planets—not life itself but the building blocks of life. In the 1990s planets were definitively discovered outside our solar system, and they are now known to be virtually ubiquitous, including a fraction in the habitable zones of their parent stars. And in the 2000s, the search for biosignatures in exoplanets became a reality, as ever more planets were discovered in abundance.

Today astrobiology is a thriving enterprise around the world. Although exobiology was a part of NASA's research efforts almost from its founding in 1958, the mid-1990s saw a rebirth as "astrobiology," which was given a much wider portfolio than the origins of life studies that had dominated exobiology. Now life was to be seen in its planetary and cosmic context to include planetary origins and evolution, the origin and evolution of higher life forms, exoplanet studies, and the search for biosignatures (Dick and Strick, 2004). In 1998, the NASA Astrobiology Institute was founded, and the first of several astrobiology roadmaps were constructed, which have served as a focus for research over the last two decades (Des Marais et al., 1998; NASA, 2015). SETI, however, was not a part of the new astrobiology program at NASA, having been terminated for petty political reasons in 1993 after one year of operation (Garber, 2014).

All of these developments did nothing to prove the existence of life beyond Earth: that was the goal still to be achieved whether in the long or short term. But adopting the Copernican and Darwinian presuppositions that Earth is not unique and that life will evolve by natural selection wherever conditions are favorable (Fry, 2015), these advances demonstrate that it would be prudent for us to think about the consequences if life were discovered, whether in microbial, complex, or intelligent form.

Indeed, a report from the World Economic Forum (2013) declared the discovery of life beyond Earth one of five X factors—emerging concerns for planet Earth of possible future importance but with unknown consequences. Giving attention to X factors, the report suggested, would lead to a more proactive approach if and when these events actually occurred, resulting in more "cognitive resilience" and perhaps preventing at least some undesirable social consequences. Such consequences could occur even if simple alien life were discovered. "Over the long term," the report argued, "the psychological and philosophical implications of the discovery could be profound. . . . The discovery of even simple life would fuel speculation about the existence of other intelligent beings and challenge many assumptions that underpin human philosophy and religion." The study of these assumptions and implications is therefore far from academic and is a worthy endeavor even if life is never discovered beyond Earth. Such studies also help to nurture a cosmic perspective sorely needed in our turbulent times.

## Early Explorations in Astrobiology and Society

Interest in astrobiology and society in its broadest sense dates back at least a quarter century to the days when NASA was planning its SETI program. In 1976–1977 when scientists first met to contemplate this program, the discussions included the possibilities of cultural evolution beyond the Earth, led by none other than the young Nobelist Joshua Lederberg, whose two-day "Workshop on Cultural Evolution," focused more specifically on "evolution of intelligent species and technological civilizations." Among the conclusions of the group, which included several scholars in the social sciences, was that "our new knowledge has changed the attitude of many specialists about the generality of cultural evolution from one of skepticism to a belief that it is a natural consequence of evolution under many environmental circumstances, given enough time" (Morrison et al., 1977, 49). This meant that cultures beyond the Earth, perhaps ending in technological civilizations capable of radio communication, were at least a possibility. A few farsighted anthropologists were even beginning to show some interest (Maruyama, 1975), an interest that has grown over the decades since (Vakoch, 2009, 2014a).

It is quite remarkable that the early practitioners of SETI were already sensitive to societal concerns. In the early 1990s, just prior to the inauguration of NASA SETI operations on the quincentennial of Columbus's first landfall in the Americas, NASA convened a series of workshops on the cultural aspects of SETI (CASETI). The intimate gathering of two dozen scholars was a model of interdisciplinary brainstorming, with astronomers, anthropologists, religious scholars, historians, several representatives from media studies, and even two diplomats. The gathering was a de facto recognition that this was a broad-based problem not to be solved by scientists alone. While the publication of the results was delayed almost a decade by the congressional cancellation of the NASA SETI program, its recommendations are still valuable for contemplating the aftermath of any successful SETI program (Billingham et al., 1999). Plans for an international conference on the subject were canceled when the NASA SETI program itself was cancelled.

The quick and untimely demise of the NASA SETI program meant that astrobiology and society discussions would be scattered and sporadic. One opportunity for a more systematic treatment of the societal aspects of astrobiology was

NASA's construction of a roadmap for astrobiology, as previously mentioned. However, although some proponents argued that astrobiology and society issues should be among the roadmap's firm goals, in the end proponents had to be content that two of the four roadmap operating principles were related to these issues, one in encouraging planetary stewardship by emphasizing planetary protection and avoiding contamination and another by recognizing "a broad societal interest in our subject," including the discovery of extraterrestrial life and engineering new life forms adapted to live on other worlds. Thus, while the roadmap and its successors served as a focus for a broad program of science research, they did not do the same for funding social sciences and humanities research.

These conditions notwithstanding, it is rather surprising that in 1999 NASA's Ames Research Center organized a workshop on the societal implications of astrobiology (Harrison and Connell, 2001). This time about 50 scholars ranging from futurists like Alvin Toffler to anthropologists, scientists, and journalists gathered to discuss the subject. Not surprisingly, the group emphasized the importance of their task: to encourage public understanding of this new science, to gauge public reaction to astrobiological discoveries, and to prepare for the future through policy decisions given "a possible sea of living worlds" (p. 34). More than a dozen recommendations were issued, including the importance of a multidisciplinary approach involving both scientists and humanists, studying the implications of a shift in our frame of reference from the Earth to a living cosmos, making "state-of-the-art preparations" (p. 37) for discovery of life, studying the ethical implications of discovering life, and implementing policy measures "to ensure the integrity of extraterrestrial life" (p. 58). They made a strong case for undertaking serious levels of research and outreach before the fact of discovery, arguing such research should be integrated into core science initiatives (as would soon be done with the Human Genome Project). "Science and society are deeply and irrevocably intertwined," they wrote, "and a mutual appreciation of the close relationship is vital to the integrity of both fields" (p. 10).

Beyond NASA several other organizations undertook initiatives on the subject. One notable meeting was sponsored by the John Templeton Foundation, which focuses on the dialogue between science and religion. The foundation convened a meeting in late 1998, only a few months after the first NASA astrobiology roadmap was constructed. Again, the meeting was interdisciplinary, including a Nobel biochemist (Christian de Duve), physicists, astronomers, theologians, one historian, and the very skeptical evolutionary biologist Richard Dawkins. The results, published as *Many Worlds: The New Universe, Extraterrestrial Life, and the Theological Implications* (Dick, 2000), read like a cauldron of nonconsensus. Over the next two decades, theological and ethical issues would become an important component of societal issues in astrobiology (Impey et al., 2013; Peters, 2013, 2014; Smith, 2014).

## Into the New Millennium

The new millennium has seen increasing, although still sporadic, interest in issues involving astrobiology and society and sponsored by a variety of organizations. In conjunction with NASA and the Templeton Foundation, in 2003 and 2004 Constance Bertka, a planetary scientist who also headed the Dialogue on Science, Ethics and Religion program at the American Association for the Advancement of Science (AAAS), convened a series of workshops at the AAAS in Washington, DC, that included ethical and theological perspectives on the origins, extent, and future of life (Bertka, 2009). Since the AAAS is the largest organization of scientists in the world, these discussions (and the Dialogue on Science, Ethics, and Religion program in general) are an indication that scientists are interested in the social impact of what they do, as well they should be. From changing definitions of life to extraterrestrial life to the future of life in the universe, these workshops proved to be a window on the many issues that need to be tackled under the umbrella of astrobiology and society. Another example is a workshop held in 2008 at the University of Arizona's Biosphere 2 artificial ecosystem facility. Like the AAAS workshops, these discussions ranged across the full spectrum of societal, cultural, and ethical issues in astrobiology (Impey et al., 2013). Yet another example is a meeting in Hven, Sweden, in 2011 sponsored by the Pufendorf Institute for Advanced Studies at Lund University. The publication of the proceedings in the scientific journal *Astrobiology* (Dick, 2012; Dunér, 2012, 2013) is another indication of interest among scientists in societal issues. Sometimes meetings have addressed more specific issues such as communication with extraterrestrial intelligence, exemplified especially in a series of volumes edited by Douglas Vakoch, who for many years held the awesome title of Director of Interstellar Communications at the SETI Institute (Vakoch, 2011, 2013, 2014a, 2014b).

Stimulating as they were, the scattered discussions of the previous decades cried out for more organization and synthesis. This was the hopeful goal of a meeting in 2009, held under the auspices of the NASA Astrobiology Institute. Some 43 invited scholars gathered at the SETI Institute to develop an "Astrobiology and Society" roadmap, fully aware of the astrobiology science process. Unlike the science roadmap, however, the societal impact roadmap (Race et al., 2012) was not officially adopted by NASA and thus has not become policy backed up by sustained funding. But the work continues at a basic level, and the process seems to be following its companion science roadmap in percolating from the bottom up with minimal funding and the hope of eventually becoming a more recognized and funded activity. That will require the two cultures to work together, and it is encouraging that the introduction to the latest 2015 Astrobiology Strategy document still lists a goal to enhance societal interest and relevance. "Astrobiology recognizes a broad societal interest in its endeavors," it states, "especially in areas such as achieving a deeper understanding of life, searching for extraterrestrial biospheres, assessing the societal implications of discovering other examples of life, and envisioning the future of life on Earth and in space" (NASA, 2015, p. xv). The document also includes as an appendix a humanities and social sciences section, the substance of which many feel should be a more integral part of the report.

Finally, NASA's establishment of the Baruch S. Blumberg NASA/Library of Congress Chair in Astrobiology in 2011 specifically to address the humanistic and societal aspects of astrobiology is a de facto recognition of the importance of these issues. This prestigious position has resulted in both individual and collective research (Grinspoon, 2016; Dick, 2015, 2018), drawing in younger scholars from a variety of disciplines and giving respectability to a field that has long been on the margins.

Thus, far from initial skepticism about a role for the social sciences and humanities in astrobiology, there is now considerable consensus that the problem of the impact of discovering life in any form is not only important but essential, and should not be left to scientists alone. The same is true of the broader aspects of astrobiology and society. When the Royal Society of London sponsored a meeting on the detection of extra-terrestrial life and the consequences for science and society in 2010 and a satellite meeting seeking a scientific and societal agenda on extra-terrestrial life, the organizers wrote:

> While scientists are obliged to assess benefits and risks that relate to their research, the political responsibility for decisions arising following the detection of extra-terrestrial life cannot and should not rest with them. Any such decision will require a broad societal dialogue and a proper political mandate. If extra-terrestrial life happens to be detected, a coordinated response that takes into account all the related sensitivities should already be in place. (Dominik and Zarnecki, 2011, 503)

My point is that these and other conferences on astrobiology and society (Table 2.1) should form the collective basis for future studies. Moreover, a few individual efforts have also concentrated on aspects of the problem. Foremost among these are psychologist Albert Harrison's (1997) volume *After Contact: The Human Response to Extraterrestrial Life* and the American diplomat Michael Michaud's (2007) *Contact with Alien Civilizations: Our Hopes and Fears about Encountering Extraterrestrials*. While some have argued that we know nothing about extraterrestrial intelligence (Billings, 2015), the anthropologist Michael Ashkenazi, one of the participants in the original CASETI workshops, has offered an answer of sorts with a large volume *What We Know About Extraterrestrial*

**Table 2.1** Twenty-Five Years of Discussions on Societal Impact of Astrobiology, 1991–2019

| Meeting | Date and Place | Sponsor | Results |
| --- | --- | --- | --- |
| Cultural Aspects of SETI (CASETI) | 1991–1992 Chaminade Conference Center, Santa Cruz, California | NASA | Billingham et al. (1999) |
| Many Worlds | November 22–24, 1998 Lyford Key, Nassau, The Bahamas | John Templeton Foundation | Dick (2000) |
| When SETI Succeeds | 1999 Hapuna Prince Big Island of Hawaii | Foundation for the Future | Tough (2000) |
| Societal Implications of Astrobiology Workshop | November 16–17, 1999 NASA Ames | NASA | Harrison et al. (2001) |
| Exploring the Origin, Extent and Future of Life | 2003 American Association for Advancement of Science Washington, DC | NASA/American Association for the Advancement of Science | Bertka (2009) |
| Astrobiology: Expanding our Views of Society and Self | May 2008 University of Arizona Biosphere 2 Institute | University of Arizona | Impey et al. (2013) |
| Astrobiology and Society | February 2009 SETI Institute | NASA Astrobiology Institute | Race et al. (2012) |
| The Detection of Extra-terrestrial Life and the Consequences for Science and Society Satellite Meeting | January 25–26, 2010 Royal Society in London, Kavli Centre, Buckinghamshire | Royal Society of London | Dominik and Zarneki (2011) |
| The History and Philosophy of Astrobiology | September 27–28, 2011, Ven, Sweden | Pufendorf Institute for Advanced Studies, Lund University, Sweden | Duner et al. (2012); Duner |
| Preparing for Discovery | September, 2014 Library of Congress | NASA/Library of Congress | Dick (2015) |
| Social and Conceptual Issues in Astrobiology 2016 | September, 2016 Clemson University | Clemson University | Smith (in press) |
| Social and Conceptual Issues in Astrobiology 2018 | April, 2018 University of Nevada, Reno | University of Nevada, Reno; Blue Marble Institute; others | Peters (2019); Smith and Abney (in press) |

*Intelligence: Foundations of Xenology* (Askenazi, 2017), arguing that we can actually infer quite a bit about extraterrestrials and therefore lay out scenarios about societal impacts. Individual efforts are also represented in a plethora of widely scattered articles, whose full extent may be measured in the 30-page bibliography of *Astrobiology, Discovery, and Societal Impact* (Dick, 2018).

## Anticipating the Future

In summary, the future for astrobiology and society studies looks bright, if not guaranteed to maintain momentum. If studies so far have been dominated by researchers in the United States, an important harbinger comes from Europe. In contrast to the NASA astrobiology roadmap, in 2017–2018 astrobiology and society became a foundational theme for the proposed European Astrobiology Institute (EAI). In contrast to the American astrobiology roadmap process in 1998, 20 years later the EAI systematically laid out the societal issues in a roadmap that bids fair to become an integral part of astrobiology in Europe (Capova and Persson, 2018). In addition to the previously mentioned Royal Society meeting, the EAI initiative has been preceded in recent years by European research on the subject of societal impacts (Dunér et al., 2012; Dunér, 2013). Although not yet fully established as of this writing, when and if the EAI is fully established, it bodes well for the astrobiology and society theme. So does the inauguration in 2017 of the Society for Conceptual Issues in Astrobiology (SoCIA), which envisions international participation, and of which this volume is a product.

This chapter is all too brief to lay out comprehensively the issues encompassed in the astrobiology and society. But the questions are legion and, potentially, Earth-shaking. Who should take the lead in preparing for discovery? What do we do if life is actually discovered, microbial or intelligent, near or far? Should national governments be in charge, international political and scientific institutions, scientists and social scientists, ethicists and theologians, or some mix thereof? How do we prevent contamination of potential microbes on Mars, Europa, Enceladus, or other habitable sites in the solar system, and (more perhaps more urgently from most Earthlings' point of view) how do we protect our planet from back contamination in the event of the discovery of microbial life? If a message is received as a result of a successful SETI program, should we answer? If so, who speaks for Earth? Should we initiate messages as part of a Messaging Extraterrestrial Intelligence (METI) program? If so what should we say, and who, if anyone, should control what is said? These questions are only the leading edge of the many decisions that will have to be made once alien life is actually discovered. And each discovery scenario will have its own unique problems and solutions.

The question is often asked why we should worry about these potential and seemingly far-out problems when we have so many actual problems on Earth. The answer is the same as for programs such as Near Earth Objects and the Human Genome: it is prudent to prepare for potential events so as to maximize the beneficial outcome that may affect all of humanity. The World Economic Forum (2013) report concluded:

> Looking forward and identifying emerging issues will help us to anticipate future challenges and adopt a more proactive approach, rather than being caught by surprise and forced into a fully reactive mode. . . . Through basic education and awareness campaigns, the general public can achieve a higher science and space literacy and cognitive resilience that would prepare them and prevent undesired social consequences of such a profound discovery and paradigm shift concerning humankind's position in the universe.

There are other reasons as well. Even if we are alone in the universe, the examination of our basic assumptions about life and intelligence, culture and civilization, and technology and communication will have been well worth it. It has been said before, but it bears repeating, that astrobiology is in many ways a search for ourselves, for our place in the universe, and for our future destiny. Our destiny will be much different if we live in the universe of Isaac Asimov, where life is human or robotic products of humans, or if we live in Arthur C. Clarke's universe, where alien life is everywhere. In either case, we need to be good stewards of our planet. But if aliens are in the mix, whether for good or ill, we will have to deal with them. The universe is what it is, not what we want it to be. Meanwhile, the presence or absence of life will be one of the greatest discoveries in the history of science.

# References

Ashkenazi, Michael. 2017. *What We Know About Extraterrestrial Intelligence: Foundations of Xenology*. Cham, Switzerland: Springer.

Bertka, Constance, ed. 2009. *Exploring the Origin, Extent, and Future of Life: Philosophical, Ethical and Theological Perspectives*. Cambridge, UK: Cambridge University Press.

Billingham, John, Roger Heyns, David Milne, et al. 1999. *Social Implications of the Detection of an Extraterrestrial Civilization*. Mountain View, CA: SETI Press.

Billings, Linda. 2015. "The Allure of Alien Life: Public and Media Framings of Extraterrestrial Life." In *The Impact of Discovering Life Beyond Earth. Cambridge*, edited by Steven J. Dick, pp. 308–323. Cambridge, UK: Cambridge University Press.

Capova, K. A., E. Persson, T. Milligan, and D. Dunér. 2018. "Astrobiology and Society in Europe Today: The White Paper on Societal Implications of Astrobiology Research in Europe." Springer International Publishing. doi:10.1007/978-3-319-96265-8

Crowe, Michael. J. 1986. *The Extraterrestrial Life Debate, 1750–1900: The Idea of a Plurality of Worlds from Kant to Lowell*. Cambridge, UK: Cambridge University Press.

Des Marais, David, J. A. Nuth, III, Louis Allamandola, et al. 2008. "The NASA Astrobiology Roadmap." *Astrobiology*, 8: 715–730.

Dick, Steven J. 1982. *Plurality of Worlds: The Origins of the Extraterrestrial Life Debate from Democritus to Kant*. Cambridge, UK: Cambridge University Press.

Dick, Steven J. 1996. *The Biological Universe: The Twentieth Century Extraterrestrial Life Debate and the Limits of Science*. Cambridge, UK: Cambridge University Press.

Dick, Steven J. 1998. *Life on Other Worlds*. Cambridge, UK: Cambridge University Press.

Dick, Steven J. 2000. *Many Worlds: The New Universe, Extraterrestrial Life and the Theological Implications*. Philadelphia: Templeton Press.

Dick, Steven J. 2012. "Critical Issues in the History, Philosophy, and Sociology of Astrobiology." *Astrobiology*, 12: 906–927.

Dick, Steven J. 2013a. "The Twentieth Century History of the Extraterrestrial Life Debate: Major Themes and Lessons Learned." In *Astrobiology, History and Society: Life Beyond Earth and the Impact of Discovery*, edited by Douglas Vakoch, pp. 133–173. Berlin: Springer-Verlag.

Dick, Steven J. 2013b. "The Societal Impact of Extraterrestrial Life: The Relevance of History and the Social Sciences." In *Astrobiology, History and Society: Life Beyond Earth and the Impact of Discovery*, edited by Douglas Vakoch, pp. 227–257. Berlin: Springer-Verlag.

Dick, Steven J. 2015. *The Impact of Discovering Life Beyond Earth*. Cambridge, UK: Cambridge University Press.

Dick, Steven J. 2018. *Astrobiology, Discovery, and Societal Impact*. Cambridge, UK: Cambridge University Press.

Dick, Steven J., and J. E. Strick. 2004. *The Living Universe: NASA and the Development of Astrobiology*. New Brunswick, NJ: Rutgers University Press.

Dominik, Martin, and J. C. Zarnecki. 2011. "The Detection of Extra-Terrestrial Life and the Consequences for Science and Society." *Philosophical Transactions of the Royal Society A*, 369.1936: 20100236. https://doi.org/10.1098/rsta.2010.0236

Dunér, David, E. Persson, and G. Holmberg, eds. 2012. *Special Issue: The History and Philosophy of Astrobiology*. *Astrobiology*, 12: 901–1016.

Dunér, David, J. Parthemore, E. Persson, and G. Holmberg, eds. 2013. *The History and Philosophy of Astrobiology: Perspectives on Extraterrestrial Life and the Human Mind*. Newcastle-upon-Tyne, UK: Cambridge Scholars.

Fry, Iris. 2015. "The Philosophy of Astrobiology: the Copernican and Darwinian Philosophical Presuppositions." In *The Impact of Discovering Life Beyond Earth*, edited by Steven J. Dick, pp. 23–37. Cambridge, UK: Cambridge University Press.

Garber, Stephen J. 2014. "A Political History of NASA's SETI Program." In *Archaeology, Anthropology and Interstellar Communication*, edited by Douglas Vakoch, pp. 23–48. Washington, DC: NASA.

Grinspoon, David. 2016. *Earth in Human Hands: Shaping our Planet's Future*. New York: Grand Central.

Guthke, Karl. S. 1990. *The Last Frontier: Imagining Other Worlds from the Copernican Revolution to Modern Science Fiction*. Ithaca, NY: Cornell University Press.

Harrison, Albert A. 1997. *After Contact: The Human Response to Extraterrestrial Life*. New York: Plenum.

Harrison, Albert A., and Kathleen Connell. 2001. *Workshop on the Societal Implications of Astrobiology.* Moffett Field: NASA Ames Research Center. Retrieved from http://www.astrosociology.org/Library/PDF/NASA-Workshop-Report-Societal-Implications-of-Astrobiology.pdf

Impey, Chris, Anna Spitz, and William Stoeger, eds. 2013. *Encountering Life in the Universe: Ethical Foundations and Social Implications of Astrobiology.* Tucson: University of Arizona Press.

Maruyama, M., and A. Harkins, eds. 1975. *Cultures Beyond the Earth: The Role of Anthropology in Outer Space.* New York: Vintage Books.

Meech, Karen. J., J. V. Keane, Michael Mumma et al. 2009. *Bioastronomy 2007: Molecules, Microbes and Extraterrestrial Life.* San Francisco: Astronomical Society of the Pacific.

Michaud, Michael A. G. 2007. *Contact with Alien Civilizations: Our Hopes and Fears about Encountering Extraterrestrials.* New York: Copernicus.

Morrison, Philip, John Billingham, and John Wolfe. 1977. *The Search for Extraterrestrial Intelligence (SETI).* Washington, DC: NASA.

NASA. 2015. *Astrobiology Strategy.* Retrieved from https://nai.nasa.gov/media/medialibrary/2016/04/NASA_Astrobiology_Strategy_2015_FINAL_041216.pdf

Peters, Ted. 2013. "Astroethics: Engaging Extraterrestrial Intelligent Life-Forms." In *Encountering Life in the Universe: Ethical Foundations and Social Implications of Astrobiology,* edited by Chris Impey, Anna Spitz, and William Stoeger, pp. 200–221. Tucson: University of Arizona Press.

Peters, Ted. 2014. "Astrotheology: A Constructive Proposal." *Zygon,* 49: 443–457.

Race, Margaret S., Kathryn Denning, Constance Bertka, et al. 2012. "Astrobiology and Society: Building an Interdisciplinary Research Community." *Astrobiology,* 12.10: 958–965.

Smith, Kelly. 2014. "Manifest Complexity: A Foundational Ethic for Astrobiology?" *Space Policy,* 30: 209–214.

Vakoch, Douglas. 2009. "Anthropological Contributions to the Search for Extraterrestrial Intelligence." In Meech et al (2009), pp. 421–427

Vakoch, Douglas. 2011. *Communication with Extraterrestrial Intelligence.* Albany, NY: SUNY Press.

Vakoch, Douglas, ed. 2013. *Astrobiology, History and Society: Life Beyond Earth and the Impact of Discovery.* Berlin: Springer-Verlag.

Vakoch, Douglas, ed. 2014a. *Archaeology, Anthropology and Interstellar Communication.* Washington, DC: NASA.

Vakoch, Douglas. 2014b. *Extraterrestrial Altruism: Evolution and Ethics in the Cosmos* Heidelberg, Germany: Springer.

Tough, Allen, ed. 2000. *When SETI Succeeds: The Impact of High-Information Contact.* Foundation for the Future, Bellevue, Washington.

World Economic Forum. 2013. *Global Risks 2013,* edited by Lee Howard. Geneva: World Economic Forum, 2013. Retrieved from http://reports.weforum.org/global-risks-2013/section-five/x-factors/#hide/img-5

# 3

# Hell Is Other Planets

## Extraterrestrial Life in the Western Theological Imagination

### Derek Malone-France

The emergence of the idea of "extraterrestrial life" (ETL) as a prominent element in Western culture is generally associated with the advent of modern science fiction. But the possibility that life (including intelligent life) might exist beyond Earth has fascinated and inspired, as well as terrified and repulsed, Western thinkers and their audiences ever since the ancient Greek Atomists hypothesized an infinite plenitude of "worlds" containing an infinite variety of life forms—thereby contradicting both the geocentric and anthropocentric character of the "single world" cosmologies of Plato and Aristotle.

Taking many different forms—some positive, some negative; some cautious, some extravagant—the *ETL hypothesis*[1] periodically reemerged as an important topic of debate (in some cases, even a preoccupation) among Western philosophers, theologians, political and cultural commentators, artists, and, eventually, scientists, at various intellectual and cultural transition points between the ancient and medieval, medieval and Renaissance, and Renaissance and modern milieus. Indeed, throughout the entire *three centuries* preceding H. G. Wells' publication of *War of the Worlds* (1897)[2]—perhaps the work today most widely associated with both the advent of modern science fiction and the inception of our cultural fascination with the possibility of ETL—the literate public in Europe and the United States had already been exposed to a more or less constant, and often quite substantial, flow of writings, along with other forms of expression of ideas, related directly and explicitly to the ETL hypothesis. And this flow included not just works by marginal figures (though there were plenty of those, to be sure) but also works by many of the most prominent and influential figures of the early- and mid-modern periods: Galileo Galilei, Michel de Montaigne, Ben Jonson, Voltaire, René Descartes, Immanuel Kant, Johannes Kepler, Bernard Le Bovier de Fontenelle, John Locke, Christiaan Huygens, Jonathan Swift, Thomas Paine, Benjamin Franklin, Alexander Pope, Ralph Waldo Emerson, G. W. F. Hegel, Ludwig Feuerbach, Mark Twain, and others.

Derek Malone-France, *Hell Is Other Planets*. In: *Social and Conceptual Issues in Astrobiology*. Edited by: Kelly C. Smith and Carlos Mariscal, Oxford University Press (2020). © Oxford University Press.
DOI: 10.1093/oso/9780190915650.001.0003

Furthermore, the positions taken by prominent thinkers in the historic ETL debates were often inextricably woven into, or bound together with, these same thinkers' perspectives and opinions on more general scientific, philosophical, and theological controversies. So, to the extent that the specific content and character of these debates is, today, largely unknown, even to many academic specialists working in areas relevant to the time periods in question, it represents an important missing element and an unseen context in relation to the scholarly and cultural stories that we have inherited and that we continue to tell ourselves about the emergence of science, the shape of many specific controversies over particular historic scientific advances, and the complex and multivalent relationship between science and religion, as they have developed alongside one another throughout Western history.

There are many books yet to be written on this too long neglected dynamic in Western cultural and intellectual history (I have cited some of the best and most notable of the relatively few serious scholarly works on the topic written so far in the notes here.)[3] In this short chapter, my aim is simply to highlight a few exemplary moments, figures, and ideas that have appeared in the course of the historic debates over ETL in the West in order to demonstrate how intimately connected to broader intellectual and cultural conversations the ETL hypothesis could be for earlier generations of thinkers—including for some major thinkers who are rarely associated with this hypothesis, either affirmatively or negatively, today. More particularly, I am interested here in the ways in which various Christian, para-Christian,[4] and eventually post-Christian thinkers in the West, beginning in the Renaissance, reacted to the conceptual possibility of ETL, with relation to core traditional Christian doctrines and the inherited anthropological assumptions associated therewith.

Not surprisingly, in each phase of the historic ETL debates, the conceptions of ETL that were put forward tended either to *strongly reflect and reinforce* or *provocatively controvert and undermine* traditional human *self-conceptions*. Thus, debates over the existence and nature of ETL have always, also, been debates over human nature, our collective identity and epistemic situation, and the significance (or insignificance) of our existence. Moreover, traditional Christianity, as constructed in the Renaissance and early modern periods, in both its Catholic and non-Catholic forms, faced unique conceptual and interpretive exigencies in relation to the possibility of ETL, precisely because of its particular metaphysical and existential construction of the divine-human relation, including especially the doctrine of the "incarnation" of God in human form and the singular dignity and significance for humanity that this doctrine has been taken to imply by *many* Christians throughout history.[5]

Some thinkers viewed these conceptual and interpretive exigencies as opportunities for the creative extension, reinterpretation, and/or transformation of traditional Christian understandings—in some cases, by way of recovery of alternative understandings from within the tradition that had been lost or left behind at some earlier point along the way. Others took their traditional understandings to be sufficient counterevidence to the ETL hypothesis, denying the possibility of life elsewhere in the cosmos on the basis that it conflicted with what they "knew" to be true, on the basis of Biblical or doctrinal authority. Still others viewed the emerging scientific plausibility of ETL as definitive counterevidence against Christianity per se. Indeed, the role played by the ETL hypothesis in the gradual development of, first, *deism* and other forms of para- or post-Christian anti-incarnationist and anti-*fideistic* forms of theism, and, then, eventually, genuine *atheism*, as culturally viable options in modern European and American society is one important example of an underappreciated historical dynamic associated with this topic.

## The Ancient Background

In their temporarily losing argument against Plato and Aristotle, the ancient Greek "Atomists" maintained the following: the fundamental "stuff" out of which the *kosmos* is constructed is an infinite proliferation of absolutely "simple" (i.e., "indivisible"), inert, and imperishable material particles (i.e., "atoms") that *move without intent* and *collide without permanent attachment*, within an infinite "void" space. The physical world is understood, essentially, to be the product of the everlastingly ongoing, random *jostling* of an infinite number of such atoms, of diverse geometric shapes, intermingling in every possible combination, across the unending expanse of the infinite void.

Contra Plato and Aristotle, therefore, the Atomists would ultimately assert that there is not merely a single "world" (i.e., orbital system of planets and stars), bounded by the heavenly spheres, centered on this little terrestrial orb of ours, and constituting the entire *kosmos*. Rather, they claimed, there is an infinite multiplicity of "worlds," spread out through infinite time, across an infinite "void" space, which is *filled to brimming* with an equally infinite plenum of atoms (and, hence, the distinction first arose between the respective notions of *kosmos* and "world"). According to the earliest known written account of the Atomists' ideas, found in Diogenes Laertius's *Lives of the Eminent Philosophers*, Leucippus was the first to formulate the basic elements of this view: "He held that the totality of things is infinite and that they all change into one another. The whole is at the same time empty and full of bodies. The worlds come into being when the bodies fall into the void and are intertwined with one another; it is their motion as their bulk increases that gives rise to the substance of the stars."[6] Leucippus's immediate successor, Democritus, went on to assert that, just as the microscopic constituents of the atomistic universe are infinite in number, so too, therefore, must be the macroscopic products of their entanglements: "The worlds are un-limited in number; they come into being and they perish."[7] And, finally, the full implications of this view were elaborated by Epicurus, who clarified that the un-limited array of worlds was not merely temporal and successive but also spatial and coexistent:

> [T]here is an unlimited number of worlds, some of them like ours, others un-like. For the atoms, being unlimited in number . . . travel to the most distant points. For atoms of this description, out of which a world might arise, or from which it might be composed, have not been used up either on one world or on a limited number of worlds, whether resembling ours or not. Hence, nothing stands in the way of an unlimited number of worlds.[8]

It is, of course, important that we not make the mistake of simply equating the ancient Atomists' conception of the multiplicity of world systems with our own, modern *astrophysical* conception of the multiplicity of solar systems, or of gal-axies. They conceived of each of the serendipitously spawned "worlds," dancing amidst the eternal chaotic whirl of atomic jostling, as an *internally closed system*. This was an infinite multiplicity of bubble-verses, so to speak, flitting across an infinite cosmic sea of atomic creation and dissolution, empirically impenetrable and unperceivable to one another.

Nevertheless, inferential reason was enough to prompt the Atomists to con-template the content of these worlds, and the principle of the *productive logic of an actual infinitude* that they formulated suggested that this content also would be infinite in form and would include, therefore, both an *infinite variety of specific forms* and *infinite replication of every such form*. Thus, in his *De Rerum Natura*, the later Roman Atomist Lucretius writes:

> Wherefore you must in like manner confess for sky and earth, for sun, moon, sea, and all else that exists, that they are not unique but of number innumer-able; since there is a deep set limit of life equally awaiting them, and they are as much made of perishable body as any kind here on earth which has so many specimens of its kind.[9]

And, so, the ETL hypothesis, the idea of other life existing beyond the Earth, in other parts of the cosmos, either near or far, came briefly into play in Western culture, only to be, for a while, preemptively set aside through the intellectual and cultural ascendance and, then, millennium-long monopolistic dominance of Platonic-Aristotelian-style, single-world cosmology.

## Rediscovery and Renaissance

A little less than three centuries after Muslim scholars, arriving in the wake of the Umayyad conquest of medieval Spain, delivered Greek and Arabic copies of *De Caelo* and other "lost" texts of Aristotle to European thinkers, igniting the "scholastic" movement in theology and philosophy, the Italian Christian-humanist Poggio recovered a copy of Lucretius's *De Rerum Natura*, bringing the ideas of the Atomists back into circulation, just in time to play a pivotal role in the European Renaissance.[10] During the 1400s and 1500s, European theologians, philosophers, and scientists *cautiously* drew on various elements of the Atomists proposals in order to poke gently at the Aristotelian consensus that Aquinas and the other Scholastics had established.

In fact, even before the infusion of atomistic thinking, late medieval and early Renaissance theologians and philosophers had debated the merits of Aristotle's conception of our geocentric "world"-system as both closed and singular. If taken in abstraction from the traditional Christian conception of divine transcendence and supremacy in relation to nature, the acceptance of Aristotle's onto-cosmological doctrine of "natural place" could well function as a justification for the assumption of a finite and geocentric universe—and, therefore, of humanity's uniqueness and singular significance. However, when juxtaposed with traditional Christian understandings of God's omnipotence and the logical implications of the doctrine of *creatio ex nihilo*,[11] the idea that any characteristic of created reality, however ontologically foundational, could represent an actual limitation on the creative capacity or freedom of God appeared to many Christian thinkers as both unscriptural and, therefore, potentially heretical. Thus, in 1277, no less authority than the influential Bishop of Paris, Etienne Tempier, proclaimed a dogmatic prohibition against a set of 219 propositions that infringed on divine sovereignty, including the claim that God could not have created other worlds. In doing so, he effectively censured Aquinas.[12]

Furthermore, during this intermediary period in the development of the ETL hypothesis, a number of important thinkers, such as William of Ockham and Nicole Oresme, used the space opened up for explicitly contra-Aristotelian reasoning by Tempier's proclamation to rebut Aristotle's arguments against a plurality of worlds on other grounds, besides that which had motivated Tempier and his fellow theological absolutists. Both Ockham and Oresme critiqued the internal logic of the Aristotelian position to arrive at carefully crafted affirmations of the *possibility* of other worlds.[13] Interestingly, although not yet informed by the Atomists' ideas, these thinkers similarly posited other

worlds as being ultimately *empirically unknowable*. So, while no longer necessarily understood as singular, our world was, nevertheless, still taken by
these late medieval and early Renaissance pluralists to be "closed"—that is,
cut off both physically and observationally from any other worlds that might
exist. Only the development of sufficiently powerful telescopes would ultimately overturn this sense of unbridgeable cosmic isolation among European
thinkers (though it is worth noting that the idea of contact with beings *on
other planets and moons of our own solar system* was not ruled out by the
closed-worlds models).

The addition of the Atomists' perspective was a lit match thrown into an already combustible theological situation. Even as the Atomists' materialist framework and at least quasi-atheistic tendencies were rejected, their thoroughgoing
*cosmogonic naturalism*, their insistence on *uniform effects from uniform causes
throughout nature*, and their elucidation of the material-mechanical productive logic implied by an *actual infinity* informed increasingly explicit contra-
Aristotelian speculations about the nature of the cosmos, speculations that were,
by this point, likewise being informed—and driven—by advances in telescopic
technology and mathematical astronomy.

On the theological and philosophical side, the question of the "plurality of
worlds"—with which the ETL hypothesis was always at least implicitly, if not explicitly, associated—was a topic of concern for such notable thinkers of this period as Montaigne,[14] Pierre Gassendi,[15] John Dunne,[16] and Leonardo da Vinci.[17]
And one major thinker of this period in particular, the tremendously influential German humanist philosopher and theologian Nicholas of Cusa, made the
ETL hypothesis an explicit component of the pluralism debate. In his celebrated
work, *Of Learned Ignorance*, Cusa, a powerful cardinal in the Catholic Church,
voiced an affirmational, anti-anthropocentric perspective on pluralism and the
idea of extraterrestrial life:

> Life, as it exists here on earth in the form of men, animals, and plants, is to be
> found, let us suppose, in a higher form in the solar and stellar regions. Rather
> than think that so many stars and parts of the heavens are uninhabited and that
> this earth of ours alone is peopled—and that with beings perhaps of an inferior
> type—we will suppose that in every region there are inhabitants, differing in
> nature and rank and all owing their origin to God, who is the center and cir-
> cumference of all stellar regions.[18]

The sort of conjunction of genuine humanism with authentic anti-
anthropocentrism represented here by Cusa is characteristic of a number of
important thinkers who engaged with the ETL hypothesis. (We will see it
again later, in the thought of Christiaan Huygens, for example.) For thinkers

like Cusa, there is a genuine sense of the *relativization* of humanity, derived from our cosmic decentering and from the implied possibility of more fully developed and/or more morally perfected forms of intelligence in the universe that they associated with this decentering. This sense of relativization, therefore, forms the basis of the anti-anthropocentric sentiment in their work.

However, for thinkers like Cusa, this relativization of humanity is *not* taken to be *cognitive* in nature. That is to say, there remains an assumption that *human cognition*—our forms and modes of reasoning and our most basic conceptual and experiential categories—*transparently relates to, and therefore accurately represents, the nature of reality* per se. This assumption, of course, follows logically from the belief, held by proto-Enlightenment thinkers of this period like Cusa, that nature was constructed by God according to a rational plan and that human reason is a teleological outcome of this plan—God wants to be known by "His" creatures and for the natural order that "He" has created to be understood by them. Accordingly, it is further assumed by thinkers like Cusa that the forms, modes, and categories of human experience and reason will be universal characteristics of intelligence throughout the cosmos. Thus, humanity's emerging rational and scientific enlightenment preserves its claim to glory in the view of these thinkers. For despite the possibility that other beings have achieved even higher realizations of the universal principles of reason than we human beings so far have, it is precisely the meaningful exercise of reason, at whatever level it is achieved, that represents the proper fulfillment of our purpose as thinking "creatures" and the proper form of worship of our "Creator," the rational architect of the cosmos.

It is not until we reach Immanuel Kant's late "critical philosophy," at the end of the 18th century, that we find a systematic consideration of cognitive relativism as an epistemic possibility that is associated explicitly with the idea of extraterrestrial intelligence, although the great French skeptic Montaigne, in the 16th century, had already suggested this as a possible implication of the plurality of worlds hypothesis[19] and Fontenelle had echoed and amplified this suggestion, with relation specifically to the ETL hypothesis, in the late 17th century (see later discussion).

Finally, as the Renaissance came to a close, one thinker threw caution to the wind. The mystically inclined Italian humanist and Dominican friar Giordano Bruno was as imaginative as he was indiscrete. Indeed, there is substantial irony in the historic treatment of Bruno—who was burned at the stake by the Catholic Inquisition in Rome in 1600 as a "martyr of science"—given the deeply unscientific character of Bruno's mind and thought.[20] While certainly well-read on the substance of much important science of his day, especially the new astronomy, Bruno's own method of reasoning and modes of justification for his ideas owed more to the ancient Hermetic tradition of occultism and alchemy and pseudo-scientific movements like medieval astrology than to the emerging tradition of Bacon, Copernicus, Galileo, and others. Nevertheless, Bruno's audacity gave voice to a maximalist, Atomist-echoing interpretation of the scope and content of the emerging telescopic picture of the cosmos, at a moment when the logic of such a view was implicitly haunting the minds of anyone contemplating the possibilities for where the new science might lead. Bruno, who claimed a visionary experience of an infinite cosmos not only populated the sun, moon, and other planets, also proclaimed that the planets and stars themselves were living, ensouled beings, propelled through infinite space by their own internal power and will.[21] In this sense, also, he recurred to the ancient Greeks, interpreting the celestial bodies as "divine" entities, as the Greeks had generally done in their own theological speculations.

Yet, however unscientific Bruno's path to the position he staked out, and however far beyond the bounds of evidentiary support, or even reason, he may have pushed certain elements of that position, it was a position that had the merit of clearly and explicitly engaging with all of the points of greatest tension, intellectually and existentially, raised by the new astronomy. Thus, Bruno's intellectual extravagance served to clarify the full potential stakes of the plurality of worlds/ETL debates of this period—as did his execution at the hands of the Catholic Church. In addition, Bruno's avid enthusiasm for the plurality of worlds and ETL hypotheses illustrates one of the most central and important dynamics that would continue to shape theological responses to these hypotheses throughout the coming centuries.

Theologically speaking, what the plurality of worlds/ETL hypothesis-related controversies of the Renaissance period most fundamentally revealed was a profound, deeply rooted, seemingly intuitional, and perhaps psycho-inclinational divide between those Christian thinkers for whom the possibility of extraterrestrial life represented an awe-inspiring expansion of their conception of the greatness of God, as opposed to those for whom this possibility represented a horror-inducing threat to their conception of the special "closeness to God" enjoyed by humanity. For the latter, a fundamental aspect and purpose of the Biblical narrative is the validation of humanity as being—despite our moral "fallen"-ness—the chosen creatures of God. In this sort of Christian mindset, the Jewish notion of God establishing a special "covenant" with a particular group of people is replaced with the claim of a special covenant for humanity, above all other creatures in nature. For a significant portion of Christians throughout history, this anthropocentric understanding of the "creation story" has formed, and continues to form, a base layer of their self-conceptions. To be told that humanity might not sit atop the "great chain of being," as the beloved apex of God's creation but, rather, in some middling position (or worse) in a cosmic chain of being stretching out beyond imagination, is to be told that a primary basis for their conceptions of the meaningfulness and worth of their existence is a lie. And, as we will continue to see as we move through the later periods in this brief historical survey, to *feel* this way about the ETL hypothesis is to have a powerful motivation to dismiss any evidence of its plausibility.

## Early Modern Fixations

As the Renaissance turned toward the early modern period, this fundamental divide between those who were drawn to and those who were repulsed by the ETL hypothesis continued to shape theological responses in ways that could often run contrary to what one might expect to be the overarching theological dynamics. For example, Protestant theologians, as a consequence of breaking theologically with the Catholic Church, were effectively liberated from the grip of Aristotle's cosmology, as a matter of institutional dogmatics, allowing them, in theory, to follow the new astronomy wherever it might take them—so long as it agreed with their readings of scripture. Furthermore, precisely insofar as Protestant thinkers wished to emphasize God's power, while frequently offering distinctly critical, even negative, theological anthropologies, which focused on the depth and intractability of human moral and existential "depravity,"[22] there would seem to have been reason for them to be more open to the idea of other worlds and other forms (perhaps even better forms) of intelligent life in the cosmos. Yet, initially, the Protestant theological approach to the ETL hypothesis was often definitively negative, precisely because seminal early Protestant theologians wished to stress the uniqueness of the connection between humanity and God that scripture established through its account of God's incarnation in human form, as Jesus of Nazareth, and the cosmic redemptive-salvific significance of the life, death, and resurrection of Jesus, as "the Christ."

One prominent example of this antipluralist tendency in early Protestant engagements with the ETL hypothesis is found in the work of Philip Melanchthon, who essentially coestablished Lutheran theology with Martin Luther. Melanchthon, whose life overlapped directly with that of Copernicus, did not see a source of common cause in an earlier figure like Tempier, who (very much like the Protestants) had been concerned with upholding both divine power and sovereignty and scriptural authority, over and against the claims of the subtle theological logicians and philosophers. Instead of embracing him as a proto-Reformationist compatriot in these regards, Melanchthon framed Tempier's 1277 proplurality proclamation as just another example of the sort of promulgation of the Catholic Church's hierarchy that the Protestants were rejecting. He identified, instead, with the earlier consensus against plurality, arguing for this view both on Aristotelian principles and on a reading of scripture focused clearly and explicitly on human uniqueness:

> We know God is a citizen of this world with us . . . we do not contrive to have him in another world, and to watch over other men also . . . the Son of God is One; our master Jesus Christ was born, died, and resurrected in this world. Nor does He manifest Himself elsewhere, nor elsewhere has HE died or resurrected. Therefore it must not be imagined that there are many worlds, because it must not be imagined that Christ died and was resurrected more often, nor must it be thought that in any other world without the knowledge of the Son of God, that men would be restored to eternal life.[23]

Ultimately, of course, the continuing advance of astronomical cosmology would overrun the peremptory theological objections of those like Melanchthon. Telescopes would continue to be improved and enlarged. Observations would continue to strengthen the evidence that other bodies—planets and moons—in our solar system were similarly constituted physical analogues of Earth and that distant stars were analogues of our sun. Simple, natural inference, therefore, necessarily and immediately raised the questions of whether these Earth-analogue bodies might also exhibit the same sorts of environmental—and, therefore, potentially biological—characteristics as our own; whether there might be hosts of other planets and moons, also, out there amongst the "other Suns" of the cosmos; and whether these other planets and moons, both here in our solar system and, perhaps, out there in other systems, might be inhabited, either by "men" and other Earth creatures or, alternatively, by more disparate forms of life quite unlike those found here on Earth.

Galileo himself insisted that, although it was certain there could not be "men," or other of the *same forms of life found on Earth* on other planets and bodies—even describing the opposing view as "damnable" in correspondence—he considered it to be an open question whether there might be forms of life *other than those found on Earth* living elsewhere in the cosmos.[24] And Johannes Kepler took an even bolder tack on the issue, writing in correspondence to a skeptic of pluralism and ETL: "I myself argue the probability by analogy. . . . I attribute humors to the stars, and regions which from the exhalations of the humors are rained upon, and living creatures, to whom this is useful."[25]

Eventually, in light of the accumulating astronomical evidence for the plurality of worlds, natural philosophers and cosmological theorists like Descartes began to draw once more on core concepts of the Atomists, now conceptually refined and admixed with their own developing scientific reductionism and physical materialism,[26] to formulate updated "mechanical" explanations of how a vast, cosmic sea of "worlds" could have come to be out of a primordial state of disordered simple material potential. And for the first time, these many "worlds" were now conceived of as spatially distributed stellar-planetary systems in something approaching the contemporary understanding thereof. Descartes's "vortices" theory of star-system formation was a direct precursor to both Immanuel Kant's and, later, Pierre-Simon Laplace's formulations of the "nebular hypothesis." Descartes mediated the tension between Atomistic mechanism and Christian transcendental supernaturalism by maintaining that the initial momentum necessary to put the mechanistic system in motion and, thereby, begin the causal process of world formation was provided by God, acting as an updated version of Aristotle's "prime mover."

In France in 1686, Bernard le Bovier de Fontenelle published *Conversations on the Plurality of Worlds*, in which the young polymath sought to convey, in an accessible way, both the scientific substance of the new astronomy and cosmology, including Descartes' version of the "vortices" theory, and the philosophical and theological implications thereof, especially those related to the ETL hypothesis.[27] Following the immense success of this work, Fontenelle would go on to become the veritable dean of French intellectual society for more than half a century, a member of both the *L'Académie française* and *L'Académie sciences*, and, indisputably, one of the greatest and most prolific translators of science for public consumption in history. In *Conversations*, Fontenelle imagines a series of dialogues between a French *philosophe* and a well-read and intelligent aristocratic lady.

Over the course of several evenings, Fontenelle's plurality and ETL advocating *philosophe* explicates the new cosmology suggested by the confluence of Copernican and Cartesian principles and seeks to reassure the disquieted yet intrigued lady, who rhetorically stands in for the dialogues' readers. In doing so, he directly takes on the intuitive divide between those who were exhilarated and those who were horrified by the thought of a vast universe hosting a plurality of worlds:

> "But," she replied, "here's a universe so large that I am lost . . . I'm nothing . . . All this immense space which holds our Sun and our planets will be merely a small piece of the universe? As many spaces as there are fixed stars? This confounds me—troubles me—terrifies me."
>
> "And as for me," I answered, "this puts me at ease. When the sky was only this blue vault, with the stars nailed to it, the universe seemed small and narrow to me; I felt oppressed by it. Now that they've given infinitely greater breadth and depth to this vault by dividing it into thousands and thousands of vortices, it seems to me that I breathe more freely, that I'm in a larger air, and certainly the universe has a completely different magnificence."[28]

While using the dialogue format to perform the sort of delicate rhetorical dance still required in France at this time, if one's proposals might be seen as running contrary to church doctrine, Fontenelle also advocates very effectively for both intellectual and theological openness to the possibility of ETL. At the same time, in the course of exploring this possibility, Fontenelle's *Conversations* exhibits one of the most fascinating, illuminating, and *morally troubling* dynamics associated with the ETL hypothesis in European thought from the late 16th century onward.

Not coincidentally, the late Renaissance and early modern astronomical clarification of our cosmological situation—which also produced more accurate mapping and navigation—overlapped fairly precisely with the age of European exploration and colonial conquest. Accordingly, considerations of the possibility of ETL during this period inevitably coincided at least conceptually and implicitly (and sometimes explicitly) with considerations of the existence and significance of the non-European "peoples" who were then being encountered in "the New World" right here on Earth. Moreover, throughout the 17th, 18th, and 19th centuries, conversations about ETL were intellectually tracked alongside the rise of modern zoology and early evolutionary thought (first in pre-Darwinian and, later, in Darwinian forms). This was the period during which—through the interaction between early taxonomic zoology, pre-Darwinian evolutionary theory, and the colonial disclosure of "other peoples"—European scholars began to develop the first *putatively scientific* conceptions of human phylogeny, on the basis of which European (and eventually American) culture would produce morally pernicious and self-serving theories of "race."

Not surprisingly, therefore, we frequently see close parallels between the sort of speculations that an author like Fontenelle makes about the potential characteristics of life forms on other worlds and the sorts of characterizations of "racial" groups that one finds in the work of the early European categorizers of humanity. In particular, early modern ETL-focused authors such as Fontenelle, just like their zoological taxonomist counterparts of that era, tended to take an important insight that was beginning to make its way into scientific naturalism—namely, that *over time* organisms are *shaped by* their environments—and to juxtapose it with caricaturing claims regarding human beings from other parts of the world and differing climates, giving a scientific-seeming veneer to bigotry. Fontenelle, for example, writes:

I'm beginning to see . . . how these Venusians are made. They resemble our Moors of Grenada, a small, black people, sunburnt and full of verve and fire, always amorous . . . But what of the inhabitants of Mercury? They're even closer to the Sun. They must be vivacious to the point of madness! I believe that they have no memory, no more than most savages; that they never think deeply on anything; that they act at random and by sudden movements, and that actually Mercury is the lunatic asylum of the Universe.[29]

Approximately fifty years later, Carl Linnaeus, the father of modern taxonomy, would characterize the supposed racial "humors" of *homo europaeus, homo asiaticus, homo americanus*, and *homo afer* as "sanguine," "melancholic," "choleric," and "phlegmatic," respectively.[30]

In addition, in later, expanded editions of *Conversations*, while seeking further to soften the blow, theologically, of the ETL hypothesis, Fontenelle takes a tack often employed during this era by colonialists, slave-traders, and their political advocates, who sought to minimize or even cancel out the moral responsibilities associated with interactions with non-European peoples. Fontenelle argues that any hypothetical extraterrestrial beings, while they might perhaps embody the sort of intelligence that sets humanity apart from other creatures here on Earth, would nevertheless not be *descendants of Adam and Eve*, and, therefore, their existence would be irrelevant to the theological doctrines of incarnation and salvation associated with Christianity.[31] In the context of the racial discourse during this period, such claims—that, for example, the indigenous people of the "West Indies," or those in Africa or Australia, were not "sons and daughters of Adam and Eve"—were commonly used to justify the denial that these people possessed eternal souls or were due any moral consideration whatsoever—in other words, to justify their dispossession and enslavement.

Fontenelle's descriptions of the imagined inhabitants of other worlds reflect both the scientific uncertainties and existential anxieties of this period during which biology was haltingly groping in the direction of evolutionary thought but still remained largely predicated on the classical idea of "fixed" natural types. Within the scope of Fontenelle's very long life, Pierre Louis Maupertuis would eventually put forward the idea of changes occurring naturally over time through the reproductive process, but it would be another century after this before Darwin would explicate the manner in which such changes occur. So it is not surprising that, on the one hand, there are passages in *Conversations*, like the one quoted earlier, in which we see ETL imagined as being very human-like, not only in character but also in physical form. Yet, at another point in the dialogues, Fontenelle writes:

I don't believe at all there are men on the Moon. Look how much the face of nature changes between here and China: other features, other shapes, other customs, and nearly other principles of reasoning. Between here and the Moon the change must be even more considerable. . . . He who would press on to the Moon assuredly would not find men there. . . . If it could be that we were rational, yet weren't men, and if besides we happened to live on the Moon, could we possibly imagine that down here on this place there were bizarre creatures who called themselves the human race?[32]

Just over a decade after Fontenelle, the science popularizer, published his coy encouragement to the pluralist-minded, the cause gained the far more assertive advocacy of one of the greatest figures of the scientific revolution, Christiaan Huygens. In *Kosmotheoros*, written at the end of his life and published posthumously in 1698, Huygens confidently champions the ETL hypothesis. Styled, like Fontenelle's *Conversations*, in a breezy, accessible way intended to engage a wide, popular readership, *Kosmotheoros* was, in fact, wildly successful and became a staple reading among well-educated Europeans during their school years for many decades. Like Fontenelle (and Cusa before him), Huygens represents the view that the existence of a multitude of other life forms, including other intelligent life forms, spread across the cosmos only enhances our sense of the grandeur both of God and of nature and that it places humanity in a more, rather than less, satisfying existential position. But unlike Fontenelle (who was following Descartes), Huygens retains a more robust sense of *teleology*—of purposiveness and ongoing divine activity—rather than, as Descartes had done, more or less reducing the role of God to the provision of initial energy input. Thus, as the German literary scholar and historian of the plurality/ETL debates, Karl Guthke, has observed:

> [Huygens'] teleology [is] anti-anthropocentric . . . it is not for the descendants of Adam that the planets and stars with their plants, animals, and so on were created . . . but for their own rationally endowed inhabitants. . . . Unlike Fontenelle . . . Huygens only very occasionally represents the creative principle that acts with intention and gives a purpose to the whole as a mechanical "Nature". . . in the overwhelming majority of instances it is God or Divine Providence whose wise ordinance Huygens emphasizes in opposition to the belief in chance of Atomists such as Democritus and Descartes.[33]

Also unlike Fontenelle, Huygens does not proffer the view that chance and dissimilarities among environments may have led to radically different forms of life and, possibly, of intelligence. Instead, still writing just far enough in advance of the development of evolutionary thinking to reasonably hold onto such a view, Huygens presumes that species are the product of a sort of universal design template and that nature is ultimately consistent in its character across stellar and planetary environments, and, therefore, God can certainly have populated other bodies with the exact same forms of life as were placed here on Earth. Indeed, Huygens imagines hosts of other "humanities," peopling other star systems, all observing the same natural order and all sharing in the same form of reason.[34]

Meanwhile, in Britain, just about thirty years after Fontenelle's *Conversations* ignited the French public's imagination regarding the ETL hypothesis, the English scientist, natural philosopher, and theologian William Derham published *Astro-theology: Or, A Demonstration of the Being and Attributes of God, from a Survey of the Heavens*. Though certainly not of Huygens' stature, when he wrote *Astro-theology*, Derham was already an esteemed Fellow of the Royal Society, having served as Boyle Lecturer, and he had made significant contributions in scientific areas including field biology, zoological/proto-evolutionary theory, and astronomy, and had produced the first relatively accurate measurement of the speed of sound. Like both Fontenelle and Huygens—whose telescope Derham actually used to make many of the original astronomical observations reported in this work, Derham sought to communicate and popularize the new astronomy and its implications. Also, like Fontenelle and Huygens, in doing so, Derham allowed his observational inferences and imagination to run wide of scientifically establishable facts (though not in a completely unchecked way). Derham also argued for the theological significance and, indeed, preferableness of what he called "the new Systeme" of astronomy, which he outlined in *Astro-theology* as being a Copernican model hybridized with an neo-Atomistic infinitude of worlds. And, yet, because he was still, like Huygens, thinking in terms of the classical neo-Aristotelian Christian notion of a plenitudinous "great chain of being" that has been arranged by God to fill creation without gap or vacuum, Derham writes:

> And now considering how accomplished the Moon, and all the other Planets are for Habitation . . . with great reason therefore the Maintainers of the new Systeme conclude those Planets, yea all the Planets of the Sun and Fixt Stars also be habitable Worlds; places as accommodated for Habitation, so stocked with proper Inhabitants. But now the next Question commonly put is what Creatures are they inhabited with? But this is a difficulty not to be resolved without a Revelation, or far better Instruments than the World hath hitherto been acquainted with.[35]

Indeed, *Astro-theology* and *Kosmotheoros* are exemplary of an entire class of ETL hypothesis–related works written during the period from the Enlightenment through the early 19th century, often by authors with genuine scientific training and expertise, in which science is mixed together with starkly pseudoscientific imaginings and often, also, with ungrounded metaphysical and mystical-theosophical speculation.[36] Furthermore, this type of hyperimaginative theological engagement with the ETL hypothesis almost certainly fed directly into the emergence and proliferation during the 19th century of demographically significant Christian or para-Christian religious movements, like Christian

Science and Mormonism, the doctrinal foundations of which included explicit affirmations of the ETL hypothesis in one form or another.

An even more extreme example of the sort of exuberant theological and imaginative response to the ETL hypothesis manifested in *Astro-theology* can be found in the later work of the great Swedish scientist and natural philosopher turned mystic-theologian Emanuel Swedenborg. Like Huygens and Derham, Swedenborg came to the ETL debates following a successful career as a scientist, engineer, and inventor. Yet, at the age of fifty-three, Swedenborg entered into a visionary-occultist phase that would last the remainder of his life and would split the opinions of those around him as to whether he would become a prophet or simply a madman. In a work with the notably long, stream-of-consciousness title, *Earths in Our Solar System which are Called Planets and Earths in the Starry Heaven, Their Inhabitants, and the Spirits and Angels There from Things Heard and Seen*, Swedenborg reports of the visions his spirit has been granted through a special divine dispensation that allowed him to experience spiritual transmigration across the cosmos while living (and, thus, supposedly acquiring just the sort of "Revelation" that Derham had said might be required to learn the actual character of e life forms of other planets):

> Inasmuch as, by the Divine mercy of the Lord, the interiors which are of my spirit have been opened in me, and it has thereby been given me to speak with spirits and angels, not only with those who are near our Earth, but also with those who are near other earths; and since I had an ardent desire to know whether there were other earths, and to know their character and the character of their inhabitants; it has been granted me by the Lord to speak and have intercourse with spirits and angels who are from other earths, with some for a day, with some for a week, with some for months; and to be instructed by them respecting the earths from and near which they were, and concerning the life, customs, and worship of their inhabitants, besides various other things there that are worthy of note.[37]

In Swedenborg's case, his theo-philosophical speculations and the emergence of a new Christian or para-Christian movement was direct and immediate: Swedenborgianism, also known today as "The New Church," is a cluster of related Christian denominational groups that were founded on the basis of his visions and his theological interpretations of the significance thereof.

Swedenborg's mystical transformation deeply disquieted many European intellectuals who were familiar with him as a sober and formidable scientific intellect, including his one-time admirer, the great Prussian philosopher, mathematical physicist, and Enlightenment rationalist par excellence Immanuel Kant—who may have derived his own early, pre-Laplacian formulation of the nebular hypothesis of stellar formation from Swedenborg. But while Kant was disturbed by Swedenborg's turn toward religious mysticism, he was by no means averse to the ETL hypothesis. In fact, probably no figure who is as foundationally important to modern Western thought across as wide a range of different disciplines spent more time considering the multidimensional philosophical and theological import of the ETL hypothesis than Kant. More specifically, as mentioned earlier, Kant was the first historically significant European thinker to systematically consider the cognitive/epistemic implications of the possibility of disparate forms of intelligence and experience.

Kant's reflections on the significance of the ETL hypothesis began at least as early as the mid-1750s, during the precritical' period in his philosophical development. In his *Universal Natural History and Theory of the Heavens*, published in 1755, Kant claims, "most of the planets are certainly inhabited."[38] Throughout the middle part of his career, as he transitioned into and then developed his critical "transcendental idealistic" perspective, he continued to mull over the implications of this presupposition, and, finally, in his late work, *Anthropology from a Pragmatic Point of View*, published in 1798, he arrives at a nuanced position regarding the nature and status of human epistemological and cognitive frameworks and categories, informed both by his critical work to this point and by his contemplation of the ETL hypothesis:

> In order to indicate a character of a certain being's species, it is necessary that it be grasped under one concept with other species known to us . . . The highest species concept may be that of *terrestrial* rational being; however, we shall not be able to name its character because we have no knowledge of *non-terrestrial* rational being that would enable us to indicate their characteristic property and so to characterize this terrestrial being among rational beings in general.—It seems, therefore, that the problem of indicating the character of the human species is absolutely insoluble, because the solution would have to be made through experience by means of the comparison of two *species* of rational being, but experience does not offer us this.[39]

So, in keeping with his critical, transcendental idealistic approach to human cognition, Kant steers us away from the sort of presupposition of universal experiential and epistemic transparency and commensurability that characterized the responses to the ETL hypothesis of so many earlier European *rationalists* (e.g., Cusa and Huygens). As Kant observes, unless and until experience provides the crucial $n = 2$ moment, in which we encounter another form of intelligence, we cannot know if it/they will experience and think as we do or, rather, in some very different way.

At the same time that Kant was contemplating what we might now label issues of "xenocognition," across the Atlantic, in America, key Enlightenment, revolutionary, and founding figures like Thomas Paine and Benjamin Franklin drew on the plurality of worlds and ETL hypotheses in formulating their respective *rejections* of Christianity. Paine, writing in *The Age of Reason*, echoed earlier European anti-incarnationist thought (especially the English deists), while bringing his own penchant for bitingly uncomplicated satirical *reductio ad absurdum* arguments into play:

> What are we to think of the Christian system of faith that forms itself upon the idea of only one world . . . From whence . . . could arise the solitary and strange conceit that the Almighty, who had millions of worlds equally dependent on His protection, should quite the care of all the rest, and come to die in our world, because, they say, one man and one woman had eaten an apple? And, on the other hand, are we to suppose that every world in the boundless creation had an Eve, an apple, a serpent, and a redeemer? In this case, the person who is irreverently called the Son of God, and sometimes God Himself, would have nothing else to do than to travel from world to world, in an endless succession of deaths, with scarcely a momentary interval of life.[40]

Meanwhile, Franklin expressed a far more unusual—but surprisingly well-sourced—view of the implications of plurality. In a short piece titled "Articles of Belief and Acts of Religion," Franklin begins by stating that, "I believe there is one Supreme most perfect Being, author and father of the gods themselves. For I believe that man is not the most perfect Being but one, but rather that there are many degrees of beings superior to him." He then goes on to contemplate the vastness of the universe that the "Supreme" divine entity has created, concluding that an entity of such infinite and incomprehensible greatness cannot be known or even meaningfully worshipped by the likes of humanity, being so far above the need or desire for the sorts of infantile and inadequate modes of engagement of which we are capable as to be essentially unaware of our existence. But then he goes on to describe a host of subordinate "created gods" and conceives for the reader "that Each has made for himself, one glorious sun, attended with a beautiful and admirable system of planets." And, Franklin says, "It is the particular wise and good God, who is the author and owner of our system, that I propose for the object of my praise and admiration."[41]

From whence could Franklin have derived such an eccentric form of *henotheism*? The answer is quite possibly from Isaac Newton, who is believed to have held and communicated some such view in the period during which Franklin was in residence in London and circulating in overlapping friendships with Newton.[42]

Over the next several generations, major American intellectual heirs of these founding figures, like Emerson and Twain, would similarly take up the ETL hypothesis and deploy its implications in tradition-critical ways. Both Emerson and Twain were directly influenced by Paine's argument in *The Age of Reason*. Emerson, in his response, seeks to soften the blow a bit, asking, in light of astronomy, "Who can be a Calvinist or who an Atheist," given the smallness of the Earth *and* the grandeur of the cosmos? Our new cosmological understanding of our situation, Emerson says, offers "not contradiction but correction" to the Christian understanding. It is true: we are no longer to be understood as the center of the universe, the main players in the cosmic drama. And yet, like Cusa and Huygens before him, Emerson, the transcendental universalist (and admirer of the late Swedenborg), claims that, "if we could carry the New Testament to the inhabitants of other worlds we might need to leave Jewish Christianity, and Roman Christianity . . ., but the moral law, justice and mercy would be at home in every climate and world where life is."[43] The acerbic Twain, on the other hand, more closely echoes the skepticism of his founding-era inspiration, Paine, writing in a letter to his wife, "How insignificant we are, with our little pygmy world!—an atom glinting with uncounted myriads of other atoms . . . & yet prating complacently of our speck as the Great World, & regarding other specks as pretty trifles made to steer our schooners by . . . Did Christ live 33 years in each of the millions & millions . . . [or] was *our* small globe the favored one of all?"[44]

## Late-Modern Reactions

As we shift toward late modernity, we find a particularly interesting and revealing thread of discourse related to the ETL hypothesis in post-Kantian German philosophy and theology. The tremendous individual/personal and collective/humanistic *self-regard* that characterized nearly all of the main thinkers and perspectives in both the Romantic and Idealist schools profoundly inflected their response to the ETL hypothesis. For example, G. W. F. Hegel, the philosophical high-priest of both self- and human-aggrandizement, dismissed the idea and import of ETL out-of-hand. In his *Encyclopedia of the Philosophical Sciences*— a work that represents, among other things, a sort of petulant tantrum against the ascendancy of Newtonian physics—Hegel asserts that the "multitude of stars in immeasurable space means nothing to Reason; this is *externality*, the *void*, the *negative infinitude* to which Reason knows itself to be superior."[45] Similarly, Hegel's contemporary and near-rival Friedrich Schelling explicitly rejected the ETL hypothesis, while essentially framing the ongoing Romantic-Idealist philosophical enterprise as a reversal of the Copernican revolution.[46] And one of Schelling's most illustrious students, the scientifically trained Norwegian-Danish theologian Henrich Steffens—sounding not unlike a latter-day Melanchthon—unabashedly expressed this theo-philosophical return to geocentricity, claiming with the characteristically ungrounded confidence of his milieu that, "[P]resent day astronomy is fast approaching the time in which our planetary system will be recognized as the most organized point in the universe . . . our earth will be recognized . . . [as the] hallowed place, on which the Lord appeared . . . as the absolute middle point of the universe.[47]

From here, the ETL hypothesis went on to play an explicit and very significant role in the development of various forms of mid- and late-19th-century *anti-theology*, which emerged, in part, as reactions against Romanticism and Idealism, in both Germany and France. Indeed, in several prominent cases, the specific thread of the ETL hypothesis not only runs continuously alongside and is tightly interwoven with the more general thread of "dialectical" reasoning that connects the Romantics and Idealists with their anti-theological successors, but it appears to have played a crucial role in the particular ways in which this dialectical reasoning was developed and deployed. For example, the ETL hypothesis, like the broader plurality of worlds debate, was still very much in the air, culturally and intellectually, for the "Young Hegelians"—the subsequent generation of thinkers who radicalized various elements of Hegel's dialectical system. This radical dialecticism was formulated in both *political* terms—as exemplified most famously, of course, in the work of Karl Marx—as well as in *religious* terms—as exemplified most influentially in the work of Ludwig Feuerbach, one of the most important progenitors of contemporary humanism and the pre-Freudian author of the psychological "projection" thesis regarding belief in God. And for Feuerbach, in particular, the ETL hypothesis was a matter of both sustained intellectual fixation and genuine existential concern over the entirety of his career.

In his earlier theological work, before he had completed his anti-theological turn, Feuerbach had concurred with Hegel and Schelling's dismissal of the ETL hypothesis. Notably, Feuerbach came to the attention of the European intellectual public—and of disapproving church and government authorities in Germany—with an (initially anonymous) publication, in which he argued against the traditional Christian doctrine of personal immortality (i.e., the claim of everlasting life after death for the individual human "self" or "subject," *qua* "immortal soul"). While the overarching focus of his argument in this work is the question of whether the human soul—understood here as the irreducible ontological locus of the sentience, subjectivity, and selfhood of the individual human being—survives the death of the body, Feuerbach spends large portions of the text engaging with and rebutting the ETL hypothesis. Sounding much like Steffens on the cosmic pride of place and singular purpose of Earth/humanity,

Feuerbach writes, "it is absolutely certain that, in all of creation, there exists but one animated and ensouled point, and . . . this point is the earth, which is the soul and purpose of the great cosmos."[48] Continuing with the Romantic-Idealistic inclination toward humanistic egoism—and exhibiting a sort of mash-up of the categories of traditional *spiritual beings* (e.g., angels and transmigrated souls) and *extraterrestrial beings* that, as we have seen, was not uncommon in theological treatments of the ETL hypothesis throughout the 17th, 18th, and 19th centuries—Feuerbach asserts that it would be "superfluous" for there to be life on other planets, claiming that, "humanity itself is the ultimate of all individual beings . . . Reason, will, freedom, science, art, and religion are the only guardian angels of humanity, are the only actually higher and more perfect beings. Infinite, everlasting life exists in these alone, but not on Saturn or Uranus or anywhere else."[49]

Yet, amidst all this terracentric and anthropocentric chest-thumping, Feuerbach also reveals a sensitivity to what he takes to be the *inverse existential import* of the possible truth of the ETL hypothesis. He comments, for example, that if there were in fact other forms of life—especially intelligent life—elsewhere in the cosmos, then *humanity would be superfluous*, and, thus, *human existence would be meaningless*.[50] Much like his Renaissance Lutheran predecessor, Melanchthon, at this early point in his reflection on the ETL hypothesis, the Romantically inclined Feuerbach did *not* view the possibility of a cosmic community of intelligent beings as a source of existential comfort ("We are not alone!"). *Nor* did he view it as a potential affirmation of the greatness of the divine ("That they, too, may know and love God!"). Instead, his attitude is that of a jealous child confronting the possibility that his parents might add to the family ("I should be enough for you!").

In spite of Feuerbach's systematic and passionate rejection of the ETL hypothesis in his earlier theological work, however, throughout his later, epoch-making, anti-theological writings, he nevertheless finds it worthwhile to explicitly reference the possibility of extraterrestrial forms of intelligence, in order to clarify and support both his projectionist thesis regarding religious consciousness and his continuing valorization of humanity as representing the "ultimate" form of intelligence in the universe. In his *Principles of Philosophy of the Future*, for example, he offers the following thought experiment:

Imagine to yourself a thinking being on a planet or a comet seeing a few paragraphs of Christian dogmatics dealing with the nature of God. What would this being conclude from these paragraphs? Perhaps the existence of a god in the sense of Christian dogmatics? No! It would infer only that there are thinking beings also on earth; it would find in the definitions of the earth inhabitants regarding their god only definitions of their own nature. For example, in the definition "God is spirit" it would find only the proof and expression of their own spirit.[51]

By the time we reach his most famous and important work, *The Essence of Christianity*, Feuerbach seems to have rethought his opposition to the ETL hypothesis almost entirely. While not quite going so far as to affirmatively endorse the existence of ETL here, he offers this very different take on the possibility (and sounds quite like an atheistic Huygens):

> There may certainly be thinking beings besides men on other planets of our solar system. But by the supposition of such beings we do not change our standing point—we extend our conceptions *quantitatively* not *qualitatively*. For as surely as on the other planets there are the same laws of motion, so surely there are the same laws of perception and thought as here. In fact, we people the other planets, not that we may place there different beings from ourselves, but more beings of our own or a similar nature.[52]

Thus, having started from the position in his early work that other intelligent life in the cosmos beyond Earth would represent a definitive *negation* of the existential significance of humanity, Feuerbach ends up deputizing our hypothetical cosmic neighbors into his argument for the universality of human cognitive categories and experiential forms. The aliens' *presumed sameness* becomes the basis for a *petitio principii* affirmation of epistemic transparency between human experience and reality per se. Thus, according to Feuerbach, we human beings can continue confidently to treat ourselves as the measure of all things, secure in the knowledge that our cognitive metrics are the product of universal forms and structures of reality and experience that are invariant throughout the cosmos.

A generation later, the leader of the international anarchist movement, the Russian philosopher and peripatetic revolutionary Mikhail Bakunin, drew on Feuerbach's psychological projectionist understanding of theology in his wide-ranging, immensely provocative—and, at many points, quite brilliant—argument for the co-implicated logic of anarchism and atheism, in *God and the State*.[53] Here, Bakunin excoriates his *theo-philosophical* predecessors, "the whole melancholy and sentimental company of poor and pallid minds who . . . established the romantic school in Germany, the Schlegels, the Tiecks, the Novalises, the Werners, the Schellings." According to Bakunin, these abstraction-obsessed theologians and philosophical theists:

> Very far from pursuing the natural order from the lower to the higher... from the relatively simple to the more complex; instead of wisely and rationally accompanying the progressive and real movement from the world called inorganic

to the world organic, vegetables, animal, and then distinctively human—from chemical matter or chemical being to living matter or living being, and from living being to thinking being—the idealists, obsessed and blinded . . . by the divine phantom which they have inherited from theology, take precisely the opposite course.[54]

According to Bakunin, "The gradual development of the material world, as well as of organic animal life and of the historically progressive intelligence of man . . . is perfectly conceivable. It is the wholly natural movement from the simple to the complex . . . a movement in conformity with all *our daily experiences*, and *consequently in conformity also with our natural logic*."[55] This is, obviously, an appeal to empirical reason (to the tribunal of "daily experiences"), over against the ungrounded a priori speculations of thinkers like Hegel, Schelling, and even Feuerbach. But it is more than that. Bakunin's understanding of what he calls "our natural logic" is framed in relation to "the *distinctive laws of our mind,* which being formed and developed only by the aid of [natural] experiences; [are], so to speak, but the mental, cerebral reproduction . . . thereof."[56] In other words, human cognition, being a natural product of natural causal processes, necessarily embodies, manifests, and reflects the underlying structures and modalities of nature, within or through which these causal processes take place.

So we have here a seeming repetition of Feuerbach's anti-Kantian claim that human experience and thought derive their form and character directly from nature itself and, therefore, reflect the natural order as it is, *an sich.* However, this seemingly neo-Idealistic claim and its implications are importantly qualified in the context of Bakunin's far more empirical and pragmatic mode of reasoning.

First, as Bakunin clearly recognizes, if the existence of humanity is not, as the Romantics and Idealists would have it, the miraculous (or at least singular) manifestation of a divinely guided evolutionary *telos* but rather the *accidentally inevitable* product of material processes driving a "wholly natural movement from the simple to the complex," then it follows that we have no reason to presume our own uniqueness. Quite the contrary, if the existence here on Earth of "life" and "intelligence" are not the result of miraculous exceptions to the universal rules of material causation but instead natural outcomes that are permitted, if not encouraged, by those rules, then it is to be expected that, in an infinite (or practically infinite) material universe, where there are likely to be many locations with similar initial system-conditions to those that characterized the early Earth, similar outcomes will have sometimes occurred. (We hear once again the echo of the Atomists—time and chance *produce* us all.)

Thus, writing just over a decade after the publication of Darwin's *Origin of Species* and reflecting an attention to both its thesis and the ETL hypothesis, Bakunin observes:

> All branches of modern science, of true and disinterested science, concur in proclaiming this grand truth, fundamental and decisive: The social world, properly speaking, the human world—in short, humanity—is nothing other than the last and supreme development—*at least on our planet and as far as we know*—the highest manifestation of animality.[57]

Along with the striking ETL-oriented caveat about humanity's supremacy embedded in this statement, Bakunin also shows a recognition of the *relativity and incompleteness* of human understanding in his discussion of the nature and aim of "science." Channeling the positivism of Auguste Comte, he writes:

> I mean by the words "absolute science," the truly universal science which would reproduce ideally, to its fullest extent and in all its infinite detail, the universe, the system of co-ordination of all the natural laws manifested by the incessant development of the world. *It is evident that such a science, the sublime object of all the efforts of the human mind, will never be fully and absolutely realized.*[58]

Though the point is never quite made fully explicit, the combination of the various principles and intuitions expressed in this work suggests that Bakunin did not assume the absolute universality of the *particular features* of the natural system as *manifested locally* on Earth, *nor, therefore, the universality or absolute applicability of the particularities of the cognitive system of humanity, as being the mental apotheosis of this particular Earthly system of material structures and dynamics.* Or, at the very least, we can say that his system does not in any way require the assumption of such universality in the way that Hegel's, Feuerbach's, and others' require it. Furthermore, there are conceptual resources internal to Bakunin's thought available to rebut both the truth and the importance of such an assumption.

It seems likely, therefore, that Bakunin believed—or, at least, would have affirmed, if prompted—that (a) insofar as "nature," in the sense of the universe at large, is a coherent totality, any particular localized component of that totality, such as Earth/humanity, will necessarily manifest *some* foundational characteristics that pervade everywhere throughout the universe,[59] but also that (b) beyond these most basic shared characteristics, there might be a wide diversity of particular characteristics associated with diverse environments (and, therefore, associated also with the organisms that evolve therein), a diversity substantive enough in its import to undercut the sort of thoroughgoing assumptions of conceptual universalism and human cognitive supra-adequacy found in the work of thinkers like Hegel, Schelling, and Feuerbach. Indeed, it is tempting to read this as an intended implication of Bakunin's use of the adjective "distinctive" in describing the "laws" governing the operations of human minds.

Thus, Bakunin seems to return to, or at least to be moving in the direction of, something like the *relativized cognitive naturalism* that Kant had arrived at in *Anthropology from a Pragmatic Point of View*.

and which the Romantics and Idealists had overturned in favor of their own return to metaphysically transparent (in Kant's terms, "transcendental realistic") conceptions of the character and form of human cognition. Accordingly, the historical debates that led from the Kantian school to the Romantic and Idealist schools, and then the various reformulations and critiques of the Romantics and Idealists that followed this transition, by thinkers like Feuerbach and Bakunin, represent very real and important precursors to contemporary debates in astrobiology, metaphysics, and philosophy of mind about the likelihood of epistemic compatibility and cognitive commensurability among intelligent species across the cosmos.

We also see, once again, in Bakunin's work, that the logic of the Atomists remains ever in play, precisely because their original set of propositions outlining the possibility of an accidental-mechanistic cosmogony of material plenitude represents the most compelling and adaptable/updateable historic alternative to teleological theistic cosmogonies (today's version par excellence being inflationary quantum mechanics/multiversalism).

Moreover, *at the very same time* that Bakunin was writing *God and the State*—in the months just before the Paris Commune of 1871—his comrade in the international revolutionary workers movement, the tremendously influential French republican-socialist, Louis Auguste Blanqui, while imprisoned at the island fortress Château du Taureau awaiting "transportation" to one of France's hellish tropical island penal colonies, took time away from his usual activist and journalistic writing to pen *L'Éternité par les Astres: Hypothèse Astronomique*.[60] In this strange and remarkable work, Blanqui essentially combines the *cosmogenic logic* of classical Atomism with the *cosmological framework* of the Laplacian "nebular hypothesis," filtered through a deeply informed, yet utterly idiosyncratic, reading of the new elemental chemistry to construct a novel—and bizarre—take on the doctrine of "eternal recurrence."

Were it a work of fiction, rather than a seemingly earnest metaphysical speculation, *L'Éternité par les Astres* might easily be mistaken for a Jorge Luis Borges story.[61] Blanqui begins by setting the cosmological scene and asserting his idiosyncratic first principles:

The universe is wholly composed of stellar systems. To create them nature has only one hundred *basic elements*[62] at its disposal. Despite the creative wealth it is able to draw from these resources and the incalculable number of combinations they allow its fecundity the result is necessarily a *finite* number, as with the elements themselves, and in order to fill its entire expanse nature must infinitely repeat each of its original or general combinations.[63]

Presumably, Blanqui's choice of "one hundred" as the supposed final number of elements that would ultimately be found to exhaustively constitute the building-blocks of nature rested on some sort of neoclassical assumption of symmetry and mathematical order in nature. (Earlier in the text, he indicates an awareness of the two widely recognized versions of the elemental or periodic table at that time, those of Lothar Meyer[64] and Dmitri Mendeleev[65], which listed fifty-five and sixty-six elements, respectively.) Blanqui continues:

> Every star, whatever it is.[66] exists therefore in infinite number in time and in space, not only in one of its aspects, but as it is in each and every second of its duration, from birth until death. All the beings spread across its surface, big or small, living or inanimate, partake in this privilege of perennity. The earth is one of these stars. Every human being is therefore eternal in every second of his existence. That which I now write in a cell in the fort of Taureau, I have written and I will write under exactly the same circumstances throughout eternity, on a table like this, with a pen like this, wearing clothing like this. And so for everyone.[67]

Here the ETL hypothesis takes on the most radically anthropocentric and anthropomorphic form possible. Blanqui turns the *metaphorical* self-projection described in Feuerbach's psychological explanation of the ETL hypothesis into a *literal* doctrine of self-replication, projecting an *infinitude of Blanquis* (and of all the rest of us) across the universe. We have met the aliens, and they are us: a starscape of human narcissism.

Yet, before holding Blanqui up for special ridicule on this account, it is important to recognize, first, that the picture of the cosmos that he articulates here is, in some ways, simply a direct implication of the positions taken before him by many others, including that great scientist Huygens, if taken to their full logical conclusions. If we imagine, as Huygens did, that the human form is, if not *imago dei* at the very least *imago universalis*, then surely in a *materially infinite* universe filled with *infinite inhabited worlds*, there must be repetition of all possible combinations of all elements of the human form and, thus, of specific "personal" instantiations of that form and, indeed—if we are really taking the concept of an actual infinity seriously—these repetitions must extend even into the fullest details of entire lived lives, entire worlds of specific lived lives, and so on. This is all of which is to say, among other things, that those today who believe in the inflationary quantum mechanical theory of infinite multiversal and dimensional branching of possibilities were not the first to dream this particular dream.

> Those curious about extraterrestrial life may smile at a mathematical conclusion that bestows upon them not only immortality but eternity. The number of our doubles is infinite in time and in space. . . . [I]s it not a consolation to know ourselves to be constantly, on millions of worlds, in the company of loved ones, who are for us here today only a memory? . . .
> At this very moment, the entire life of our planet, from birth until death, is being replicated, day by day, on a myriad of brother stars. . . . Eternity imperturbably replays the same spectacle, *ad infinitum*.[68]

And, second, we must recognize how this Blanquian position was *carried forward*, immediately and directly, in an only very slightly modified form, by a major thinker who is still taken quite seriously in the philosophical literature today. *L'Éternité par les Astres* was published in 1872, precisely ten years before Friedrich Nietzsche would first mention the idea of "eternal recurrence," with which he is now so closely associated, in *The Gay Science*. And there is good reason to believe that Nietzsche was directly influenced by Blanqui's formulation of the doctrine when he wrote this work. Friedrich Albert Lange had discussed Blanqui's *L'Éternité* in his revised and expanded 1872–1873 edition of his *History of Materialism and Critique of Its Present Importance*,[69] a work that Nietzsche studied carefully (and which also included substantial and laudatory discussion of the Greek Atomists). Of course, Nietzsche's much more famous formulation of the doctrine of recurrence focuses exclusively on the *temporal* dimension of the everlasting "cosmic process" by which we are endlessly duplicated and reduplicated. He writes:

Whoever thou mayest be, beloved stranger, whom I meet here for the first time, avail thyself of this happy hour and of the stillness around us, and above us, and let me tell thee something of the thought which has suddenly risen before me like a star which would fain shed down its rays upon thee and every one, as befits the nature of light . . . Fellow man! Your whole life, like a sandglass, will always be reversed and will ever run out again, —a long minute of time will elapse until all those conditions out of which you were evolved return in the wheel of the cosmic process.[70]

But the echo of Blanqui seems quite distinct in both the logic and melancholic character of Nietzsche's formulation. Indeed, even Nietzsche's choice of metaphor here—"the thought which has suddenly risen before me like a star"—could be taken as a subtle nod toward Blanqui's *L'Éternité*.

So, as can be seen here, the ETL hypothesis is a genuinely important thread that runs through, and sometimes even plays a determinative role in, the development of and relationships between some of the most important philosophical, theological, and anti-theological perspectives on human existence and understanding at-play during the 19th century—a mostly invisible thread, in terms of the standard contemporary treatments of these perspectives.

## Postmodern Revival

In our own time, we see a proliferation of new forms of ETL-related theological and quasi-theological imaginings. In the academic study of religion, we now recognize an entire category of "UFO religions," which includes groups like the Unarians, the Raëlians, the Solarians, and, most significantly, the Scientologists.[71] In addition, the so-called simulation theory of human existence, derived from the work of the philosopher Nick Bostrom and popularized by the industrialist Elon Musk, should be also understood as representing, among other things, a quasi-religious notion of ETL.[72] According to the theory, super-advanced intelligent beings have constructed our entire reality, including our planet and its history, in the equivalent of a digital simulation environment—making our programmers (that is, our *creators*), by definition, extraterrestrial. And in a fascinating and revealing irony, this 21st-century, science and technology based update of the ETL hypothesis actually (though, I take it, unintentionally) implies a *return to prescientific supernaturalism*. For, precisely insofar as our hypothesized simulation creators—our digital gods— have constructed the rules of our simulated reality, presumably they have the power to suspend, alter, or violate these rules. At any point, we must presume, on this supposedly plausible scenario, the program may be rewritten, updated, deleted—*silicon supernaturalism*.

At the same time, Christian thinkers are once again grappling with the ETL hypothesis, now with the fresh urgency of concrete possibility, as the field of astrobiology moves into mainstream science, and the search for ETL looks more and more like one that will inevitably yield actual results. So, for example, the Vatican funds extensive and ongoing inquiry into both the science and the theological and philosophical implications of astrobiology. And the Pope's own astronomer, Brother Guy Consolmagno, S.J., has published a book titled, *Would You Baptize an Extraterrestrial?*[73]

In short, the ETL hypothesis is alive and well and continues to provoke fundamental and profound questions about human nature, our collective identity and epistemic situation, and the significance (or insignificance) of our existence, just as it has since the ancient Atomists raised the question two and a half millennia ago.

# Notes

1. Not to be confused with the contemporary "Extraterrestrial Hypothesis" (ETH)—terminology that generally refers to the belief that the best abductive explanation available for at least some unidentified flying objects (UFOs) is the existence of extraterrestrial space craft visiting Earth. I am *not* addressing this UFO-oriented hypothesis in this chapter.

2. H. G. Wells, *The War of the Worlds* (New York: Harper and Brothers Publishers, 1898).

3. The lion's share of the existing scholarship on this topic has been carried out over the last three and a half decades by two researchers, the University of Notre Dame–based philosopher and intellectual historian Michael Crowe, who has focused on the modern period, and former historian of NASA and Baruch S. Blumberg NASA/Library of Congress Chair in Astrobiology, Steven J. Dick, who has focused on the ancient to early modern period. (See notes below for specific citations.)

4. By "para-Christian," I mean theological, quasi-theological, and/or philosophical perspectives that emerged as offshoots from or reactions to, or as otherwise fundamentally shaped by conceptual and cultural proximity to, Christianity and which, therefore, share basic anthropological, metaphysical, and/or moral principles and assumptions with Christianity—for example, 17th-century English deism, 19th-century German philosophical idealism.

5. As the great 20th-century theologian and chemist Arthur Peacocke has put it in our time: "Christians have to ask themselves (and sceptics will certainly ask them), What can the cosmic significance possibly be of the localized, terrestrial event of the existence of the historical Jesus? Does not the mere possibility of extraterrestrial life render nonsensical all the superlative claims made by the Christian church about his significance? Would ET, Alpha-Arcturians, Martians, et al., need an incarnation and all it is supposed to accomplish, as much as *homo sapiens* on planet Earth? Only a contemporary theology that can cope convincingly with such questions can hope to be credible today"; "The Challenge and Stimulus of the Epic of Evolution to Theology," in *Many Worlds: The New Universe, Extraterrestrial Life, and the Theological Implications*, ed. Steve Dick (Philadelphia: Templeton Foundation Press, 2000), 103. See, also Ernan McMullen's "Life and Intelligence Far from Earth: Formulating Theological Issues," in the same collection (*Many Worlds*, 151–175).

6. Diogenes Laertius, *Lives of the Eminent Philosophers*, trans. Pamela Mensch, ed. James Miller (Oxford: Oxford University Press, 2018), 450.

7. Ibid., 457.

8. Ibid., 509.

9. Lucretius, *De Rerum Natura (On the Nature of Things)*, trans. W. H. D. Rowe (Cambridge, MA: Loeb Classical Library, 1924), Bk 2, Lines 1052–1066.

10. See Dick, *Plurality*, 44.

11. Christianity—unlike (pre-Hellenized) Judaism—framed the creation of the world by God as occurring in a metaphysical vacuum. Rather than shaping order out of a primordial potentiality, God, according to classic Christian doctrine, brought both the substance and order of the cosmos into being, by a simple, transcendent, supernatural

*fiat*, thus, *creating something out of nothing* (and overcoming the classic Greek naturalistic principle *ex nihilo nihil fit*). The implication, in relation to the framing of the traditional Abrahamic theological notion of God's "omnipotence," is that divine power is *absolute* in every meaningful sense; that the rules of nature, indeed, even the metaphysical characteristics and structures of reality, per se, are the product of a *divine command*; and, therefore, the metaphysical and natural orders are subject to divine interruption, supersession, or negation.

12. See Dick, *Plurality*, 28, 36.

13. Ibid., 30–37.

14. Michel de Montaigne, *Essays*, in *The Complete Works of Montaigne*, trans. Donald Frame (Palo Alto, CA: Stanford University Press, 1958).

15. Pierre Gassendi, *Animadversiones in decim librum Diogenes Laertii, qui est de vita, moribus, placitisque Epicuri* (Lyon: Lugduni, 1649).

16. John Donne, *The Complete English Poems*, ed. C. A. Patrides (New York: Everyman's Library, 1991).

17. Leonrado da Vinci, *The Notebooks of Leonardo da Vinci*, ed. Jean Paul Richter (New York: Dover Books, 1970).

18. *Of Learned Ignorance*, trans. Germain Heron (New Haven, CT: Yale University Press, 1954), 114–115.

19. Montaigne, *Complete Works*, 390.

20. It is also worth noting that much contemporary scholarship suggests that Bruno's sentence had little if anything to do with his advocacy for Copernican-style cosmology—his rejection of the Church's doctrines regarding Jesus's divinity and "co-eternality" with God, the Trinity, and Hell were more than sufficient reason for the Inquisition to condemn him.

21. See, for example, Giordano Bruno, *The Ash Wednesday Supper: A New Translation*, trans. Hilary Gatti (Toronto: University of Toronto Press, 2018), 123.

22. See, for example: Jean Calvin, *Institutes of the Christian Religion*, trans. Ford Lewis Battles, ed. John T. McNeill (Louisville: Westminster John Knox Press, 1960), 250–254 and *passim*.

23. Translated quote from Dick, *Plurality*, 88–89.

24. See Dick, *Plurality*, 86.

25. Ibid., 73.

26. Descartes was, of course, a "dualist," affirming the ontological reality of both "matter" and "mind" or "soul," but precisely because his metaphysics was reductionistic with regard to the physical universe—as distinct from the transcendent, eternal realities of God and the human soul—he was a pure materialist with regard to physical explanations, excepting only the first causal activity of the divine in imbuing initial motion into the material continuum of the primal physical universe.

27. Bernard le Bovier de Fontenelle, *Conversations on the Plurality of Worlds*, trans. H. A. Hargreaves (Berkley: University of California Press, 1990).

28. Ibid., 63.

29. Ibid., 49.

30. Carl Linnæus, *Systema naturæ, Editio decima* (Stockhom: Salvius, 1758).

31. Fontenelle, *Conversations*, 6.

32. Ibid., 32.

33. Karl Guthke, *The Last Frontier: Imagining Other Worlds from the Copernican Revolution to Modern Science Fiction*, trans. Helen Atkins (Ithaca, NY: Cornell University Press, 1990), 239–240.

34. Ibid., 240.

35. William Derham, *Astro-theology, Or a Demonstration of the Being and Attributes of God from a Survey of the Heavens—Corrected Fourth Edition* (London: W. J. Innys, 1721), xlviii–xlix.

36. Guthke provides insightful discussions of this particular genre of ETL writing throughout *Last Frontier*.

37. Emanuel Swedenborg, *Earths in Our Solar System which are Called Planets and Earths in the Starry Heaven, Their Inhabitants, and the Spirits and Angels There from Things Heard and Seen* (London: Morrison & Gibb, Ltd., 1962), 1.

38. Immanuel Kant, *Universal Natural History and Theory of the Heavens*, trans. Olaf Reinhardt, in *Natural Science*, ed. Eric Watkins (Cambridge: Cambridge University Press, 2012), 297.

39. Immanuel Kant, *Anthropology from a Pragmatic Point of View*, trans. Robert Louden (Cambridge: Cambridge University Press, 2006), 225.

40. Thomas Paine, *The Complete Writings of Thomas Paine*, ed. Philip S. Foner (New York: Citadel Press, 1945), 1:504.

41. Benjamin Franklin, "Articles of Belief and Acts of Religion," in *Works of Benjamin Franklin* (Boston: Hilliard, Gray, and Company, 1840), 2:1–3.

42. See Crowe, *Sourcebook*, 207.

43. Ralph Waldo Emerson, *The Journals and Miscellaneous Notebooks of Ralph Waldo Emerson, Vol. 4, 1832-43*, ed. Alfred R. Ferguson (Cambridge, MA: Harvard University Press, 1964), 177.

44. Mark Twain, *The Love Letters of Mark Twain*, ed. Dixon Wechter (New York: Harper and Brothers, 1949), 133.

45. *Hegel's Philosophy of Nature: Being Part Two of the Encyclopedia of the Philosophical Sciences*, trans. and ed. A. V Miller and J. N. Findlay (Oxford: Oxford University Press, 1970), 62 (emphasis added).

46. *Ideas for a Philosophy of Nature: as Introduction to the Study of this Science*, trans. E. E. Harris and P. Heath (Cambridge: Cambridge University Press, 1988).

47. Henrich Steffens, *Christliche Religionsphilosophie* (Breslau, 1839), 205–206, translated and quoted in Crowe, *Debate*, 260.

48. Ludwig Feuerbach, *Thoughts on Death and Immortality*, trans. James A. Massey (Berkley: University of California Press, 1980), 62.

49. Ibid., 71.

50. Ibid., 60.

51. Ludwig Feuerbach, *Principles of the Philosophy of the Future*, trans. Manfred Vogel (Indianapolis: Hackett, 1986), 9–10.

52. Ludwig Feuerbach, *The Essence of Christianity*, trans. George Elliot (New York: Prometheus Books, 1989), 11.

53. Mikhail Bakunin, *God and the State*, trans. Benjamin R. Tucker (New York: Dover Books, 1970).

54. Ibid., 14.

55. Ibid., 13.

56. Ibid., 14 (emphasis added).

57. Ibid., 9 (emphasis added).

58. Ibid., 34 (emphasis added).

59. If nothing were universal across reality, in what sense would reality be a coherent "totality"?

60. Louis August Blanqui, *L'Éternité par les Astres: Hypothèse Astronomique* (Paris: Librarie Germer Baillière, 1872).

61. And, in fact, Borges not only knew Blanqui's *L'Éternité*, he acknowledged its influence on his writing.

62. Blanqui speaks in an Atomistic idiom here, using the term "corps simples" (literally, "simple bodies"). But the idea at work is very clearly—both from the proximate context and the fuller text—his understanding of the newly periodicized chemical elements (see later discussion). Despite having somehow found time to absorb something of the cutting-edge scientific discussions around the construction of the periodic table of elements underway at that time, Blanqui was still thinking of them, as the Atomists thought of their simple material atoms, as representing something like *shapes* or *forms* that combine in very mechanical ways.

63. Blanqui, *L'Éternité, Résumé, Passage 1* (my translation).

64. Lothar Meyer, *Die Modernen Theorin der Chemie und ihr Bedeutung für die chemische Statik* (Breslau: Maruschke & Berendt, 1864).

65. Dmitri Mendeleev, "Ueber die Beziehungen der Eigenschaften zu den Atomgewichten der Elemente," in *Zeitschrift für Chemie*, 12, 405–406.

66. At this time, it remained common for the term "star" to be used in a way neutral to the distinction between what we now term "stars" and "planets." Indeed, real controversies still existed about these classifications even among astronomers during Blanqui's life, controversies that would be finally settled only by more powerful telescopes.

67. Blanqui, *L'Éternité, Résumé, Passages 2–3* (my translation).

68. Ibid., *Passages 5–12*.

69. Friedrich Albert Lange, *Geschichte des Materialismus und Kritik seiner Bedeutung in der Gegenwart* (Iserlohn: J. Baedeker, 1873).

70. Friedrich Nietzsche, *The Complete Works of Friedrich Nietzsche*, ed. Oscar Levy (Edinburgh and London: T. N. Foulis, 1909–1913), Vol. 16, "Eternal Recurrence," Passages 24–25.

71. See, for example: "UFO Religions," John A. Saliba, in *Encyclopedia of Religion*, ed. Lindsay Jones. Vol. 14. 2nd ed. (Detroit, MI: Macmillan Reference USA, 2005), 9432–9436.

72. See Elon Musk, "Are We in a Simulation" (https://www.youtube.com/watch?v=xBKRuI2zHp0); Mike Wall, "We're Probably Living in a Simulation, Elon Musk Says," Space.com (https://www.space.com/41749-elon-musk-living-in-simulation-rogan-podcast.html); Olivia Solon, "Is Our World a Simulation? Why Some

Scientists Say It's More Likely Than Not," *The Guardian*, October 11, 2016 (https://www.theguardian.com/technology/2016/oct/11/simulated-world-elon-musk-the-matrix).

73. Guy Consolmagno and Paul Mueller, *Would You Baptize an Extraterrestrial?. . . and Other Questions from the Astronomers' In-box at the Vatican Observatory* (New York: Penguin Random House, 2018).

# PART II

# WHAT IS LIFE?

# 4

# Three Lives and Astrobiology

*Lucas Mix*

## Introduction

Astrobiologists study the origin, extent, and future of life in the universe.[1] The *Oxford English Dictionary* and other references label it a branch of biology, while the *American Heritage Dictionary* provides a broader but still natural science focused definition: "The scientific study of the possible origin, distribution, evolution, and future of life in the universe, including that on Earth, using a combination of methods from biology, chemistry, and astronomy." Nonetheless, increasing commentary on humanities, social sciences, and social implications in astrobiology suggest that it pushes the boundaries of natural science (Bertka, 2009; Ćirković, 2012; Impey et al., 2013; Vakoch, 2013; Mix, 2018b; Vainio, 2018).

The word "life" has numerous meanings, making it valuable to ask about the scope of astrobiology, both in terms of subject matter and methodology. I distinguish three basic life-concepts (Table 4.1). They correspond to traditional categories of vegetables, animals, and rational animals (Mix, 2018a). Each addresses a different "life" or aspect of "life" present in common, historical, and philosophical usage. Biological life ($life_1$) shares traits with all cellular life on Earth (archaea, eubacteria, and eukarya). Internal or conscious life ($life_2$) shares subjective interiority with humans. Rational life ($life_3$) shares intellect with all minds that can distinguish truth from non-truth.

## Why We Need Three Lives

As with many other topics in astrobiology, the same words have different meanings in different contexts and some translation will be necessary.[2] When working across disciplines, it can be easy to miss linguistic subtleties. For example, the term "organic chemistry" is a verbal fossil. Prior to the 19th century, many believed that life was the only source of complex carbon compounds (Mix 2009, p. 27; *OED*). The terms organic (meaning biological) and organic (meaning with carbon-carbon bonds) are equivocal; they have different meanings, despite having a common origin. Linguists would call them cognates, biologists

Lucas Mix, *Three Lives and Astrobiology*. In: *Social and Conceptual Issues in Astrobiology*. Edited by: Kelly C. Smith and Carlos Mariscal, Oxford University Press (2020). © Oxford University Press. DOI: 10.1093/oso/9780190915650.001.0004

**Table 4.1.** Three Lives

---

Life$_1$—Biological Life
> *sharing "life" with all cellular Earth life*
> *e.g., organisms such as bacteria, ferns, and humans*
> *problem cases: viruses, self-replicating programs (artificial life?)*
> *traditional category: vegetable life (or vegetables plus animals if they are distinct)*

Life$_2$—Internal Life
> *sharing interior "life" with all conscious beings*
> *e.g., humans and other subjects with will or sentience*
> *problem cases: trees, fish, machine learning algorithms (artificial consciousness?)*
> *traditional category: animal life*

Life$_3$—Rational Life
> *sharing rational "life" with all rational beings*
> *e.g., minds capable of distinguishing truth from non-truth*
> *problem cases: human newborns, humans with dementia, artificial general*
> *    intelligence*
> *traditional category: rational life*

---

homologs. The various versions of life face similar issues. Astrobiologists in all fields can benefit from a better understanding of what the general public—and various scholars—are looking for, when they are looking for "life."

When giving public talks on astrobiology, I have been surprised by how many people connect the search for life with the search for intelligence. Astrobiology as natural science has focused on life$_1$. The Search for Extraterrestrial Intelligence (SETI) has focused on life$_3$. Meanwhile, most ethical systems depend, critically, on ideas about will and suffering connected to life$_2$.

The three lives possess different origins, extents, and futures. They are generally, though not universally, understood to be stacked. Life$_1$ arises in a non-living universe. Life$_2$ arises in the context of life$_1$, and life$_3$ in the context of life$_2$. Most ethicists have attributed more value and agency to living things at the "higher levels." The typology helps to clarify what we are looking for. It reveals how traditional ideas sneak into modern debates. It also identifies three "hard problems of life" relating to the origin and extent of biological organization, consciousness, and reason. The Drake equation, the Fermi paradox, and the anthropic principle provide concrete examples in astrobiology.

## Life$_1$—Biological or Organismic Life

From before the time of Plato, Western scholars recognized a category of things that behave differently (Claus, 1981). They commonly associated living things with causal and structural features. Causally, their behavior showed apparent purpose and freedom that could not be explained using basic physical rules. Structurally, their parts appeared to work together; hence, they were "organisms." Ancient philosophers debated about borderline cases (i.e., magnets, stars, planets, angels), but most thought along still familiar lines. Following Plato and Aristotle, they considered all of them ensouled and would have used a term like "psychology" for the study of all living things, what we now call biology (Goetz and Taliaferro, 2011). This usage persisted through the Middle Ages, and, despite terminological debates, scholars thought of living things as having something special to explain their motion and structure.[3]

Modern conceptions of $life_1$ recognize the same basic distinctions. Since the early 20th century, most have focused on how evolution and/or metabolism explain both causality and structure (Luisi, 1998; Trifinov, 2011; Mix, 2015). We may be unable to objectively differentiate living things from their products in this regard (Mix, 2014). Nonetheless, the basic category remains intuitive despite long debate (e.g., Cleland, 2012; Shields, 2012). How should it be bounded? Can definitions can be natural, or even useful? Bacteria, trees, dogs, and so on share features that make them a meaningful group, worthy of a common discipline (i.e., biology), and sufficiently interesting to look for beyond Earth.

I call $life_1$ a constellation of meaning or life-concept. It requires a basic hypothesis with three parts.

1. A distinction can be made between known living things (i.e., archaea, eubacteria, and eukaryotes) and other known things (e.g., rocks and clouds).[4]
2. That distinction aids in understanding known living things.
3. There exist unknown things best understood as living.

If the hypothesis is true, then the set of living things, known and unknown, constitutes $life_1$. The hypothesis may prove false. It is, however, necessary for astrobiology. A search for life requires some concept of life that shapes methodology and defines success.

Almost all biological theories or definitions of life (e.g., replicators, regulators) require $life_1$. They attempt to identify a distinction that makes the hypothesis true. Astrobiology does not require either a correct definition or a consensus definition of life in this narrow sense. It does require a meaningful target for the search.

## Origin and Extent of $Life_1$

Astrobiologists associate the origin of $life_1$ with the origin of evolving populations (Hays et al., 2015). They take for granted the existence of a pre-biotic universe, such that the historical origin of $life_1$ is a meaningful question. Once, there were no organisms; now there are: how did they arise? The origin of individual organisms becomes a less interesting question, largely because the transition from one generation to the next is so well understood, mechanically. Organisms fuse and divide. We see debate around the proper definition of biological individuality (e.g., Godfrey-Smith, 2009, pp. 69–145; Clarke, 2011) but, for any given definition (currently popular), the origin of individuals is straightforward. The origin of the first evolving population remains an active area of research.

Life$_1$, including both organisms and their products, extends unambiguously throughout the atmosphere and upper crust of Earth. We have evidence for life$_1$-like chemistry on Mars but, as yet, no clear evidence of organisms, past or present.

I will not argue that biologists have a privileged place with regard to the definition of life. Life is too broad a term. Biologists have a clear idea of life$_1$ grounded in biochemistry shared by archaea, eubacteria, and eukyarotes. It matches closely with premodern concepts of vegetable life, which included all embodied living things, not just the kingdom Plantae (Mix, 2018a, pp. 213–224). Our society largely defers to biologists and their definitions with regard to the life of bacteria and other organisms, with the potential exception of animals. Even then, there seems to be consensus that animal bodies reflect life$_1$. Those who wish to defend a different understanding of biological life, if they wish to be understood, must articulate how their understanding differs from the biologists' orthodoxy. Meanwhile, two other uses of the word life should also be spelled out.

## Life$_2$—Internal, Subjective, or Conscious Life

A second life-concept arises from introspection. Human life has often been characterized by an interior locus, as both patient of sensation and agent in the world. Both require a distinction between inside and outside. The external environment shapes the internal subject while the internal subject shapes the external environment. Aristotle referred to these processes as sensation and willed motion (Aristotle, 1986, 2.6; Shields, 2007, pp. 277, 294). Those faculties defined "animals" in the Middle Ages and provided for the vegetable/animal dichotomy still in common use.

Life$_2$ should not be confused with modern animals. Linnaeus (1758, pp. 6–8) used traditional notions of sensation and will to differentiate animals from other organisms. He created a term still in use: kingdom Animalia. By the 20th century, however, biologists had redefined it as a narrow range of multicellular eukaryotes. The term *animal* suggests breath, but also *anima*, a soul or mind. I avoid the term here, as neither Aristotle's category nor the kingdom Animalia corresponds to modern views of life$_2$. Subjectivity and consciousness come closer.

Augustine introduced an internal subjective self. He started with a belief in personal existence, then asked, "If I am wrong in this, who is wrong?"[5] From this, he argued that we have an interior life distinct from the world and unknowable to others. His argument had a profound influence on Western ideas about personal identity and consciousness.[6] His "inner life" came to be seen as logically necessary for selfhood, though many considered it a black box (Matthews, 2000, p. 135). From the 17th century onward, it grew in popularity, especially when contrasted with objective sensory experience and mechanical physics (Martin and Barresi, 2006, pp. 120–122). For Augustine and Descartes, the observing self was essentially and by definition distinct from the observable universe.

Recent work in psychology and neuroscience attempts to bridge the gap, but the language of interiority remains. Attempting to explain human decision-making as brain function, Koechline and Summerfield (2007) define executive function or "executive control [as] the ability to select actions or thoughts in relation to internal goals" (p. 229). This concept of executive function was developed in the early and mid-20th century as a way of speaking about the bridge from mechanical life$_1$ to intentional or attentional life$_2$. Psychologist Roy Baumeister (2008) refers to "inner processes." Even the most mechanical of metaphors—the brain as computer—invokes some sense of input and output.

Having identified the boundary, we can easily identify such processes, but the boundary is precisely the issue at stake. David Chalmers (1995) sums it up succinctly as the "hard problem of consciousness" (p. 201). The experience of sensation ("qualia," Nagel, 1974) and chosen action ("standard agency," Hornsby, 2004) require an interior that aligns with neither the spatial interiority of physical aggregates nor the biological interiority of organisms.

Life$_2$ requires a basic hypothesis with three parts.

1. A distinction can be made between known conscious things (i.e., humans) and other known things (e.g., rocks and clouds).[7]
2. That distinction aids in understanding known conscious things.
3. There exist unknown things best understood as conscious.

The hypothesis may prove false. If it is true, then the set of conscious things, known and unknown, constitutes life$_2$. The existence of known conscious things does not preclude the possibility of known things that are not known to be conscious. Descartes provides a rare example of life$_2$ that includes humans and no other animals (Voss, 2000). Most versions include chimpanzees and other animals among the known conscious things.

Most people do not believe that unicorns exist. They, nonetheless, could tell you a great deal about unicorns. The concept has a history and a common usage, which makes the statement "unicorns do not exist" meaningful. Astrobiologists know this, because they have had to look for things that may or may not be there. They have had to turn hypotheses about Martians into concrete experiments. Consciousness, qualia, and standard agency may be illusions. Naming internal life does not reify it. To the contrary; it will be a necessary step in having real discussions about whether or not it exists.

No one discipline has a privileged place in defining $life_2$, though psychology, philosophy, and theology all have a large stake. Internal life enjoys a deep history and pervasive usage. The subjectivity of Augustine and Descartes, along with the subjective personhood of Boethius and Locke, invokes a metaphysical interiority that makes $life_2$ distinct.

One may wish to redefine words like "sensation," "will," and "consciousness." I suspect this will be necessary to advance our understanding. It will require reworking the entire Cartesian frame for thinking about thinking.[8] Those who wish to be understood, will need to articulate how their vision differs from common and historical conceptions.

## Origin and Extent of Life$_2$

$Life_2$ interiority distinguishes an individual subject from its environment and from other subjects.[9] Thus, life at large and a singular life may have different boundaries. Ethical discussions about the beginning of life (vis-à-vis abortion) and the end of life (vis-à-vis euthanasia) revolve around discrete $life_2$ individuals. $Life_1$ organisms need not have such well-defined boundaries (Mix, 2018a, pp. 181–182, 225–238). In the last century, it has been common to view consciousness as something that emerges within animals only at a certain stage in embryonic development and may fail before the body does. *Every* instance of internal life represents $life_2$ arising from non-$life_2$.

$Life_2$ may not be a helpful category. Nietzsche and others used individuation as a way to distinguish animals ($life_2$) from vegetables ($life_1$). He went on to critique the $life_2$ hypothesis. Individuation may not be a useful concept; if it is necessary for $life_2$, $life_2$ may not be a useful concept. Darwin viewed sensation and willed motion as adaptations. They arise from the interaction between $life_1$ populations and the environment (Mix, 2018a, pp. 199–205). Numerous philosophers have attempted to eliminate $life_2$, or make it an emergent property of $life_1$.

$Life_2$ may be a helpful category, even if it does not arise from $life_1$. Some suggest a continuity of consciousness with matter called panpsychism (Goff et al., 2017). The constituents of matter may have constituents of consciousness at every level (pan-protopsychism; e.g., Lucretius, *De Rerum Natura*; Mathews, 2003). Alternatively, everything in the physical universe may have its own interiority (e.g., Leibniz, *Monadology*). Or we could say that all existence participates in a larger cosmic $life_2$.

These ambiguities lead me to believe that $life_2$ will not be useful in the context of astrobiology. Nonetheless, we regularly choose to treat some entities as though they had sensitivity and agency, as though they suffer and take responsibility. For example, Peter Singer (1979, pp. 50–51) argues that we must respect beings that suffer and beings with interests. Indeed, it is hard to imagine an ethical system that does not make these distinctions. Such (at least relative) sensitivity and agency have been observed across the kingdom Animalia and even kingdom Plantae (Chamovitz, 2012; Wohlleben, 2016). Several scholars have questioned whether we should be more mindful of plants as moral patients, capable of suffering (e.g., Hall, 2011; Marder, 2014; Marder and Irigaray, 2016).

Such broad-frame questions about the categories of life make astrobiology an ideal launching point for discussions about biological ethics. In considering life-as-we-do-not-know-it, we have an opportunity to reimagine life as we encounter it on a daily basis. This openness to change helps stimulate popular support for astrobiology science and attracts a broad range of scholars from the humanities. It also makes it doubly important to understand and articulate the categories we currently use.

If astrobiologists—as natural scientists or not—are looking for $life_2$, then it is worth asking what examples of $life_2$ we wish to use as a starting point. Should we include only humans? Should we include complex plants, such as oak trees? It is also worth asking whether we expect instances of $life_2$ to be a subset of instances of $life_1$. Schneider (2015) has asked whether consciousness need be present in intelligent aliens and computers. MacLennan (2009), likewise, has asked about robot suffering.

## A Note on Human Life

Ideas about individuality in human $life_1$ and human $life_2$ overlap, but they are not the same. Human $life_1$ represents a continuum running from the first examples of *Homo sapiens* through the present day. It must include placentas and cancers as well as more extreme examples. Cellular metabolism can continue for four to ten minutes after bodily death (Vass, 2007). Tissues and cell lines can be cultured indefinitely as made famous in the HeLa cell line (Skloot, 2010). We can easily imagine human $life_2$ ending early, or at least dissociating from a related $life_1$ body in the case of brain death. Similarly, we can imagine it arriving (or associating) late in embryonic development. Conjoined twins can share a single metabolism and genome but maintain distinct consciousnesses.

Locke distinguished between humans and persons.[10] I commend this distinction. Reduction or reconciliation may be possible, but it will benefit from a clear statement of how our languages and explanations depend upon each other. More importantly, it is not clear to me that we should too quickly infer the same relationship when observing other examples of life$_1$ and life$_2$. Neither may exhibit the individuality or agency we attribute to ourselves.

## Life$_3$—Rational Life

Historically, scholars have associated life$_2$ with animals, or at least higher animals (for some value of "higher"). An additional distinction was reserved for humans. Plato and Aristotle felt that neither life$_1$ nor life$_2$ sufficed to explain reasoning. They believed that ideas were immaterial and eternal and could only be held by an immaterial mind (Martin and Barresi, 2006, pp. 19–22).[11] Sensation and preference were animal faculties. Only a mind could judge whether a sensation was accurate or a preference just. For two thousand years, their followers viewed the intellect as a veridical faculty capable of making these determinations through participation in ideal reality: the *logos* and, later, the mind of God. The intellect provided infallible, immaterial perception, distinct from the fallible, material perception of the body.

Belief in rational certainty began to erode in the high Middle Ages (Osler, 1994). Aquinas and Descartes held on to it, while Ockham, Hume, and the empiricists rejected it. Among natural scientists and much of the general public, "reason" became a term for subjectively processing data. Descartes removed the human mind from the physical world to make the physical world simple enough to understand and to provide a repository for genuine knowledge. He preserved life$_3$ and genuine knowledge by making them independent of life$_1$ and physics. This makes life$_1$ explanations of life$_3$ extremely problematic. Several scholars have suggested such explanations undermine our epistemology, making all conclusions inescapably subjective (e.g., Schrödinger, 1992, p. 119; Plantinga, 2011, pp. 309–316; Nagel, 2012, p. 75). Plantinga provides a telling quote from Darwin:

> But then with me the horrid doubt always arises whether the convictions of man's mind, which has been developed from the mind of the lower animals, are of any value or at all trustworthy. Would any one trust in the convictions of a monkey's mind, if there are any convictions in such a mind? (Darwin, 1881)

I am satisfied, from the perspective of natural science, with refined subjective data processing. Natural science can provide strong confidence. This view of "reason," however, is continuous with emotion. We possess a range of fallible heuristics, some of which are more logical and more energy intensive (Kahneman, 2011). This must be distinguished from stronger claims many still wish to make, not only in religion and philosophy but in science. As with life$_2$, life$_3$ may be illusory, but proponents of this position must make it clear that they depart from common and historical usage. If by "reason" and "knowledge" they mean the subjective beliefs of individuals, it will be important to make this clear. Likewise, claims to reason and knowledge that transcend empiricism should be transparent in how they intend to do so. Many will wish the epistemic certainty of life$_3$ but are unwilling to make the ontological commitment. The step—from consciousness to traditional intelligence—requires a label so that we can meaningfully discuss our position on it, for or against.

Standard accounts of agency as well as traditional Christian accounts of will require more than interiority; they require a mind that judges. The agent chooses, using both imagination and intention. The moral actor follows or neglects conscience and intellect. Thus, life$_3$ invokes more than life$_2$. Standard agency is usually reserved to life$_3$ individuals, while the parallel account for life$_2$ individuals gets labeled as "primitive agency" (Jones, 2017, p. xiv; Burge, 2009), "minimal agency" (Barandiarian et al., 2009), or something similar.[12]

Like life$_2$ subjects, life$_3$ minds require a locus with clear boundaries. Notions of truth or falsehood commonly refer to discrete propositions present in discrete minds (see, e.g., McGrath, 2012). It also raises the questions of whether propositions might exist outside of the minds we so frequently associate with life$_1$ brains. Books and computers hold many informational strings, but can they be called propositions, much less true propositions, without a mind to interpret them?

Life$_3$ requires a basic hypothesis with three parts.

1. A distinction can be made between known rational things (i.e., minds) and other known things (e.g., rocks and clouds).[13]
2. That distinction aids in understanding known rational things.
3. There exist unknown things best understood as rational.

The hypothesis may prove false. If it is true, then the set of rational things, known and unknown, constitutes life$_3$.

## Origin and Extent of Life$_3$

Life$_3$ brings us into areas related to the anthropic principle. What does the existence of life$_{1-3}$ tell us about the cosmos? Historically, many have argued that an ordered universe requires an orderer, an intelligible universe requires a pre-existent intelligence.[14] Prior to the Enlightenment, many took it for granted that life$_{1-3}$ arose through participation in cosmic life (Ruse, 2010). Cartesian dualism removed order and mind from the physical world, making a nonphysical designer and a temporal origin of life necessary. In using a machine metaphor, Descartes and contemporaries invoked a mechanic to design and wind up the mechanism (Canguilhem, 2008, p. 86). Intelligent design arguments—arguing from the apparent design of living things to a creator—arose during an awkward period when life$_{1-3}$ were viewed as needing transcendent order but the physical universe was not (roughly 1650–1850).

Early approaches to life$_3$ worked from a premise that *logos* or the cosmic mind exists in eternity. To the extent it was created (in Jewish, Christian, and Muslim theology), it was the eternal foundation for a temporal physical universe.[15] Individual minds arose through inheritance of souls (*traducianism*) or, more popularly, through the immediate intervention of the divine *logos*, "specially creating" souls as needed. This view dominated in European thought through the 14th century. Several Medieval scholars considered the possibility that "mind" was eternal, with individual minds participating temporally, making life$_3$ minds less distinct than life$_2$ subjects.[16]

Following Descartes, Enlightenment thinkers were more inclined to think dualistically, with life$_1$ arising mechanically, life$_3$ arising eternally or by special creation, and life$_2$ floating awkwardly in the middle. The temporal origin of life$_3$ became a popular question after Darwin provided a foundation for biological (life$_1$) explanations of "reason." That reason, however, following the skepticism of Hume, Locke, and others, is only debatably life$_3$. Consequently, most modern thinkers see life$_3$ as pervasive in space and time (e.g., monotheistic *logos*, German idealism, or panpsychism) or illusory. Less ontologically freighted accounts of "reason," in line with life$_2$, have largely taken over in the latter case.

I have argued elsewhere that we use "life" to mean roughly like us and "intelligent" to mean very like us (Mix, 2009, p. 288). As with internal life, a serious search for life$_3$ or some other concept of intelligence will require greater clarity about the examples we argue from, the traits that interest us, and the tests we hope to use.

## Life$_3$ as Tool Use

One recent approach has been to identify intelligence with tool use at some level of complexity. For the past 50 years, SETI researchers have identified intelligence with the ability to build a radio telescope. In the *Star Trek* universe, Starfleet identifies interstellar spaceflight as the marker of a civilization sufficiently advanced to contact. Kardashev (1964) proposed ranking civilizations by how much energy they could harness (planetary scale, stellar scale, or galactic scale). While technological markers strike me as somewhat arbitrary, they are clear and empirically tractable. They seem reasonable potential replacements for traditional intelligence.

One caveat should be added. Many view intelligent life as "higher" in some evaluative sense. This may refer to dignity, freedom, or access to truth. Ancient and Medieval views on immaterial intellect gave humans special status. They possessed an eternal aspect lacking in all other organisms. They participated in the mind of God, giving them access to truth and morality in a unique way. Intellect ($life_3$) was fundamentally distinct from bodily sensation ($life_2$). If we redefine "intelligence" as tool use, intelligent beings lose this special status. It is unclear to me why it would entail greater dignity, better access to truth, better morality, or even evolutionary advancement. Alternatively, we might use tool use as indicator of $life_3$, without claiming that it gives us the critical distinction. In this case, we must look more closely at what it is meant to indicate.

## Artifice

Use of the term "artifice" reveals important distinctions between the three life-concepts. I briefly discuss artificial selection as well as artificial life, consciousness, and intelligence.

In *Origin of Species*, Charles Darwin argues from artificial selection—in the intentional breeding of animals and plants by humans—to natural selection—in changes of populations due to inheritance, variation, and differential survival. Some evolutionary biologists, myself included, have taken this to mean that artificial selection is a special case of natural selection in which the activity of humans is involved. Others seem devoted to the idea that there is something categorically different. For them, artifice involves a $life_3$ imposition on normal causality. Any time that artifice is invoked as "unnatural," a distinction is being made between $life_1$ and $life_{2/3}$. Human agency or intellect has interfered with the normal course of biology.

If artificial selection means nothing more than one species selectively encouraging another for self-interest, then there is no reason to distinguish human selection of grasses for food from ant selection of fungus for food. Both have one species breeding another. In this case, ants have been practicing agriculture for 45 to 65 million years, far longer than humans (Mueller et al., 2001). Such a broad definition of artifice, however, would spread artificial selection throughout the tree of life, including the manipulation of insects by endosymbiotic bacteria (e.g., Wohlbachia). Humans must possess special consciousness ($life_2$) or rational agency (life$_3$) to differentiate human-mediated selection from selection mediated by other species.

This broader "artifice" fails to do the type of work either popularizers of science or ethicists wish it to do. Some researchers have proposed a new geologic epoch, the Anthropocene, to recognize the profound impact of human activity on geologic processes (Monastersky, 2015). Human impact on the atmosphere, geosphere, hydrosphere, and biosphere has been dramatic, and yet it is unclear why this should be distinguished from the dramatic impact of cyanobacteria or land plants, both of which had incredibly broad effects historically (Payne et al., 2011; Willis and McElwain, 2014, p. 59).

In ethics, many arguments against "playing God" or "messing with the laws of nature" imply a distinction between human action and natural action. While artificial selection may have been a necessary concept in the construction of evolutionary theory, and while it may be a useful short-hand when speaking about human influence on non-human survival, it does not currently reflect a different explanatory regime in evolutionary biology.

In the case of artificial life, life-concepts play a role for both creator and created. Humans have crafted entities within or at the borders of life$_1$, life$_2$, and life$_3$. Planned pregnancies involve intentional human action. If life$_2$ or life$_3$ categories do meaningful work in "artifice," then planned pregnancies result in artificial life. The seeming oddness of that statement shows how deeply life-concepts shape modern speech.

We can also speak of *in silico* creations. Self-replicating programs and other forms of artificial life share properties with common organisms. Machine learning algorithms modify themselves. If this requires self-recognition, they could be considered artificial life$_2$ or artificial consciousness. Hypothetical general artificial intelligence could meet the criterial for artificial life$_3$.

## Hierarchy and Value

Both science and theology reveal reasons to be skeptical of the classical hierarchy of life. Modern phylogenetic trees of life displace humans from a position "above" other organisms. Rather we are one of many branches on a very bushy tree, each equally distant in time and evolution from the root, the last universal common ancestor. Likewise, rejections of progress in evolutionary biology suggest that all organisms adapt to their environment, but cannot—qua life$_1$—be said to progress in any absolute sense (Gould, 1996; Ruse, 2009). Similarly, Christian theology in the last fifty years has moved toward an emphasis on God-centered rather than human-centered pictures of creation (Gustafson, 1994; Mix, 2016). These are, of course only two examples of recent literature across fields that questions the idea that humans are the "highest" or even the most "complex" examples of life. Within biology, Maynard-Smith and Szathmáry (1995, pp. 4–5) suggest a "fallacy of progress" and highlight the divergent trajectories of many species. Perhaps other organisms are as specialized as we are but in different ways. Perhaps human life$_{2-3}$ are paralleled by dolphin life$_{2'-3'}$, different manifestations of the same traits, or even completely different specializations, invisible to human perception. Or perhaps they share the same life$_{2-3}$ in ways we are only now beginning to appreciate.

More importantly, we should ask whether the aspects of life$_2$ and life$_3$ will always be found in the context of life$_1$. Could consciousness and reason arise without biology? Could they outlast it? There is a common, though by no means universal, belief that intelligence is a natural and common product of biology. We should seriously consider the possibility that life$_3$ arises infrequently, may not be adaptive on long time scales, or may come and go. Other adaptations may be more successful elsewhere.

## The Hard Problems of Life

Chalmers (1995) proposed the hard problem of consciousness. Using the three life-concepts here presented, I want to propose that there are at least three hard problems related to life. Each problem reflects a gap between explanatory regimes. The *hard problem of biological organization* deals with the shift from physical to biological explanation and asks how function relates to physics.[17] The *hard problem of consciousness* deals with the shift from biological to subjective explanation and asks how suffering and agency relate to physics and biology. The *hard problem of intellect* deals with the shift from subjective to rational explanation and asks how experience becomes genuine knowledge. In each case, we face serious questions about how we treat identity and causality. In each case, explanations in one frame appear insufficient to other frames. In addressing the gaps, we must attend closely to see whether we have denied that something exists (e.g., mind is an illusion), modified the rules of explanation (e.g., added teleology to mechanical science), or genuinely provided a reduction.

Scholars have developed distinct modes of reasoning about each life. Moving between them can cause controversies, particularly in astrobiology. John Stuart Mill (1885, p. 8) distinguished two uses of the word "nature." One contrasts the physical world with the supernatural. The other contrasts nature and artifice, the world with and without human agency. In the second use, artifice indicates some version of life$_2$ or life$_3$. Genetic conditioning reflects nature, while social conditioning reflects nurture. Animal behavior is nature, but human behavior is artifice. Dussault (2016) discusses these and suggests alternative, non-dualistic approaches to "nature."

The subject matter is contentious because we are invested in the answers. Awareness about relationships among the three lives and between these lives and our explanatory frameworks will be essential to creating meaningful communication across fields. It can also aid communication with the public, whose training may vary widely. The three lives warrant further and far more specific investigation. Here I can only sketch out two areas where the typology is particularly important for astrobiologists.

## The Drake Equation and the Fermi Paradox

Calculations on the probability of detecting life frequently rest on implicit assumptions about $life_{1-3}$, and their relationships. The Drake equation estimates the number of civilizations with which we might communicate.

$$N = R_* f_p n_e f_l f_i f_c L$$

The right side of the equation includes three sets of terms. The first three cover the frequency of habitable planets.[18] The second three cover the frequency of life.[19] The last term covers the duration of technological civilizations, turning a rate into a number. Drake has been clear that the equation represents research priorities, rather than a useful estimator. Nonetheless, it reveals implicit use of the traditional hierarchy.

First, the equation presumes that life ($life_1$) and intelligence ($life_{2/3}$) and technological intelligence are meaningful categories based on useful distinctions. Second, it presumes that intelligence only arises in the context of life. Third, the presence of only one term for duration suggests that these transitions occur in only one direction.

In recent decades, we have learned much about the frequency of habitable planets. We have learned less about the frequency of life. Papers on the probability of finding life usually attend closely to the former without addressing the latter or how much they influence the final estimate (Mix, 2018b). Sandberg and colleagues (2018) look at uncertainty across estimates for all seven variables. Uncertainty in $f_l$ dominates other factors dramatically. Using probability distributions in place of point estimates, they find it unsurprising that we have not detected an alien civilization. The Fermi paradox (or observation) of an apparently lifeless universe is consistent with available data. This does not mean life is rare. It simply means we lack sufficient data to make estimates.

## The Anthropic Principle

Discussions of the anthropic principle, and how our existence shapes our understanding, bring the relationship of $life_1$ to $life_{2/3}$ into focus.[20] Weak forms of the anthropic principle note observation selection effects or observer bias. How does the fact that we exist constrain what we can observe? Stronger forms state that $life_{1-3}$ must arise due to fundamental parameters of the universe.

Many authors (e.g., Kärkkäinen, 2015, pp. 140–143) have attempted to frame the weak anthropic principle in terms of biology alone, a "biopic principle," conflating $life_1$ with $life_{2/3}$. And yet, $life_2$ subjectivity or $life_3$ judgment drive the argument. Carter's (1974) original formulation and most later approaches (e.g., Wheeler, 1988; Bostrom, 2013) invoke "observers." Observers require a point of view and an interior subject of sensation.[21] Several authors go even further. Tipler's (1982) final anthropic principle requires "intelligent information processing," using $life_3$ language. Vainio (2018) speaks of "embodied conscious agents," arguably pulling in all three life-concepts. As with the Drake equation, these authors jump too quickly from the frequency of precursors (habitable planets and carbon chemistry) to the frequency of life. If we grant that observers must be carbon-based, this will constrain our answers. Once again, though, the steps that actually involve life have not been addressed. Here ambiguity about the meaning and origin of $life_2$ and $life_3$ compound the problem.

Many scholars hold an a priori commitment to the Copernican principle, or the principle of mediocrity, which claims we (the observers) have no privileged place in the universe. Problems arise once we start to ask what we are being compared to (Ćirković, 2012, p. 56). Typical with regard to what reference class? Organic chemistry does not occur uniformly in space; it clusters at planets and moons. $Life_1$ does not occur uniformly among locations with organic chemistry. Even on Earth, life is more abundant in places with moderate energy flux: not too hot and not too cold. The idea of a habitable zone depends upon this nonuniformity. We still do not know what other nonuniformities may be involved (e.g., tides, water, UV shielding, etc.) in the origins of $life_{1-3}$. Earth observers may not be typical in any interesting sense.

## Conclusion

What are we looking for when we look for life? The three lives represent common conceptions of biological life ($life_1$), internal or conscious life ($life_2$), and rational or intelligent life (life$_3$). Each reflects a specific hypothesis about how we understand the universe. Each gives us an object for search and a phenomenon to explain. Each has its own origin, extent, and future. Their history in Western thought shapes our expectations and priorities in astrobiology.

Life$_1$ has the clearest grounding and boundaries, defined by the modern field of biology. Many concrete questions exist at the boundary of physics and biology (and chemistry, planetary science, etc.). These can largely be addressed with the language and methodology of the natural sciences. Life$_{2-3}$ have their own languages, developed over centuries of discussion. When astrobiology moves into these areas, the rules of discussion come into question. Attending to the differences can clarify expectations. It can also connect researchers to the longer history of discussion. Not all searchers for "intelligent life" search for the same thing.

The three lives may not be meaningful categories. They may not be related along traditional lines. Often that ambiguity can be set aside. We have sufficient consensus to keep working on individual projects. Sometimes these issues critically effect our conclusions. This is especially true for assessing the probability of various lives arising and for understanding the relationship of humans to their biological, chemical, and astronomical environment. Looking to the future, the three lives do not tell us how life, consciousness, and intelligence are related, but, like the Drake equation, they can identify the areas that need to be explored.

## Acknowledgments

Thank you to the Center of Theological Inquiry and the members in residence 2015–2016 for rigorous discussion of the meanings of life. Thank you to Rika Anderson, William Brown, Luis Campos, Carlos Mariscal, Susan Schneider, Frederick Simmons, Kelly Smith, and William Werpehowski for comments.

## Notes

1. Bertka (2009) summarizing the three basic questions that begin the NASA Astrobiology Roadmaps: How does life begin and evolve? Does life exist elsewhere in the Universe? What is the future of life on Earth and beyond? (Morrison and Schmidt, 1999; Des Marais et al., 2003, p. 219; Des Marais et al., 2008, p. 715; Hays et al., 2015).

2. Some of the more obvious examples include "terrestrial" in geology and biology, "organic" in biology and chemistry, and "metal" in chemistry and astronomy.

3. A number of scholars, often following Stoic or Hebrew precedents, gave plants a "plant-nature" and reserved soul to animals. They nonetheless attributed something beyond—but rarely contrary to—basic physics. See Mix (2018a) for a detailed history.

4. This distinction need not involve a formal definition, a natural kind, or distinct boundaries; see Mix (2015). It does require examples of both life and non-life.

5. For example, *City of God*, 11.26.

6. Augustine's argument from doubt, or *Dubito*, influenced Ibn Rushd's *Flying Man* and Descartes' *Cogito*. Together they shaped Christian, Muslim, and Jewish anthropology.

7. This distinction requires examples of both conscious and non-conscious things.

8. Dennett (1991), Mathews (2003), and Millikan (2017), just to name three prominent examples, make clear that they are reworking the Cartesian frame with attendant ontology and epistemology. Following Schrödinger (1992) and Foucault (1994), I wonder about the extent to which such a move would destabilize the paradigm of natural science, but that is a much bigger question.

9. Some authors, particularly in Buddhist and Hindu philosophies, speak of a continuous consciousness. Western thinkers after the Enlightenment have favored discrete subjective minds.

10. *Essay Concerning Human Understanding*, 2:27.26.

11. Debate persists about the extent to which Aristotle's rational soul was immaterial. For discussion see Johnson (2005, pp. 171–172). European scholars consistently interpreted it as necessarily immaterial by the beginning of the Common Era. It began to be questioned again in the late 20th century.

12. Several Medieval scholars suggested intermediate levels of will and agency (Ivry, 2012). Descartes simplified matters by claiming that humans have both life$_2$ and life$_3$, while all other creatures have neither (Voss, 2000; Lennon, 2000). This clear demarcation between humans and non-humans persisted until the 20th century.

13. This distinction requires examples of both rational and non-rational things.

14. For example, Pierre Gassendi (Osler 1994, pp. 36–77) and Darwin (1960). Notably, they viewed the universe in this way and did not see organisms providing special evidence.

15. Leading examples are the dual creation accounts of Origin, Augustine, and Aquinas.

16. For example, the Agent Intellect of Alexander of Aphrodesias and Ibn Sînâ (Martin and Barresi, 2006).

17. This is more general statement of the "hard problem of life" described by Walker and Davies (2017). They identify the relevant distinction for life$_1$ as locally encoded information with causal efficacy.

18. The rate of suitable star formation ($R_*$), the fraction of such stars with planets ($f_p$), the fraction of such planetary systems with a habitable planet ($n_e$).

19. The fraction of habitable planets where life arises ($f_l$), the fraction of such biospheres in which intelligence arises ($f_i$), the fraction of intelligent species with radio transmissions ($f_c$).

20. Readers looking for a summary of the anthropic principle in astrobiology are encouraged to read Scharf (2014) for the state of the art in science. For philosophical concerns, see Mix (2009, pp. 58–65) and Ćirković (2012, pp. 56–85).

21. This need not require a Cartesian theater or any substantial "viewer." It does require a conceptual distinction between the world as it is and the world as it is experienced.

# References

Aristotle. *De Anima (On the Soul)*, translated by Hugh Lawson-Tancred. London: Penguin, 1986.

"Astro-, comb. form." *Oxford English Dictionary Online*. Oxford: Oxford University Press, June 2018.

"Astrobiology." *American Heritage Dictionary of the English Language*, 5th ed. Boston: Mifflin Harcourt, 2018.

Barandiaran, Xabier E., Ezequiel Di Paolo, and Marieke Rohde. "Defining Agency: Individuality, Normativity, Asymmetry, and Spatio-Temporality in Action." *Adaptive Behavior* 17, no. 5 (2009): 367–386.

Baumeister, R. F. Free Will in Scientific Psychology. *Perspectives on Psychological Science* 3 (2008): 14–19.

Baumeister, R. F., E. J. Masicampo, and K. D. Vohs. "Do Conscious Thoughts Cause Behavior?" *Annual Review of Psychology* 62 (2011): 331–361.

Bertka, Constance M., ed. *Exploring the Origin, Extent, and Future of Life*. Cambridge, UK: Cambridge University Press, 2009.

Bostrom, Nick. *Anthropic Bias: Observation Selection Effects in Science and Philosophy*. New York: Routledge, 2013.

Burge, Tyler. "Primitive Agency and Natural Norms." *Philosophy and Phenomenological Research* 79, no. 2 (2009): 251–278.

Canguilhem, Georges. *Knowledge of Life*. New York: Fordham University Press, 2008.

Carter, Brandon. "Large Number Coincidences and the Anthropic Principle in Cosmology." In *Confrontation of Cosmological Theory with Observational Data*, edited by M. S. Longair, 291–298. Dordrecht: Reidel, 1974.

Chalmers, D. J. Facing up to the problem of consciousness. *Journal of Consciousness Studies* 2, no. 3 (1995): 200–219.

Chamovitz, Daniel. *What a Plant Knows: A Field Guide to the Senses of Your Garden and Beyond*. London: Farrar, Straus and Giroux, 2012.

Ćirković, Milan M. *The Astrobiological Landscape: Philosophical Foundations of the Study of Cosmic Life*. Cambridge, UK: Cambridge University Press, 2012.

Clarke, Ellen. "The Problem of Biological Individuality." *Biological Theory* 5, no. 4 (2010): 312–325.

Claus, David B. *Toward the Soul: An Inquiry into the Meaning of ψυχή before Plato*. New Haven, CT: Yale University Press, 1981.

Cleland, Carol E. "Life Without Definitions." *Synthese* 185, no. 1 (2012): 125–144.

Darwin, Charles. Letter to Asa Gray, May 22, 1860. Darwin Correspondence Project. Retrieved from https://www.darwinproject.ac.uk/letter/DCP-LETT-2814.xml.

Darwin, Charles. Letter to William Graham, July 3, 1881. Darwin Correspondence Project. Retrieved from https://www.darwinproject.ac.uk/letter/DCP-LETT-13230.xml.

Dennett, Daniel. *Consciousness Explained*. Boston: Little, Brown, 1991.

Des Marais, David J., Louis J. Allamandola, Steven A. Benner, et al. "The NASA Astrobiology Roadmap." *Astrobiology* 3, no. 2 (2003): 219–235.

Des Marais, David J., Joseph A. Nuth III, Louis J. Allamandola, et al. "The NASA Astrobiology Roadmap." *Astrobiology* 8, no. 4 (2008): 715–730.

Dussault, Antoine C. "Ecological Nature: A Non-Dualistic Concept for Rethinking Humankind's Place in the World." *Ethics & the Environment* 21, no. 1 (2016): 1–37.

Foucault, Michel. *The Order of Things; An Archaeology of the Human Sciences*. New York: Vintage, 1994.

Godfrey-Smith, Peter. *Darwinian Populations and Natural Selection*. New York: Oxford University Press, 2009.

Goetz, Stewart, and Charles Taliaferro. *A Brief History of the Soul*. Malden, MA: Wiley-Blackwell, 2011.

Goff, Philip, William Seager, and Sean Allen-Hermanson. "Panpsychism." In *Stanford Encyclopedia of Philosophy*, Winter 2017 ed. Stanford University, 1997–. Retrieved from https://plato.stanford.edu/archives/win2017/entries/panpsychism/.

Gould, Stephen Jay. *Full House: The Spread of Excellence from Plato to Darwin*. New York: Harmony Books, 1996.

Gustafson, H. M. *Sense of the Divine: The Natural Environment from a Theocentric Perspective*. Cleveland: Pilgrim, 1994.

Hall, Matthew. *Plants as Persons: A Philosophical Botany*. Albany, NY: SUNY Press, 2011.

Hays, L., L. Archenbach, J. Bailey, et al. *NASA Astrobiology Strategy*. National Aeronautics and Space Administration, 2015.

Hornsby, Jennifer. "Agency and Actions." In *Agency and Action*, edited by John Hyman and Helen Steward, 1–24. Cambridge, UK: Cambridge University Press, 2004.

Impey, Christopher, A. H. Spitz, and William R. Stoeger, eds. *Encountering Life in the Universe*. Tucson: University of Arizona, 2013.

Ivry, Alfred. "Arabic and Islamic Psychology and Philosophy of Mind." In *Stanford Encyclopedia of Philosophy*, Summer 2012 ed. Stanford University, 1997–. Retrieved from https://plato.stanford.edu/archives/sum2012/entries/arabic-islamic-mind/.

Johnson, Monte Ransome. *Aristotle on Teleology*. New York: Oxford University Press, 2005.

Jones, Derek M. *The Biological Foundations of Action*. History and Philosophy of Biology. New York: Routledge, 2017.

Kahneman, Daniel. *Thinking, Fast and Slow*. New York: Farrar, Straus and Giroux, 2011.

Kardashev, Nikolai S. "Transmission of Information by Extraterrestrial Civilizations." *Soviet Astronomy* 8 (1964): 217.

Kärkkäinen, Veli-Matti. *Creation and Humanity: A Constructive Christian Theology for the Pluralistic World*, Vol. 3. Grand Rapids, MI: Eerdmans, 2015.

Koechlin, E., and C. Summerfield. "An Information Theoretical Approach to Prefrontal Executive Function." *Trends in Cognitive Sciences* 11, no. 6 (2007): 229–235.

Libet, B., C. A. Gleason, E. W. Wright, and D. K. Pearl. "Time of Conscious Intention to Act In relation to Onset of Cerebral Activity (Readiness-Potential)." *Brain* 106, no. 3 (1983): 623–642.

Linnaeus, Carolus. *Systema Naturae*. Stockholm: Laurentius Salvus, 1758.

Lennon, Thomas M. "Bayle and Late Seventeenth-Century Thought." In *Psyche and Soma: Physicians and Metaphysicians on the Mind-Body Problem from Antiquity to Enlightenment*, edited by John P. Wright and Paul Potter, 197–215. Oxford: Clarendon, 2000.

Luisi, Pier Luigi. "About Various Definitions of Life." *Origins of Life and Evolution of the Biosphere* 28, no. 4–6 (1998): 613–622.

MacLennan, B. J. "Robots React, But Can They Feel? A Protophenomenological Analysis." In *Handbook of Research on Synthetic Emotions and Sociable Robotics: New Applications in Affective Computing and Artificial Intelligence*, edited by J. Vallverdú and Casacuberta, D., 133–153. New York: Information Science Reference, 2009.

Marder, Michael. *The Philosopher's Plant: An Intellectual Herbarium*. New York: Columbia University Press, 2014.

Marder, Michael, and Luce Irigaray. *Through Vegetal Being: Two Philosophical Perspectives*. New York: Columbia University Press, 2016.

Martin, Raymond, and John Barresi. *The Rise and Fall of Soul and Self: An Intellectual History of Personal Identity*. New York: Columbia University Press, 2006.

Mathews, Freya. *For Love of Matter*. Albany, NY: State University of New York Press, 2003.

Matthews, Gareth. "Internalist Reasoning in Augustine for Mind-Body Dualism." In *Psyche and Soma: Physicians and Metaphysicians on the Mind-Body Problem from Antiquity to Enlightenment*, edited by John P. Wright and Paul Potter, 133–145. Oxford: Clarendon, 2000.

Maynard Smith, John, and Eörs Szathmáry. *Major Transitions in Evolution*. New York: Oxford University Press, 1995.

McGrath, M. (2012) Propositions. In *The Stanford Encyclopedia of Philosophy*, Spring 2014 Edition, edited by E. W. Zalta. Retrieved from http://plato.stanford.edu/archives/spr2014/entries/propositions/.

Millikan, Ruth Garrett. *Beyond Concepts: Unicepts, Language, and Natural Information*. Oxford: Oxford University Press, 2017.

Mill, John Stuart. *Nature, the Utility of Religion, and Theism*, 3rd ed. London: Longman, Green, 1885.

Mix, Lucas John. *Life in Space: Astrobiology for Everyone*. Cambridge, MA: Harvard University Press, 2009.

Mix, Lucas J. "Proper Activity, Preference, and the Meaning of Life." *Philosophy & Theory in Biology* 6 (2014).

Mix, Lucas John. "Defending Definitions of Life." *Astrobiology* 15, no. 1 (2015): 15–19.

Mix, Lucas John. "Life-Value Narratives and the Impact of Astrobiology on Christian Ethics." *Zygon* 51, no. 2 (2016): 520–535.

Mix, Lucas John. *Life Concepts from Aristotle to Darwin: On Vegetable Souls*. New York: Palgrave MacMillan, 2018a.

Mix, Lucas John. "Philosophy and Data in Astrobiology." *International Journal of Astrobiology* 17, no. 2 (2018b): 189–200.

Monastersky, Richard. "Anthropocene: The Human Age." *Nature* 519, no. 7542 (2015): 144–147.

Morrison, D., and G. K. Schmidt. *NASA Astrobiology Roadmap*. Moffett Field, CA: Ames Research Center, 1999.

Mueller, Ulrich G., Ted R. Schultz, Cameron R. Currie, Rachelle M. M. Adams, and David Malloch. "The Origin of the Attine Ant-Fungus Mutualism." *The Quarterly Review of Biology* 76, no. 2 (2001): 169–197.

Nagel, Thomas. "What Is It Like To Be a Bat?" *The Philosophical Review* 83, no. 4 (1974): 435–450.

Nagel, Thomas. *Mind and Cosmos: Why the Materialist Neo-Darwinian Conception of Nature Is Almost Certainly False.* New York: Oxford University Press, 2012.

*New Oxford Annotated Bible: New Revised Standard Version with the Apocrypha.* Edited by Michael D. Coogan. New York: Oxford University Press, 2001.

"Organic, adj. and n." *Oxford English Dictionary Online.* Oxford: Oxford University Press, June 2018.

Osler, Margaret J. *Divine Will and the Mechanical Philosophy.* Cambridge: Cambridge University Press, 1994.

Payne, Jonathan L., Craig R. McClain, Alison G. Boyer, et al. "The Evolutionary Consequences of Oxygenic Photosynthesis: A Body Size Perspective." *Photosynthesis Research* 107, no. 1 (2011): 37–57.

Plantinga, Alvin. *Where the Conflict Really Lies: Science, Religion, and Naturalism.* New York: Oxford University Press, 2011.

Ratzinger, C. J., and A. Bovone. "Instruction on Respect for Human Life in Its Origin and on the Dignity of Procreation." Congregation for the Doctrine of the Faith, the Feast of the Chair of St. Peter, the Apostle. Rome, February 2, 1987.

Ruse, Michael. *Monad to Man: The Concept of Progress in Evolutionary Biology.* Cambridge, MA: Harvard University Press, 2009.

Ruse, Michael. *Science and Spirituality: Making Room for Faith in the Age of Science.* New York: Cambridge University Press, 2010.

Sandberg, Anders, Eric Drexler, and Toby Ord. "Dissolving the Fermi Paradox." *arXiv preprint arXiv:1806.02404* (2018).

Scharf, Caleb. *The Copernicus Complex: Our Cosmic Significance in a Universe of Planets and Probabilities.* New York: Farrar, Straus and Giroux, 2014.

Schneider, Susan. "Alien Minds." In *Impact of Discovering Life beyond Earth*, edited by Steven J. Dick, 189–206. Cambridge, MA: Cambridge University Press, 2015.

Schrödinger, Erwin. *What Is Life? The Physical Aspect of the Living Cell; With, Mind and Matter & Autobiographical Sketches.* Cambridge, UK: Cambridge University Press, 1992.

Shields, Christopher. *Aristotle.* New York: Routledge, 2007.

Shields, Christopher. "The Dialectic of Life." *Synthese* 185, no. 1 (2012): 103–124.

Singer, Peter. *Practical Ethics.* Cambridge, UK: Cambridge University Press, 1979.

Skloot, R. *The Immortal Life of Henrietta Lacks.* New York: Crown Publishers, 2010.

Tillich, Paul. *Systematic Theology, Volume III.* Chicago: University of Chicago Press, 1963.

Tipler, Frank J. "Anthropic-Principle Arguments Against Steady-State Cosmological Theories." *The Observatory* 102 (1982): 36–39.

Trifonov, Edward N. "Vocabulary of Definitions of Life Suggests a Definition." *Journal of Biomolecular Structure and Dynamics* 29, no. 2 (2011): 259–266.

Vainio, Olli-Pekka. *Cosmology in Theological Perspective.* Grand Rapids, MI: Baker Academic, 2018.

Vakoch, Douglas A. *Astrobiology, History, and Society.* Berlin: Springer-Verlag, 2013.

Vass, A. "After a Person's pulse and Breathing Stop, How Much Later Does All Cellular Metabolism Stop?" *Scientific American*, July 16, 2007.

Voss, Stephen. "Descartes: Heart and Soul." In *Psyche and Soma: Physicians and Metaphysicians on the Mind-Body Problem from Antiquity to Enlightenment*, edited by John P. Wright and Paul Potter, 173–196. Oxford: Clarendon, 2000.

Walker, Sara Imari, and Paul C. W. Davies. "The 'Hard Problem' of Life." In *From Matter to Life: Information and Causality*, edited by S. I. Walker, P .C. W. Davies and G. F. R. Ellis, 19–36. Cambridge: Cambridge University Press, 2017.

Wheeler, John Archibald. "World as System Self-Synthesized by Quantum Networking." In *Probability in the Sciences*, 103–129. Dordrecht: Springer, 1988.

Willis, K., and J. McElwain. *The Evolution of Plants*, 2nd ed. New York: Oxford University Press, 2014.

Wohlleben, Peter. *The Hidden Life of Trees: What They Feel, How They Communicate— Discoveries from a Secret World*. Vancouver, BC: Greystone Books, 2016.

# 5

# Dimensions of Life Definitions

*Emily C. Parke*

## Life on Mars?

In early 2018 NASA announced that it had found organic matter on Mars (Potter, 2018). Specifically, researchers drilled into Martian rocks and heated them at high temperatures, catalyzing the release of organic molecules trapped there in 3.5-billion-year-old mudslides (Eigenbrode et al., 2018). This finding was presented in major media outlets as an enticing step forward in the search for life on Mars (e.g., Chang, 2018; Sample, 2018).

Organic matter had been found on Mars before, just not in anything near these concentrations (Eigenbrode et al., 2018). Also, these organic molecules are of a sort that could have been produced by life but could also be produced abiotically, by purely chemical processes. In other words, if life with the biochemistry of Earth life were around, then we would expect to find these sorts of molecules. But finding these sorts of molecules does not mean that life is, or even was, around.

Claims varied about the bearing of this finding on the search for life. Here are some representative examples:

> Are there signs of life on Mars? . . . We don't know, but these results tell us we are on the right track. (Potter 2018)

> Whether it holds a record of ancient life, is the food for extant life, or has existed in the absence of life, organic matter in martian materials holds chemical clues to planetary conditions and processes. (Eigenbrode et al., 2018)

> It's not a direct indicator that life may have existed on Mars. (Tamblyn, 2018)

This variation is understandable, regarding whether and to what extent this finding points to life. First, there are further empirical and theoretical questions to address about the significance of such a finding. Second, there is no consensus on what life is in the first place. This chapter discusses three ways that answers to

Emily C. Parke, *Dimensions of Life Definitions.* In: *Social and Conceptual Issues in Astrobiology.* Edited by: Kelly C. Smith and Carlos Mariscal, Oxford University Press (2020). © Oxford University Press.
DOI: 10.1093/oso/9780190915650.001.0005

the question "What is life?" can vary and how this variation might bear on life-detection efforts in astrobiology.

A natural place to start in thinking about what this recent discovery can tell us about life is the "NASA definition" of life: "Life is a self-sustaining chemical system capable of Darwinian evolution." This definition was proposed several decades ago by Carl Sagan and popularized in subsequent discussion (Joyce et al., 1994) and is endorsed by NASA in the context of its life-detection efforts.[1]

There is no straightforward evidential link between the 2018 finding on Mars and the NASA definition of life. Of course, there is much more to the search for life than this definition. There are more fine-grained background assumptions at stake that do not figure directly into this "official" definition of life: for example, about what life is, what life does, and what life requires. These sorts of assumptions drive the search for organic matter of the sort NASA recently discovered, along with its tentative designation as a small but real mark of success in the search for life on Mars. What this recent finding does is raise the likelihood that there could have been life forms on Mars that are biochemically akin to life on Earth. That is significant, even if it does not bear much, at face value, on the closest thing to a candidate definition of life at stake here.

This is not just NASA's issue. Anyone looking for life in the universe is in a bit of a tough spot: It cannot be assumed that life elsewhere would resemble life as we know it, in even the most basic ways. So we need some way to recognize it that abstracts away from life as we know it. Many people have thought that coming up with a definition of life is a natural and appealing way to achieve this.

## Defining Life and Why It Matters for Astrobiology

Here we run into a puzzle that philosophers and scientists have been arguing about for a long time. In the Western tradition, accounts of the debate about defining life tend to trace it back several millennia to Aristotle, who distinguished living from nonliving things in terms of functions like reproduction and nutrition. Since then, hundreds of others have weighed in, from early modern natural philosophers to contemporary philosophers and scientists. Some have attempted to define life in terms of a single property, like evolution or metabolism; others give a list of properties. Some definitions emphasize biochemical particulars; others are wholly functional. When someone proposes definitional criteria for life, someone else raises a counterexample (e.g., if metabolism is sufficient for life, then candle flames arguably qualify as living; if reproduction is necessary for life, then sterile hybrids like mules are problematically disqualified).

Other authors have surveyed extensively the landscape of proposed definitions of life and their various borderline cases and counterexamples (e.g., Sagan, 1970;

Bedau, 1998; Luisi, 1998; Pályi et al., 2002; Popa, 2004; Oliver and Perry, 2006; Cleland and Chyba, 2007; Bedau and Cleland, 2010; Trifonov, 2011; Mix, 2015; Mariscal and Doolittle, 2018). I will not repeat their efforts here. The key points for the purpose of this chapter are that this landscape is substantial (with as many as 100+ definitions of life; Popa, 2004; Trifonov, 2011), proposed definitions of life vary widely, and there is no consensus on a definition among astrobiologists, let alone across disciplines with a stake in the matter.[2]

Some authors have suggested that addressing the question "What is life?" is fundamental or even necessary for biology (Cleland and Chyba, 2007; Farnsworth et al., 2013; Mix, 2015). But while some biologists are interested in that question, biologists in general do not need an answer to it. The subject matter of biology, the living world and its phenomena, is clear enough without one. Borderline cases for defining life do not cause empirical or theoretical problems for most biologists; for example, the lack of consensus on whether viruses are ultimately nonliving or living does not impede microbiologists' study of bacteria and their viruses.

In contrast to biologists in general, astrobiologists cannot take the status of their subject matter as living or nonliving for granted. Its status as such is often precisely what is at stake. There are at least two reasons to think astrobiologists need an understanding of what counts as life. The first is to set search criteria for finding "life as we don't know it" in the universe. The second is to set success conditions conducive to agreement about when life has been found and when it has not.[3]

In addition to particular cases like the recent Mars finding discussed at the beginning of this chapter, the meaning of 'life' figures into a broader agenda in astrobiology: looking for biosignatures. There are various ways to spell out what a biosignature is. Here are four sample characterizations:

An observable feature of a planet, such as its atmospheric composition, that our *present* models cannot reproduce when including the abiotic physical and chemical processes we know about. (Léger et al., 2011; emphasis original)

Distinctive suites of durable textural, mineralogical, or chemical indicators of life. (Campbell, 2017)

A feature whose presence or abundance requires a biological origin. (Des Marais et al., 2001)

Evidence that life exists or existed. (Benner, 2010)

There is a spectrum here from understandings of biosignatures that require some particular chemical or material assumptions about life to those that do not. The first two rely on assumptions about biochemistry or the kinds of material traces life leaves; the latter two need not (note that on Benner's characterization of a biosignature, artifacts qualify: If we found technology or artificial intelligence elsewhere in the solar system, we could be pretty sure that life is or was there to create it; Benner, 2010).

Some understanding of what counts as life is built into these understandings of what counts as a biosignature—or, at least, into their use in any efforts to identify biosignatures. Of course, biosignatures could be conceptualized, and candidate ones assessed, without a definition or even a hypothesis about life beyond life as we know it. We could just take everything we know about life on Earth—its biochemical constraints, its effects on the atmosphere, the microscopic and macroscopic phenomena and artifacts it produces, and so forth—and use that collection of observations to guide the search for biosignatures. But many people invested in finding life beyond Earth want more than that. They want to be in a position to find signatures of life that might be markedly unlike life as we know it, perhaps even at the basic biochemical level.

This desire for grounds to identify "life as we don't know it" has been a key driver in the search for a definition of life (see discussion in Cleland, 2012). Beyond the debate about which definition of life is *the* definition, in recent years there has also been a meta-debate taking place about the whole project itself. In particular, the proliferation of 100+ proposed candidate definitions of life has led some philosophers and scientists to dismiss the project of trying to define life as pointless or hopeless. This definition skepticism comes in at least three flavors (see discussion in Smith, 2018). Carol Cleland has argued that there are too many problems facing the project of strictly defining life, including the sorts of counterexamples and borderline cases mentioned earlier. She says what we need instead is a broader theory of life, but scientists are not in a position to formulate such a theory, because our knowledge of life is based on a single sample: it all descended from the same common ancestor and therefore shares key fundamental properties (Cleland and Chyba, 2007; Cleland, 2012). Edouard Machery (2012) has argued that scientists use an abundance of different definitions of life, there is no reason to think they will converge on a single, unanimous one, and the project of defining life (as a scientific theoretical concept, anyway) is pointless. Jack Szostak (2012) has argued that science can proceed just fine without worrying about the definition of life, at least in the context of origin of life research.

In response to these and other pessimists about defining life, several recent discussions have proposed that we should be *pluralists* about life. That is, we should accept that multiple, even conflicting definitions of life can coexist, suited to different research agendas. A pluralist position denies that the aim of defining life is to settle on a single definition. In any case, only a subset of the 100+ definitions of life referred to earlier are proper definitions in the strict sense that philosophers like to talk about: proposing necessary and sufficient conditions intended for unanimous acceptance. In practice, many so-called definitions of life are less strict. They are better understood as *working characterizations* of life: conceptual frameworks used to guide and make sense of research in a given context, tailored to the agenda of a particular group or field with a stake in understanding life. This is in line with Cleland's (2012) suggestion that astrobiologists search for life with "tentative criteria," rather than strict definitions, in mind. For further discussion of this idea that definitions of life are often more operational or provisional, and that it is not such bad thing to have more than one of them, see (Oliver and Perry, 2006; Griesemer, 2015; Mix, 2015; Bich and Green, 2018).

The combined views of the pessimists and the pluralists point to a grim outlook for settling on a unanimous definition of life, at least today. The positions of pluralism about life, and the more relaxed understanding of what qualifies as a definition, are controversial but popular in the current literature on life—I will not argue for them at length here. For the rest of this chapter I assume these two positions and ask: How do the *ways* life is defined bear on scientific practice in astrobiology? In particular, how can the ways life is defined affect how astrobiologists understand search criteria and success conditions for finding life in the universe?

## Three Ways to Define Life

Existing discussions of ways to define life have focused on the content of different definitions, categorizing them into clusters such as evolutionary versus thermodynamic versus metabolic definitions of life. This is often for the purpose of assessing the relative merits and implications of different definitional categories, and endorsing one category as fundamental (Sagan, 1970; Pályi et al., 2004; Popa, 2004; Kompanichenko, 2008; Trifonov, 2011; Mix, 2015). Here I discuss something different: three ways to categorize definitions that vary in their strategies or commitments regarding *how* life is defined. These abstract away from the specific content or focal feature(s) of definitions of life. There are at least three such dimensions along which definitions of life can vary: treating living/nonliving as a dichotomy or a matter of degree, defining living individuals or living collectives, and defining life materially or functionally.[4] I discuss each in turn and suggest how variation in each dimension can affect the role of the concept of life in setting search criteria or success conditions in astrobiology.

## Living/Nonliving: Dichotomy or Matter of Degree?

Most proposed definitions of life offer a way to distinguish between two categories, living and nonliving. In particular, they specify criteria for drawing a line that will include what is living and exclude what is nonliving. A paradigm example of this is Maturana and Varela's (1973) "All that is living must be based on autopoiesis, and if a system is discovered to be autopoietic, that system is defined as living, i.e., it must correspond to the definition of minimal life" (cited in Popa, 2004). An alternative is to treat the difference between living and nonliving explicitly as a continuum or a matter of degree. Several recent accounts have done this: they specify a few features as definitional of life, but life comes in degrees. Christophe Malaterre's (2010) account specifies five features (individuation, replication, variation, metabolism, and coupling of components) and allows systems to instantiate each of those features to varying degrees. Mark Bedau's (2012) account specifies three features (a container, program, and metabolism) and allows for a spectrum of Boolean combinations of those features and their relationships of mutual support and integration, shading from nonliving to living. On these matter-of-degree accounts, there is no matter of fact about a clear line between what is nonliving and living, just a gradual scale.

Treating nonliving/living as a dichotomy versus a matter of degree will make a difference to the sorts of claims that can be made about life in the universe. Much of the debate about life has treated the matter of defining life as if the aim is to demarcate two kinds or categories, living and nonliving. A clean divide between what is nonliving and what is living might seem naturally appealing in the context of finding life in the universe. It would be nice to have unambiguous yes-or-no answers when candidate life forms are identified. It would be nice—once the empirical details are sorted out, which is of course no small matter—to be able to say in a given case that life-detection missions like NASA's had either found life, or had not.

On the other hand, understanding life as a matter of degree better reflects the truth of the matter in thinking about the origin of life, on Earth or anywhere. It makes sense for the search for life on other planets to understand life in a way that explicitly captures the possibility of discovering not only living or nonliving things but also intermediaries: minimal or marginal cases of life, where life is understood as something that comes in degrees. Astrobiologists are not looking only for living organisms or communities. They are looking for evidence that life could have once existed elsewhere in the universe or could be in the process of emerging elsewhere. To the extent that the latter constitutes part of the agenda of searching for life, it makes sense to conceptualize life as a matter of degree.

If living/nonliving is understood as a matter of degree rather than two dichotomous categories, the possibility space for interpreting life-detection findings, like the ones discussed at the outset of this chapter, can be understood differently. Discussions have tended to frame that possibility space in terms of three options: life was found, life was not found, or the results are ambiguous between biotic and abiotic explanations. A fourth option—which is consistent with the third option but backgrounded in discussions treating living/nonliving as a dichotomy—could be that there are systems on Mars (for example) best understood as transitional between paradigm cases of nonlife and paradigm cases of life, as one would have found on Earth sometime in the window between roughly 4.3 and 3.8 billion years ago. A clearer shared understanding of the in-between status of such not fully living (but not nonliving) forms would be an ideal starting point for detecting them.

## Living Individuals or Collectives?

Many existing definitions of life propose grounds to separate living organisms, like lizards and lactobacilli, from nonliving things like liposomes and laptops. This is a natural way to understand the question "What is life?"—in terms of what makes living *individuals* living. Some definitions of life, however, are based on properties that by definition apply only to *collectives* of individuals or collectives of individuals over time. An example of the former sort of property is variation; an example of the latter sort is evolution. These appear in many popular definitions of life, including the NASA definition discussed earlier. When variation or evolution are cited as definitional features of life, the bearer of those features is a population, not an individual. So while the question is often interpreted as asking what distinguishes a living thing (organism) from a nonliving thing, a number of authors treat living populations (e.g., Bedau, 1998; Smith, 2018) or even the whole biosphere (Lovelock, 1979; Feinberg and Shapiro, 1980) as the focal unit in characterizing life, rather than individual organisms.

There is no principled reason to prefer defining living individuals to defining living collectives (or vice versa), nor is there any principled problem with including features of both levels of biological organization in the same definition of life. Plenty of definitions do this. But there are some potential conceptual complications, in the context of searching for life in the universe.

The features emphasized in definitions of life are not always the same as the features targeted by life-detection tools and techniques. Furthermore, they are often framed at a different level (that of individual organisms versus collectives of them). Biologists and astrobiologists use different tools and timescales to look for *evidence of evolutionary processes* in nature, as opposed to *evidence of living things*, like microbes. Regarding tools, the 1976 Mars Viking missions involved techniques by which researchers could in principle identify traces of microbes and their metabolic activity; the same is true for bids for sites for the upcoming 2020 Rover missions. These approaches to finding life in the universe prioritize looking for evidence of microbial life forms, not (direct) evidence of evolution or variation. There are other ways to look for evolution and variation. With our familiar microbial Earth life, this is typically done by looking for genetic markers (ribosomal RNA sequences) and their change over time, or their differences with respect to one another or to a putative common ancestor. It is not obvious how these methods would translate to looking for life as we don't know it, given their reliance on particulars not only of DNA-based biochemistry but of conserved genome regions common to familiar life.

Regarding time scales, if life is understood purely in terms of features of individual organisms and we are searching for one, we could in principle find one (or evidence of one) in an instant. Finding evolutionary processes, or direct evidence of them, has to be done over time (but see Benner [2010] for a suggested way around this, discussed later). So variation in this feature of how life is defined could make an empirical and theoretical difference. Specifying criteria for finding living individuals, versus living evolving populations, can influence the kinds of signatures of life astrobiologists look for, the tools they use to look for them, and the timescales on which they can do so. Of course, the projects of searching for living individuals and evolving populations are compatible and can be pursued concurrently. But they can be conceptually and practically distinguished in ways that matter for astrobiology.

## Material or Functional?

Some definitions of life specify material particulars like biochemistry, carbon-based biochemistry, or nucleic acids. For example, Perrett (1952) defines life as "a potentially self-perpetuating system of linked organic reactions, catalyzed stepwise and almost isothermally by complex and specific organic catalysts which are themselves produced by the system." Other definitions are purely functional and invite a variety of material or even digital instantiations of life as we don't know it. An example is "life is self-reproduction with variation" (Trifonov, 2011), which could be instantiated by familiar carbon-based life, silicon-based life, self-replicating computer programs, and any number of other systems, chemical or otherwise. How material or functional a definition is can come in degrees: for example, Perrett's "linked organic reactions" specifies a more fine-grained material feature than the NASA definition's criterion that life is a "chemical system." Furthermore, many definitions of life combine material and functional elements.

Purely functional definitions leave more room for finding life in the universe that does not resemble our current sample of life in even its most basic biochemical aspects. On the other hand, they give less guidance about what to look for. Many definitions of life combine functional with loosely material elements—they are based on a coarse-grained material understanding of life with reference to chemistry but without more fine-grained assumptions about molecular specifics like nucleic acids. These include, for example, the NASA definition, and Pace's (2011) definition of life as "a self-replicating, evolving system expected to be based on organic chemistry."

The more material details a definition of life specifies, the more readily it allows for direct specific claims about biosignatures: what specifically to search for or whether a particular finding qualifies as life or a sign of life. Purely functional definitions will not do this directly. But they can provide a template for more specific guidance on searching for life, once the appropriate details are filled in. Benner (2010) exemplifies this sort of reasoning, connecting a relatively functional definition of life to more specific claims about biochemical signatures of life. He takes the NASA definition ("life is a self-sustaining chemical system capable of Darwinian evolution") as a starting point and supplements it with the assumption that chemical systems capable of Darwinian evolution must be based on polyelectrolites (molecules with repeating charges in their backbones). He argues that we should focus the search for life on polyelectrolites, and evidence of polyelectrolites would qualify as evidence of life under that definition.

## Conclusion

I have discussed three dimensions on which accounts of life can vary and suggested that variation along each of these dimensions can affect how the concept of life figures in to theoretical and empirical efforts in astrobiology: in forming search criteria for life in the universe, understanding success conditions, and communicating about them. In summary:

- **Living/nonliving as a dichotomy or a matter of degree**: Dichotomous understandings of living/nonliving might seem conducive to clearer answers about the status of discoveries in the universe; matter-of-degree understandings are conducive to clearly recognizing transitional entities as such.
- **Living individuals or collectives**: Focusing on features of individual life forms (such as self-reproduction and metabolism), versus features of collectives of individuals (such as variation and evolution), can influence both conceptual search criteria for life and the tools and techniques involved.

- **Material versus functional understandings of life:** More material understandings of life give clearer guidance without supplementation with auxiliary assumptions. More functional understandings leave more room for interpretation and enable searching for "life as we don't know it" at even the most basic chemical level.

The second and third dimensions, especially, are not straightforward either/ or choices. For example, some definitions of life specify only properties of individuals, some specify only properties of collectives, and others explicitly address both (e.g., Ruiz-Mirazo et al., 2004). Definitions can be specifically material, permissively functional, or in between. So these three dimensions do not give us a simple possibility space of eight ways to define life. Rather, we can think of them as axes along which approaches to defining life can be assessed or compared.

The aim of this short chapter is not to make value judgments about the superiority of any particular way to define life along any of these three dimensions— at least not in a wholesale way across astrobiology, let alone across fields with a stake in the concept of life (see footnote 2). Rather, this is a call for clearer communication about which understanding of life is at stake in the context of a given project or finding—and, specifically, how that understanding relates to search criteria and success conditions for finding life or signs of life. Explicitly recognizing a plurality of ways to define life in the sense discussed here (in addition to the range of focal features regarded as definitional of life) is a starting point for clearer discussion of the background assumptions at stake in life-detection efforts.

## Acknowledgments

I am grateful to Kathy Campbell, Carlos Mariscal, Kelly Smith and an anonymous reviewer for helpful feedback on drafts of this chapter.

## Notes

1. For example at astrobiology.nasa.gov/research/life-detection/about/; accessed March 2019.
2. These other disciplines include research on the origin of life, bottom-up synthetic biology, artificial life, and environmental ethics (see discussion in Machery, 2012; Bich and Green, 2017; Parke, in prep.).

3. Research on the origin of life is another important part of the agenda of astrobiology. For the purpose of this chapter I focus in particular on the role of defining life in the search for life in the universe.

4. These are three important ones; I think there are others as well. For example, other relevant dimensions include (a) pragmatic versus theoretical definitions of life and (b) whether life is treated as a (natural) kind, as is the norm, or not (for arguments that life is not a natural kind see Hermida, 2017; Mariscal and Doolittle, 2018).

# References

Bedau M (1998) Four puzzles about life. *Artificial Life*, 4(2), 125–140.

Bedau M (2012) A functional account of degrees of minimal chemical life. *Synthese*, 185(1), 73–88.

Bedau M & Cleland CE (Eds.) (2010) *The nature of life: classical and contemporary perspectives from philosophy and science*. Cambridge, UK: Cambridge University Press.

Benner SA (2010) Defining life. *Astrobiology*, 10, 1021–1030.

Bich L & Green S (2018) Is defining life pointless? Operational definitions at the frontiers of biology. *Synthese*, 195, 3919–3946.

Campbell K (2017, November 2) Hot springs and earliest life on land. *A History of Stones*. Retrieved from www.stonehistorian.space/2017/11/hot-springs-and-earliest-life-on-land.html

Chang K (2018, June 7) Life on Mars? Rover's latest discovery puts it 'on the table'. *The New York Times*. Retrieved from www.nytimes.com/2018/06/07/science/mars-nasa-life.html

Cleland CE (2012) Life without definitions. *Synthese*, 185, 125–144.

Cleland CE & Chyba CF (2007) Does "life" have a definition? In W. T. Sullivan & J. A. Baross (Eds.), *Planets and life: the emerging science of astrobiology*. Cambridge, UK: Cambridge University Press.

Des Marais DJ, Harwit MO, Jucks KW, Kasting JF, Lin DN, Lunine JI, . . . Woolf NJ (2002) Remote sensing of planetary properties and biosignatures on extrasolar terrestrial planets. *Astrobiology*, 2(2), 153–181.

Eigenbrode JL, Summons RE, Steele A, Frissinet C, Millan M, Navarro-Gonzalez R, . . . Coll P (2018) Organic matter preserved in 3-billion-year-old mudstones at Gale crater, Mars. *Science*, 360(6393), 1096–1101.

Farnsworth KD, Nelson J & Gershenson C (2013) Living is information processing: from molecules to global systems. *Acta Biotheoretica*, 61(2), 203–222.

Feinberg G & Shapiro R (1980) *Life beyond Earth: the intelligent earthling's guide to life in the universe*. New York, NY: William Morrow.

Griesemer J (2015) The enduring value of Gánti's chemoton model and life criteria: heuristic pursuit of exact theoretical biology. *Journal of Theoretical Biology*, 381, 23–28.

Hermida M (2016) Life on Earth is an individual. *Theory in Biosciences*, 135, 37–44.

Joyce GF, Deamer DW & Fleischaker G (1994) *Origins of life: the central concepts*. Boston, MA: Jones and Bartlett.

Kompanichenko V (2008) Three stages of the origin of life process: bifurcation, stabilization and inversion, *International Journal of Astrobiology*, 7(1), 27–46.

Léger A, Fontecave M, Labeyrie A, Samuel B, Demangeon O & Valencia D (2011) Is the presence of oxygen on an exoplanet a reliable biosignature? *Astrobiology*, 11(4), 335–41.

Lovelock J (1979) *Gaia: a new look at life on Earth*. Oxford, UK: Oxford University Press.

Luisi PL (1998) About various definitions of life. *Origins of Life and Evolution of the Biosphere*, 28, 613–622.

Machery E (2012) Why I stopped worrying about the definition of life . . . and why you should as well. *Synthese*, 185(1), 145–164.

Malaterre C (2010) Lifeness signatures and the roots of the tree of life. *Biology and Philosophy*, 25, 643–658.

Mariscal C & Doolittle F (2018) Life and Life only: a radical alternative to life definitionism. *Synthese*. Retrieved from https://doi.org/10.1007/s11229-018-1852-2

Mix LJ (2015) Defending definitions of life. *Astrobiology*, 15(1), 15–19.

Oliver JD & Perry RS (2006) Definitely life but not definitively. *Origins of Life and Evolution of Biospheres*, 36(5–6), 515–521.

Pace NR (2001) The universal nature of biochemistry. *Proceedings of the National Academy of Sciences of the United States of America*, 98(3), 805.

Pályi G, Zucchi C & Caglioti L (Eds.) (2002) *Fundamentals of life*. Paris: Elsevier.

Parke EC (in preparation) Finding, explaining, and engineering life. Manuscript in preparation.

Perret M (1952) Biochemistry and bacteria. *New Biology*, 12, 68.

Popa R (2004) *Between necessity and probability: searching for the definition and origin of life*. Berlin: Springer.

Potter S (2018) NASA finds ancient organic material, mysterious methane on Mars [NASA press release]. Retrieved from www.nasa.gov/press-release/nasa-finds-ancient-organic-material-mysterious-methane-on-mars

Ruiz-Mirazo K, Peretó J & Moreno A (2004) A universal definition of life: autonomy and open-ended evolution. *Origins of Life and Evolution of the Biosphere*, 34, 323–346.

Sagan C (1970) Life. In *Encyclopaedia Britannica* (pp. 1083–1083A). Chicago: Encyclopaedia Britannica.

Sample I (2018, June 7) NASA Mars rover finds organic matter in ancient lake bed. *The Guardian*. Retrieved from www.theguardian.com/science/2018/jun/07/nasa-mars-rover-finds-organic-matter-in-ancient-lake-bed

Smith K (2018) Life as adaptive capacity: bringing new life to an old debate. *Biological Theory*, 13(2), 76–92.

Szostak JW (2012) Attempts to define life do not help to understand the origin of life. *Journal of Biomolecular Structure and Dynamics*, 29(4), 599–600.

Tamblyn T (2018, June 8) NASA's Mars rover has found organic matter on the planet's surface. *Huffington Post*. Retrieved from www.huffingtonpost.co.uk/entry/nasas-mars-rover-has-found-organic-matter-on-the-planets-surface_uk_5b1a4e2ce4b0bbb7a0db7486

Trifonov EN (2011) Vocabulary of definitions of life suggests a definition. *Journal of Biomolecular Structure and Dynamics*, 29(2), 259–266.

# 6

# Meaning of the Living State

*Cole Mathis*

## Introduction

Astrobiology is the study of life in the universe [1]. However, in spite of rigorous debate, the astrobiology community does not have an agreed upon definition of life [2, 3]. To make progress in the face of this conceptual issue, astrobiologists focus on specific properties of living systems, such as replication or cellular respiration [4, 5]. This has allowed researchers to make progress in limited domains, such as characterizing the emergence of Darwinian evolution or quantifying the detectability of biosignatures [6, 7]. Unfortunately, without a consistent definition of life, there is no clear way to integrate the progress from these domains into a better understanding of life in the universe or its origin on Earth. Here I elaborate on the emerging concept of the *living state,* which may provide a framework to enable such integration.

References to a *living state* can be found throughout origin of life and astrobiology science [8, 9, 10, 11, 12, 13]. For different authors, the *living state* often has different meanings and connotations associated with it. For some, this term appears to be a convenient linguistic tool, used to describe the phenomena associated with biology [10, 9]. For others, this concept is intended to characterize life as a unique class of nonequilibrium processes [8, 13].

Perhaps the earliest mention of the *living state* was by the Nobel Laureate biochemist Albert Szent-Gyorgyi. In 1941, he wrote two very similar manuscripts, one for *Nature* and one for *Science* [11, 12]. In both he argues that to make progress biochemists must probe the submolecular structure of biomolecules [11, 12]. In particular, he drew inspiration from the electronic properties of crystals and semiconductors which were just becoming clear thanks to advances in statistical and condensed matter physics [11, 14]. Szent-Gyorgyi was struck by the collective behavior of electrons in semiconductors and hypothesized that similar principles were at play in the function of biomolecules [11, 14, 15]. He suggested that the deepest mysteries in biochemistry would only be explained by appealing to submolecular considerations. He went on to posit that certain features of the *living state* may be consequences of quantum mechanical laws [11, 12]. For Szent-Gyorgyi, the *living state* could be distinguished from

Cole Mathis, *Meaning of the Living State.* In: *Social and Conceptual Issues in Astrobiology.* Edited by: Kelly C. Smith and Carlos Mariscal, Oxford University Press (2020). © Oxford University Press. DOI: 10.1093/oso/9780190915650.001.0006 nonliving states based on the collective behavior of electrons. Interestingly, he rejected this idea later in this career, but it has recently seen renewed interest from other researchers [16, 17].

Since this first use by Albert Szent-Gyorgyi, the term the has been used by many more authors [8, 9, 10, 13]. Most of these authors use the *living state* when discussing the origin of life on Earth. These authors chose to investigate the origin of the *living state* rather than the origin of living cells, or organisms. The adoption of this term may be due to the realization that the "atoms" of biology cannot exist in isolation, physically or conceptually [18]. The description of living systems requires a specification of a macroscopic (or at least mesoscopic) system, which not only contains individual components (such as cells or organism) but also the nature of their interactions and their environment [18]. Therefore, the *living state* is used to refer to the essential features of biological processes that are not strictly contained within individual objects but rather manifest in the interactions between objects.

This use of the term can be found in a review of the progress on the RNA world hypothesis by Higgs and Lehman [9]. The RNA world posits that RNA played a crucial role in origin and early evolution of life on Earth [19]. In an RNA world scenario, RNA molecules are assumed to have been, at some point, the primary information carrying molecules required for primitive genetics, as well as the primary enzymatic molecules required for primitive metabolisms. Higgs and Lehman describe the evidence for an RNA world as well as the processes which would be required for it to exist. They report progress on RNA nucleotide synthesis, describe various models of RNA polymerization, and explore the concept of molecular cooperation [9]. In that review the authors define the *living state* to mean a state of the world in which the processes of enzymatic nucleotide synthesis, polymerization, and recombination are coordinated in a such a way that RNA molecules are reliably and robustly produced. This *living state* is contrasted to the *dead state* where all those processes may exist in an uncoordinated or unorganized manner (see specifically box 3) [9]. Thus these authors use the *living state* to identify the global scale organization necessary for the persistence of the RNA world.

Other researchers have used the concept of the *living state* to explicitly place biological phenomena within the epistemological scope of statistical physics [8, 20]. Within this framework biological phenomena at a given scale of organization (say, the cell) are explained and understood by appealing to the statistical properties of the dynamics of the smaller scales *and* larger scales. This is analogous to how distinct states of matter are understood by appealing to the statistical properties of atoms, with the important distinction that statistical physicists have historically not included constraints from larger levels of organization, which are essential in determining the properties of living systems. This conception of the *living state* may enable astrobiologists to integrate progress from different disciplinary perspectives into a quantitative theory of life. Living systems are influenced by many different processes, such as geological, geochemical, atmospheric, and astronomical processes [21, 22, 23, 24]. Understanding biological organization through the lens of the *living state* does not attempt to reduce all of these processes to physics but rather generalizes the approach of statistical physics to accommodate the diversity of phenomena seen in the biosphere. To understand how the tools of statistical physics can be used in this way, it is important to understand the history of that field.

## A Brief Synopsis of Statistical Physics

The goal of statistical physics is to reconcile the microscopic behavior of atoms or molecules with the macroscopic properties of materials. In the late 19th century the foundations of statistical mechanics were developed by Ludwig Boltzmann, Josiah Willard Gibbs, and James Clerk Maxwell [25]. At the time, the laws of thermodynamics were still being established but the primacy of thermodynamic descriptions of natural and artificial systems were widely accepted [25]. By contrast, there were still debates about the legitimacy of atomic theory [25]. Boltzmann's goal was to advance atomic theory by showing it was consistent with the known laws of thermodynamics [25]. To that end, Boltzmann calculated the average properties of particles interacting according to Newtonian mechanics. By taking the limit where the number of particles gets very large, Boltzmann proved that his formalism reproduced the second law of thermodynamics. In essence he demonstrated that the second law of thermodynamics was a statistically guaranteed consequence of Newton's laws of motion applied to a very large number of particles. This was the first explicit demonstration that a macroscopic theory (thermodynamics) could emerge from coarse-graining (in this case by averaging) a microscopic theory (Newtonian Mechanics).

The emergence of a macroscopic theory from a microscopic theory can be understood from the example of the ideal gas law. Gases are composed of a very large number of molecules. Each one of those molecules obeys Newton's laws of motion and therefore can be described by its velocity and position. If the number of particles in the gas is N, the number of parameters required to describe the gas using Newtons laws would be 6N, because each molecule has components of its velocity in three dimensions, similarly for its position. For any large number of particles the information required to describe the dynamical properties of a gas could become huge. However, it turns out that as a larger and larger number of particles are considered, the statistical properties of the gas become highly constrained [26]. These statistical constraints guarantee that the system will have certain features [26]. In the case of gases, those features are the pressure, temperature, and volume of the gas. In the thermodynamic limit, where the number of molecules goes to an arbitrarily large number, these features completely characterize the entire gas system [27].

By the early 20th century, atomic theory was widely accepted, thanks in part to Boltzmann and the development of quantum mechanics [14]. Around this time, research in statistical physics became organized around the concept of phase transitions [14, 28]. Some examples of phase transitions are the familiar phenomena of the melting of solid materials and the evaporation of liquids. Prior to the development of statistical physics, certain features of phase transitions were well understood experimentally [14]. For example, it was well known that pure metals had very specific melting points, thanks to the many industrial uses of metallurgy. However, experimental and theoretical interests in phase transitions were reinvigorated in the 1930s thanks to the discovery of superfluid helium and superconducting metals [14]. While the foundations of statistical mechanics and thermochemistry provided by Boltzmann and Gibbs had demonstrated that microscopic laws of motion acting on Newtonian particles could give rise to the macroscopic properties of materials, the study of phase transitions attempted to understand how the same microscopic laws applied to the same particles could give rise to such a diversity of macroscopic phenomena [29]. How was it that water molecules, subjected to the same microscopic laws of physics, could collectively exhibit the properties of a solid, liquid, or gas? The empirical facts provided by new phases of matter would elude theoretical explanation for most of the century [14, 30].

Quantum mechanics had provided a description of single (or few) electrons and their interaction with hydrogen nuclei, but these new phases of matter presented novel patterns in large systems with many electrons. These phenomena were some of the first examples of collective behavior in physics [30]. Understanding these processes required a set of theoretical tools known as the Renormalization Group (RG) [30]. The RG was developed simultaneously in statistical physics and quantum field theory [28, 30]. Initially these techniques were implemented in an ad hoc manner to deal with infinities that emerged in quantum field theories. However, the subsequent formalization of RG thanks to Freeman Dyson and later Kenneth Wilson demonstrated that RG techniques need not be ad hoc. The modern understanding of the RG is that it represents a set of tools to describe how different theories transform into each other when viewed from different perspectives [28, 29, 30].

Distinct states of matter emerge from similar microscopic systems because the collective behavior of the microscopic parts changes as larger and larger systems are considered [29, 30]. For example, the key difference between steam and liquid water is that individual molecules in steam have velocities seemingly independent from one another, whereas in the liquid state they are strongly correlated. This difference is not obvious at the microscopic scale. When observing a single molecule, whether in the gas or liquid, its motion will be correlated with the other molecules nearby due to intermolecular forces. However, as we consider more particles, the effect of this correlation tends towards zero in the gas because particles rarely interact in gases due to their low density. Meanwhile, in the liquid, with its higher density and therefore the higher interaction frequency, the effect of these correlations tends to increase. This qualitative difference (between zero and non-zero correlation) emerges as a consequence of quantitative differences in the microscopic dynamics and is responsible for the different macroscopic properties of the two phases [29]. In the study of phase transitions, these qualitative differences are usually tracked using *order parameters,* which are macroscopic properties that distinguish between different states. Often order parameters will take on a value of zero in one phase and a non-zero value in the other [28, 30].

The history of statistical mechanics is a story of reconciling different descriptions of nature. Equilibrium statistical mechanics was successful because Boltzmann demonstrated that the laws of thermodynamics emerge as a consequence of the dynamics of many-particle systems [25, 29]. Those properties, which are statistically guaranteed by the microscopic dynamics, end up defining thermal states at the macroscopic scale [26, 30]. The renormalization group demonstrated how systems with similar microscopic dynamics can result in different macroscopic states by formalizing how descriptions of those microscopic dynamics change as they are probed at different sizes or scales [29]. In summary, as a scientific enterprise, statistical physics in the 20th century provided answers to two very general questions [29]: (a) What are statistically guaranteed consequences of a given set of dynamics? and (b) under what circumstances do those consequences change? As a conceptual framework, the *living state* attempts to leverage these theoretical advances to integrate progress from many different fields into a coherent theory of living systems.

## Life as a State of Matter

Using the theoretical approach of statistical physics to investigate biological phenomena provides an opportunity to reconceptualize our understanding of biology. The notion of the *living state* emerges in the attempt to realize that theoretical approach. The *living state* is defined by the collection of all statistically guaranteed properties associated with the biosphere, in the same way that the gaseous state is defined by the pressure, temperature, and volume of the container. The framework of the *living state* does not necessarily propose a definition of life but rather a description of the features of life on Earth that are relevant at the large scale (both space and time). Currently there is no scientific consensus around which statistical properties must be associated with the biosphere; however, some suggestions include the topological properties of biochemical networks [31], interactions between the biosphere and abiotic surface processes [32], and the flow of electrons through organic matter [8]. This prospective has led many researchers to reevaluate established empirical data, as in [20], and it has led to new scientific questions [31, 33].

In the study of thermal states, the relevant properties emerge as the number of particles approaches $10^{23}$ (one mole). It is still not clear how to determine the appropriately large scale at which the relevant features of the *living state* emerge. Biologists study living systems at a number of different length and energy scales, from the molecular to the ecological. Recent advances in DNA sequencing, metagenomic analysis, and information sciences have enabled scientists to develop databases that span all of these scales [31]. These global databases have opened the possibility of studying life on Earth at the scale of the entire biosphere [8, 31, 32, 34, 35]. These studies have led some authors to suggest that the relevant features of the *living state* only emerge at the scale of the entire planet.

Viewing biological phenomena as a planetary scale processes represents a radical departure from many traditional perspectives in biology [36]. For example, Falkowski et al. argue that one of the most important features of the *living state* is the way in which it facilitates global scale cycling of material and energy by interacting with geological and atmospheric processes, which occur at a scale much larger than individual cells or populations [32]. They argue that these processes emerge not due to the dynamics of individual organism or even species. Instead they suggest that horizontal gene transfer is one of the key dynamical processes that statistically guarantees those features of the biosphere [32]. Prioritizing the role of horizontal gene transfer stands in stark contrast to most work in biology, which emphasizes the role of evolutionary dynamics by vertical descent in shaping the relevant features of living systems [36, 37].

While some researchers have suggested that the defining characteristics of the *living state* emerge at the scale of the biosphere, others (including myself) have suggested that the defining features of the *living state* emerge at many scales, not just one [31, 38]. We recently demonstrated this concept using biochemical reaction networks, which were constructed using genomic data [31]. We analyzed over 28,000 networks across three different scales of organization. We used individual genomes to construct networks for organisms, metagenomes to construct networks for ecosystems, and every known biochemical reaction to construct a network for the entire biosphere. By comparing the statistical features of these networks, we found that they shared certain properties that could not be explained simply by the shared rules of biochemistry (which are determined according to mass balance) [31]. These features appeared in genomes from different evolutionary domains, in metagenomes from different environments, and in the biosphere as a whole. The ubiquity of these features suggest that there may be underlying dynamical laws out of which these emerge as a statistical guarantee.

As the essential features of the *living state* are better characterized and understood they will help inform our understanding of the origin of life. In the context of the *living state,* the origin of life has a natural interpretation as a phase transition [8, 33]. Just as in thermal states where the laws of physics are the same for molecules in a gas or a liquid, the laws of organic chemistry are the same for carbon in the *living state* or in the nonliving state, but the macroscopic consequences of those laws are very different. Understanding how these macroscopic differences manifest will require identifying the relevant order parameters for distinguishing the living and nonliving states.

Contemporary research in the origin of life suggests a few candidate order parameters [8, 33]. Smith and Morowitz have argued extensively that the origin of life on Earth emerged as a response to planetary scale disequilibria [8]. This chemical disequilibria is due to the extremely different oxidation states of the Earth's mantle and atmosphere, where the relatively reduced mantle is much richer in electrons than the relatively oxidized atmosphere. They argue the biosphere dissipates this disequilibria by facilitating the flow of electrons from reduced sources in the mantle to oxidized sinks in the ocean and atmosphere. Accordingly, they suggest that the flow of electrons through organic carbon may be a key order parameter for the *living state* [8]. This conclusion is remarkably similar to Szent-Gyorgyi's original hypothesis that the collective behavior of electrons is responsible for the *living state.*

In my own work I have demonstrated that the origin of lifelike properties may be effectively tracked using information theoretic quantities [33]. We developed a chemical kinetic model of primitive replicators that are strongly coupled to a dynamic environment. In that model we observed two stable states, which dynamically emerged. In first state, labeled the non-life state, few replicators exist, and they are not selected according to their fitness. By contrast, the life state is dominated by replicators that were dynamically selected according to their fitness. To characterize the relationship between replicators and their environment we employed mutual information, which is a nonlinear measure of correlations. We saw that the transition from the non-life to the life state was tracked by these correlations, consistent with the idea that the *living state* is characterized more by the relationship between individual components rather than the components themselves [33].

Both features of the biosphere discussed here—life's interface with geochemical processes and the universal features of biochemical networks—may be independent of the particular of the details of terrestrial biochemistry. Any living system would be expected to interface with its planetary environment, and the universal features of biochemical networks cannot be explained by their shared biochemistry. Similarly, the two candidate order parameters for the living state discussed here do not require specific information about life on Earth. Accordingly, these features should be of great interest to astrobiologists who seek to understand life as it could be, not life as it is on Earth. The key to understanding the relevance of these features to biological organization lies in viewing life as a state of matter that manifests at many scales, not just at the scale of individual cells or organisms. By adopting this framework, astrobiologists can exploit the powerful theoretical tools and techniques of statistical physics to develop a theory that explains the interactions between the many biological and abiotic scales organization that characterize life on Earth, and (potentially) elsewhere in the universe.

## Conclusion

The *living state* is defined by the collection of statistically guaranteed properties associated with the biosphere. This concept emerges when scientists attempt to apply theoretical concepts from the field of statistical physics to characterize biological systems. Adopting this prospective leads to new scientific hypotheses regarding the nature of life on Earth as well as its origins. These new research directions assume that "life" is a phenomena that manifests at a macroscopic scale and attempt to identify the key parameters characterizing that phenomena. These features may be independent of particulars of Earth life's chemistry and would therefore be useful in guiding searches for life beyond our planet. Thus, the concept of the *living state* may prove fundamental in the future of astrobiology.

## References

[1] C. Scharf, N. Virgo, H. J. Cleaves, M. Aono, N. Aubert-Kato, A. Aydinoglu, A. Barahona, L. M. Barge, S. A. Benner, M. Biehl, et al., "A strategy for origins of life research," *Astrobiology*, vol. 15, no. 12, pp. 1031–1042, 2015.
[2] L. J. Mix, "Defending definitions of life," *Astrobiology*, vol. 15, no. 1, pp. 15–19, 2015.
[3] S. A. Tsokolov, "Why is the definition of life so elusive? Epistemological considerations," *Astrobiology*, vol. 9, no. 4, pp. 401–412, 2009.
[4] M. Eigen, "Natural selection: a phase transition?," *Biophysical chemistry*, vol. 85, no. 2–3, pp. 101–123, 2000.
[5] N. Lane, J. F. Allen, and W. Martin, "How did LUCA make a living? chemiosmosis in the origin of life," *BioEssays*, vol. 32, no. 4, pp. 271–280, 2010.

[6] M. A. Nowak and H. Ohtsuki, "Prevolutionary dynamics and the origin of evolution," *Proceedings of the National Academy of Sciences of the United States of America*, vol. 105, no. 39, pp. 14924–14927, 2008.

[7] S. I. Walker, W. Bains, L. Cronin, S. DasSarma, S. Danielache, S. Domagal-Goldman, B. Kacar, N. Y. Kiang, A. Lenardic, C. T. Reinhard, et al., "Exoplanet biosignatures: future directions," *Astrobiology*, vol. 18, no. 6, pp. 779–824, 2018.

[8] E. Smith and H. J. Morowitz, *The Origin and Nature of Life on Earth: The Emergence of the Fourth Geosphere*. Cambridge, UK: Cambridge University Press, 2016.

[9] P. G. Higgs and N. Lehman, "The RNA world: molecular cooperation at the origins of life," *Nature Reviews Genetics*, vol. 16, no. 1, pp. 7–17, 2015.

[10] L. Cronin and S. I. Walker, "Beyond prebiotic chemistry," *Science*, vol. 352, no. 6290, pp. 1174–1175, 2016.

[11] A. Szent-Gyorgyi, "The study of energy–levels in biochemistry," *Nature*, vol. 148, no. 3745, p. 157, 1941.

[12] A. Szent-Gyorgyi, "Towards a new biochemistry?" *Science*, vol. 93, no. 2426, pp. 609–611, 1941.

[13] A. Pross, "Toward a general theory of evolution: extending Darwinian theory to inanimate matter," *Journal of Systems Chemistry*, vol. 2, no. 1, p. 1, 2011.

[14] L. Hoddeson, E. Braun, J. Teichmann, and S. Weart, *Out of the crystal maze: chapters from the history of solid state physics*. Oxford, UK: Oxford University Press, 1992.

[15] A. Szent-Gyorgyi, "The living state and cancer," *Proceedings of the National Academy of Sciences of the United States of America*, vol. 74, no. 7, pp. 2844–2847, 1977.

[16] A. Szent-Gyorgyi, *Introduction to a submolecular biology*. New York, NY: Academic Press, 1960.

[17] G. Vattay, D. Salahub, I. Csabai, A. Nassimi, and S. A. Kaufmann, "Quantum criticality at the origin of life," *Journal of Physics: Conference Series*, vol. 626, p. 012023, 2015.

[18] D. J. Nicholson, "Biological atomism and cell theory," *Studies in History and Philosophy of Science Part C: Studies in History and Philosophy of Biological and Biomedical Sciences*, vol. 41, no. 3, pp. 202–211, 2010.

[19] J. F. Atkins, R. F. Gesteland, and T. R. Cech, eds., *The RNA World* (3rd ed.). Plainview, NY: Cold Spring Harbor, 2005.

[20] E. Smith and H. J. Morowitz, "Universality in intermediary metabolism," *Proceedings of the National Academy of Sciences of the United States of America*, vol. 101, no. 36, pp. 13168–13173, 2004.

[21] G. A. Dolby, S. E. Bennett, A. Lira-Noriega, B. T. Wilder, and A. Munguía-Vega, "Assessing the geological and climatic forcing of biodiversity and evolution surrounding the gulf of California," *Journal of the Southwest*, vol. 57, no. 2, pp. 391–455, 2015.

[22] E. L. Shock and E. S. Boyd, "Principles of geobiochemistry," *Elements*, vol. 11, no. 6, pp. 395–401, 2015.

[23] T. T. Huynh and C. J. Poulsen, "Rising atmospheric $CO_2$ as a possible trigger for the end-Triassic mass extinction," *Palaeogeography, Palaeoclimatology, Palaeoecology*, vol. 217, no. 3-4, pp. 223–242, 2005.

[24] C. H. Lineweaver, Y. Fenner, and B. K. Gibson, "The Galactic habitable zone and the age distribution of complex life in the Milky Way," *Science*, vol. 303, no. 5654, pp. 59–62, 2004.

[25] C. Cercignani, *Ludwig Boltzmann: the man who trusted atoms*. Oxford, UK: Oxford University Press, 1998.

[26] H. Touchette, "The large deviation approach to statistical mechanics," *Physics Reports*, vol. 478, no. 1, pp. 1–69, 2009.

[27] E. Smith, "Large-deviation principles, stochastic effective actions, path entropies, and the structure and meaning of thermodynamic descriptions," *Reports on Progress in Physics*, vol. 74, no. 4, p. 046601, 2011.

[28] N. Goldenfeld, *Lectures on phase transitions and the renormalization group*. Advanced Book Program. Reading, MA: Addison-Wesley, 1992.

[29] L. P. Kadanoff, *Statistical physics: statics, dynamics and renormalization*. Singapore: World Scientific, 2000.

[30] L. P. Kadanoff, "More is the same; phase transitions and mean field theories," *Journal of Statistical Physics*, vol. 137, no. 5–6, p. 777, 2009.

[31] H. Kim, H. B. Smith, C. Mathis, J. Raymond, and S. I. Walker, "Universal scaling across biochemical networks on Earth," *Science Advances*, vol. 5, no. 1, p. eaau0149, 2019.

[32] P. G. Falkowski, T. Fenchel, and E. F. Delong, "The microbial engines that drive Earth's biogeochemical cycles," *Science*, vol. 320, no. 5879, pp. 1034–1039, 2008.

[33] C. Mathis, T. Bhattacharya, and S. I. Walker, "The emergence of life as a first-order phase transition," *Astrobiology*, vol. 17, no. 3, pp. 266–276, 2017.

[34] R. Braakman, M. J. Follows, and S. W. Chisholm, "Metabolic evolution and the self-organization of ecosystems," *Proceedings of the National Academy of Sciences of the United States of America*, vol. 114, no. 15, pp. E3091–E3100, 2017.

[35] Y. M. Bar-On, R. Phillips, and R. Milo, "The biomass distribution on Earth," *Proceedings of the National Academy of Sciences of the United States of America*, p. 201711842, 2018.

[36] P. Godfrey-Smith, *Philosophy of biology*. Princeton, NJ: Princeton University Press, 2013.

[37] D. C. Krakauer, J. P. Collins, D. Erwin, J. C. Flack, W. Fontana, M. D. Laubichler, S. J. Prohaska, G. B. West, and P. F. Stadler, "The challenges and scope of theoretical biology," *Journal of Theoretical Biology*, vol. 276, no. 1, pp. 269–276, 2011.

[38] S. I. Walker, "Origins of life: a problem for physics, a key issues review," *Reports on Progress in Physics*, vol. 80, no. 9, p. 092601, 2017.

# 7

# Life as It Could Be

*Luis Campos*

*Life as it could be* could mean many different things. Life as it could have been? As it could be now, as yet undetected or unknown, here on Earth or elsewhere? As it could be at some point yet to come? As *we* might make it? "Life as it could be" incorporates all these possibilities—all of these many ways to think about the limits of life, ranging from the astrobiological to visions for the engineering of life here on Earth.

At the first Blumberg astrobiology symposium held at the Library of Congress in 2013, an invitation was made to think about "a world in which biology proliferates, in which we have a lot of different biologies." What if what nature has provided here on Earth is not the only biology possible? one speaker asked his fellow attendees. What if at this particular moment in history we might be facing "not the total depletion of biodiversity" but find ourselves instead to "be at the edge of a new Cambrian explosion of biodiversity"? These questions were framed from within the discourse of astrobiology but could just as easily have been presented by synthetic biologists interested in envisioning biology as kind of technology platform generating endless forms most beautiful (and useful).

Pushing the engineering of life past traditional limits in molecular biology and expanding the envelope of life to forms never before extant, synthetic biologists are now beginning to design experimental ways of getting at what astrobiologists have long suspected: that the life we know here on Earth is but a subset of vast combinatorial possibilities in the universe. "Most of biotechnology has yet to be imagined, let alone made true," the synthetic biologist Drew Endy of Stanford University said at the Synthetic Biology 7.0 meeting in Singapore in 2016. What might it be like, he asked, if we were able to "reimplement life in a manner of our own choosing?"[1] What if we moved beyond the attributes "we inherited from the continent of natural lineages? You can imagine arriving on a new continent where we are totally unconstrained by the lineages that preexist in nature," he conjectured. "Instead of just imagining the world as it exists and as we inherit it from nature, I think it is becoming increasingly important that we understand how to *imagine* worlds that might be, how we would choose to design and construct them."

Luis Campos, *Life as It Could Be*. In: *Social and Conceptual Issues in Astrobiology*. Edited by: Kelly C. Smith and Carlos Mariscal, Oxford University Press (2020). © Oxford University Press. DOI: 10.1093/oso/9780190915650.001.0007

New Cambrian explosions, new continents, and untold worlds of biodiversity—such metaphors, attempting to conceive the previously inconceivable, resonate between the future engineered possibilities of our world and speculations about possible biologies on habitable others. But these resonances are not merely happenstance moments—in fact, there is a curious and compelling deeper history interlinking scientific speculation about new forms of life elsewhere in the universe with visions for the human-directed engineering of new forms of life on Earth. For decades already, the astrobiological and the synthetic biological have mutually inspired each other and even overlapped in fascinating genealogical ways. A brief survey of some of the more prominent points of intersection over the last century will better illustrate this claim, which has heretofore often been missing from the narratives that both astrobiologists and synthetic biologists have respectively told about their own fields. I then conclude by showing that such overlaps are not merely metaphorical, discursive, and conceptual in nature but interweave science and science fiction and even play a central role in one of the great stories of scientific mentorship of the 20th century.

The science fiction novelist Arthur C. Clarke once claimed that "The best book ever written about the future opens with these words: 'There are two futures, the future of desire and the future of fate, and man's reason has never learned to separate them.'" These opening lines come from a book published in 1929 by the Irish X-ray crystallographer J. D. Bernal and present one famous early starting point for the crossroads of astrobiology and synthetic biology. Bernal's book spoke of many things, even including visions of transgenerational human spaceflight to far-off worlds and the transhuman evolutionary modifications such travel might entail over generations. Titled *The World, The Flesh and the Devil*—a sort of literary invocation of familiar Biblical themes—Bernal sought to explore and attack "the three enemies of the rational soul": the world, as in leaving the planet; the flesh, that we should cause "interference in a highly unnatural manner with our germ plasm"; and the devil—that we need to deal with our "desires and fears . . . imagination and stupidities." The extension of life from Earth to elsewhere, and the concomitant biotechnologies of the self that would be necessary to do so, are a constant thread running through his thought and challenged any easy distinction between what nature might provide (what we might call the nature of fate) and an engineered aesthetic (the nature of desire):

> The new life would be more plastic, more directly controllable and at the same time more variable and more permanent than that produced by the triumphant opportunism of nature. . . . Such a change would be as important as that in which life first appeared on the earth's surface and might be as gradual and imperceptible. . . . The need to determine the desirable form of the humanly-controlled universe . . . is nothing more nor less than art.[2]

Our envisioned synthetic futures of life in space are as cosmic in their significance for Bernal as the origin of life on earth itself.

A few years later, the celebrated English naturalist Julian Huxley picked up on how nature seemed constantly to be engineering itself: "Evolution is one long sermon on the text of the infinite plasticity of living matter," he wrote in 1931.[3] And even some twenty years later, in 1953, he returned to the theme to highlight again the role of human ingenuity in evolution: "the destiny of man on earth . . . is to be the agent of the world processes of evolution, the sole agent capable of leading it to new heights, and enabling it to realize new possibilities . . . to ever-fresh realizations of new possibilities for living substance . . . leading life into regions of new evolutionary opportunity." Huxley noted that "man finds himself in the unexpected position of business manager for the cosmic process of evolution" and that we "are the agents of further evolution, and that there can be no action higher or more noble than raising the inherent possibilities of life."[4]

By the 1950s, understanding the cosmic process of evolution frequently entailed studying the origin of life on the early Earth, a common touchstone for the forerunners of both astrobiology and synthetic biology. For many inspired by the famous Urey-Miller experiment in 1953 (which zapped gases with electricity to produce amino acids), attempts to experimentally study the origin of life on Earth meant trying also to understand how life might be synthesized in the test tube. Indeed, attempting to synthesize life (or its immediate antecedents) from scratch was a key experimental method for investigating the potentialities and prospects of life. And at midcentury, such efforts were international in scope: "The date of the Moscow International Symposium 'on the origin of life on earth'" in 1957, noted one scientist, "will probably be remembered in later years as the time when it finally became respectable in scientific circles to admit a more than ideological interest in the problem of how to make life in the laboratory."[5] Indeed, so clear was the shared goal of studying possible habitats for life's emergence and even potentially making a primordial form that within a short time one lab at NASA Ames had secretaries answering the phones saying, "life synthesis?"[6] (The lab was later renamed "chemical evolution.")

But not only were the study of the origin of life and the effort to make life in the test tube related quests at midcentury: such efforts were frequently framed within a larger cosmic context, and with consideration of how life might have emerged on *other* planets as well.[7] Already by midcentury, the forerunners of astrobiology and synthetic biology were conceptually joined at the hip. The construction of life by humans or by nature—"life synthesis" or "chemical evolution"—was of deep interest.

Even as the American microbial geneticist and Nobel laureate Joshua Lederberg coined the term "exobiology" in 1959, he also envisioned the field's history intertwined with the earthly biology with which he was more familiar.

Linking the terrestrial with the heavenly, while now a commonplace for astro-biology, grew out of efforts to understand fundamental problems: "The prime questions of exobiology, life beyond the earth, concern molecular biology," Lederberg noted. He wondered: "Do the Martian organisms use DNA and amino acids as we do, or are there other solutions to the basic problem of the architec-ture of evolution?"[8] What other multiple biologies were possible? (How might life be?)

"The age of synthesis is in its infancy, but is clearly discernible," the biol-ogist James Danielli noted in 1972, tying together the astrobiological with the engineered. "Life synthesis techniques make it possible to explore all possible combinations of genes which are viable." Struck by the astronomical possibilities of genomical combinations, he envisioned that number of novel genomes "is so huge that only a minute proportion can ever have existed, at any time, on earth" and he quoted others who asserted that

> the evolutionary process has not sorted through most of the possible genomes so that those most efficient in a particular environment have had the oppor-tunity to exist. It necessarily follows that, even considered from the point of view of ability to compete in a given environment, most of the more efficient organisms do not exist. Even more so, most of the organisms which *could* exist to fulfill the special demands of civilization do not exist now, but can be brought into being by using various combinations of techniques for life synthesis. Without going into detail, life synthesis techniques make it possible to explore all possible combinations of genes which are viable.[9]

Space might hold astronomical possibilities for life, but life held astronomical possibilities within itself.

These synthetic biological futures mirrored other industrial futures Danielli envisioned at midcentury, and brought still other synthetic futures nigh:

> The present mechanical-style computer industry may be displaced by biological-style computer design . . . increased automation and control of in-dustry by biological-style computers may accompany a disappearance of the classical need to work . . . the goals of society may become, on the one hand, much more decisively the experience of the inner world, of the total range of possible experience, and, on the other hand, the intensive exploration of the galaxy.

The futures of inner synthesis and of outer space were therefore never far apart: life synthesis, would lead to nothing less than a utopian workplace with time for the thoughtful consideration and exploration of the galaxy. While Bernal worried about overcoming the devils of our desires and fears, Danielli foresaw cosmic frontiers for industry and leisure alike. He also even theorized how scientists might design organisms for Mars and thought the process would be similar in conceptual terms to developing organisms for the treatment of sewage. The ultimate challenge was simply to learn how to "take components from different organisms and put them together and get a working cell. Eventually we hope to be able to set boundaries to what components can be mixed together." Whether for practical purposes on Earth, or for extraterrestrial purposes on Mars, what is clear is that even in the years before the actual invention of recombinant DNA techniques that we now call "genetic engineering," prominent visions of "life as it could be" incorporated simultaneously astrobiological and synthetic biological dimensions.

It is also striking that the emerging concerns in astrobiology over back- and forward-contamination ("planetary protection" in today's lingo) came to have a second life in synthetic biology. The same language of contamination and containment—and of evolutionary or laboratory barriers—traveled across both fields, aided clearly in part by *The Andromeda Strain*, a popular new sci-fi thriller written by Michael Crichton in 1969. Through fiction and discussion of speculative futures, exobiological concerns about contamination came to intersect with more terrestrial concerns about the potential hazards of genetically engineered forms of life newly constructed in the laboratory on Earth.[10] In fact, Crichton's novel was explicitly invoked by molecular biologists who participated in the famous 1975 Asilomar conference proposing laboratory safeguards for recombinant DNA work. These sorts of cases suggest that we would do well to seek to understand the unexpected and sometimes unruly cultural narratives that condition and may even deeply structure the development of new sciences into "life as it could be." Sometimes, fictions matter.

Time and again over the past century, the astronomically large possibilities for life, of systems of heredity and their mutations, and of the niches they might exist in have overlapped and resonated with the astronomically large numbers of worlds of possibility revealed by astronomy itself. The statistics and thinking about the one routinely informed the statistics and thinking about the other: even as astronomy framed the space of possible celestial abodes for life, the combinatorial possibilities of synthesized life became routinely framed in astrobiological ways.

Although the terms used have changed, the question of the mutual relations between astrobiology and synthetic biology has long been a focus of concern. One 1962 National Academies report saw "cosmobiology, man into space, and environmental biology" as fundamentally related endeavors.[11] While the search for extraterrestrial life came first in importance, and the "immense task for the biological engineer of putting man into space adequately protected" came second, the report noted that "[i]n a real sense they are three aspects of one and the same undertaking—a general extension of the scope of nature that the biologist can bring within his grasp."[12]

This history of moments of mutual inspiration has only been briefly sketched here, and more research will undoubtedly bring other connections to light. But it is clear enough that, even in more recent years, astrobiology and synthetic biology continue to mutually inspire each other. The first man to have his genome sequenced, J. Craig Venter, often talks about beaming back digital DNA sequences from Mars. But he has also described how such a proposal, an outgrowth of his interest in minimal genomes and "minimal life," was inspired by discussions in the mid-1990s about the possibility of nanobacteria found in Martian meteorites. One could observe and wonder how small life could be morphologically, Venter said, or one could experiment to figure it out genetically, experimentally. In other words, wondering about putative "life as it could be" from Mars was a direct inspiration to wondering about "life as it could be" down here—and not how we might find it in nature but how we might *make* it to be. This is doubtless only one of many moments of astrobiological inspiration behind more recent synthetic biological work—and vice versa.

But of the many rich histories of "life as it could be" remaining to be told, there is one that qualifies as perhaps the most striking, highlighting in deeply personal

**Figure 7.1.** "The red thread slowly weaves its way upwards." A birthday card sent from Carl Sagan to H. J. Muller in 1955.
*Source:* Courtesy, The Lilly Library, Indiana University, Bloomington, Indiana.

and biographical ways the shared roots and mutual overlap of astrobiological and synthetic biological concerns. This story begins in the middle of the century, when one of the foremost scientists of his generation, a visionary for the future of evolution under human control—Hermann J. Muller—had won the Nobel Prize. Some years later, he received a letter from a college student who would one day become one of the most famous astrobiologists of all time: Carl Sagan. "Astrobiology is coming of age," Sagan wrote to Muller, on September 15, 1958.[13] Sagan was twenty-three; Muller was sixty-eight.

Sagan needs no introduction. And Muller, in the history of biology, is a legendarily significant and fascinating figure. In the early years of the 20th century, Muller worked in the fly laboratory at Columbia University, one of the epicenters of classical genetics, and made a number of fundamental discoveries. Muller's early interest in the uses of the new radioactive element radium eventually led him to the use of x-rays to produce mutations in the hereditary substance. He succeeded beyond his wildest dreams in 1926 and essentially inaugurated the modern study of radiation genetics.[14] With the development and use of the atomic bomb, his work seemed even more important and urgent: Muller's insights into the effects of ionizing radiation on the hereditary material were regularly sought out, and he received the Nobel Prize for his work in 1946, not coincidentally after the bombings of Hiroshima and Nagasaki. His fame only increased in a Cold War world gaining new awareness of the dangers of fallout from nuclear testing. For how long would our world remain habitable? Muller wondered, at just the moment that pioneering astrobiologists began to ask such questions of other worlds.

It was during a celebratory lecture at Indiana University in 1947, one attended by over 4,000 people, that Muller publicly tied together his life's work and indicated an interest not only in the "possibilities" open to humanity for "remaking the earth and we can now be sure, adventuring upon other planets," he said, "but the biological ones of remolding the life forms around us . . ."[15] Muller was not simply an expert geneticist who sought to understand how nature had made things: time and again, he actively sought to explore what he called "the wondrous potentialities of development thus disclosed in life forms," "the basic possibilities of living things,"[16] and the prospects of life as it could be. If we find a way to put our wishes "in harmony with biological possibilities," he concluded, "the world of plants and animals should be increasingly ours to remold as we choose."[17] Biology held nothing but potential.

At a symposium on the origin of life held at Yeshiva University in 1959 centrally concerned with exobiology, Muller said:

Surely we will get to Mars within the next generation, to be conservative, if we do not have nuclear war. Surely, considering the inhospitality of Martian conditions, we will find that life has not evolved nearly as far or become nearly as abundant or diverse, as on the earth.

Will it too have its basis in some kind of genetic nucleic acids? If I could be put into a deep freeze and thereby live until that day, I would wager being dropped in boiling oil if it were not composed of nucleic acid. Beyond that anyone would be rash to predict the chemistry, morphology and physiology of that life on its upper levels. . . . This will be one of the most fascinating fields for the biologist, biochemist and geneticists of one or two generations hence, if not of some of you. Let us not forget here that our solar system is one speck in many billions in our own galaxy . . . and there are many billions of other galaxies.[18]

Muller was fascinated by the biological possibilities of these billions and billions of potential abodes for life. But why stop with billions? "Our own sun is not 'the only pebble on the beach,'" he noted: "it is probably a very gross underestimate to say that many trillions of planets besides our own are at this moment serving as the abode of life."[19]

"In many other worlds besides ours living matter must co-exist," he wrote, "and that in the star-bespattered abysses of the cosmos there must surely dwell a myriad exotic forms. . . . How much we may some day be able to learn of this life, though super-telescopes and other scientific telesenses, without coming into direct contact with it, is at present an unanswerable question."[20] In a 1961 piece titled, "Life Forms to Be Expected Elsewhere than on Earth," Muller even wondered if "our Earth in turn may be poor in life compared with what exists on some of the even more favoured planets of other stars."[21] Muller was fascinated with endless prospects for life as it could be: "What a comparative study all this would make and how enlightening."[22]

The outer astrobiological possibilities were mirrored by inner genetic ones for Muller: "We may bear in mind that we too contain, *within* each one of us, a veritable universe. . . . in addition to the outer world we have an inner world to understand and to administer, a world no less intricate, and no less directly important to us all."[23] There were "new opportunities" having to do with "the strange winds that are now blowing in from the atom, the gene, and outer space."[24] "We transitional creatures," he wrote, "must not shrink from our destiny or fear it" but must work "in functional alliance with our genes"[25] so that "evolution will become, for the first time, a conscious process": "That will be the highest form of freedom that man, or life, can have."[26] Life unlimited.

Muller even predicted that the time would come when even our "machines" will be "more like living organisms,"[27] a theme Danielli would later develop further. But all of this was but a prelude: "Even those, however, seem but tame beside the problem that challenges all the daring of the race—the problem of extraterrestrial projection." By this he meant that "man, as a species is at the beginning of an adventure whose epic strains may ring down the corridors of the universe." Evoking Bernal's own tale of the escape of earthly life into space, Muller argued that "man need no longer remain confined to his planet of origin" and that "[s]oon he will venture into the cosmos and his jobs of external creation will have begun in earnest."[28] In this way, as in Huxley's vision of a "business manager for the cosmic process of evolution," the astrobiological and the synthetic were again intertwined, with the sublime language of a higher calling:

> It is up to us to do our bit . . . and to use what we know constructively. . . . Our reward will be that of helping man to gain the highest freedom possible: the finding of endless worlds both outside and inside himself, and the privilege of engaging in endless creation.[29]

Apart from "reaching Mars and in probing into the nature of its organisms," Muller noted, which would otherwise be "the most exciting story in the exploration of life that has ever happened to man," the most exciting development was what is "going on right now in those laboratories of ours where biochemists and geneticists are disentangling the warp and woof of which our own earthly life is composed."[30] Lecturing under the title "Man's Responsibility for His Genetic Heritage," and only a month after the launch of the first Soviet satellite, Muller cautioned of the need for international cooperation in this endeavor: "The world cannot afford to allow to individual countries their separate genetic sputniks!"[31]

It was in the heart of this period of Muller's greatest thinking about the cosmos and life as it could be—the prospects both for astrobiology *and* synthetic biology—that he encountered the young Carl Sagan. The record appears to begin around 1952, when a seventeen-year-old Carl Sagan wrote to Muller, describing a radio program "Ad Astra" he was hosting on the local station WUCB-Chicago. Muller replied to Sagan's questions and concluded his letter with a generous flourish: "You are welcome to take anything you wish from the above, and to condense or paraphrase."[32]

A warm mentoring relationship was quickly established, and the two even went together to a science fiction congress in Chicago in 1952. (Sometimes, indeed, fictions matter.) Two years later, Sagan sent Muller a science fiction story of his own, which Muller found "very amusing."[33] Not only did Sagan end up sending Muller a copy of his undergraduate thesis that year, but he even dedicated it to him, "To H. J. Muller, with sincere appreciation." The following year, Muller went to an American Institute of Biological Sciences meeting in East Lansing in 1955 to hear Sagan present, and that same year Sagan followed up on a conversation and wrote to Muller to say: "It is sad to think that there may be no biologists on the first expedition to Mars."[34] Sagan the budding planetary scientist and astrobiologist and Muller the geneticist and biological engineer: the two men talked over the course of years about the limits of life, what life forms might be found elsewhere, and how life might adapt to new environments beyond the Earth. And Mars was a common thread in Muller and Sagan's conversations. On April 26, 1956, one of them even signed off, "Take my greetings along to the Martians." (The American geneticist Jim Crow would later remark that "if it were possible to send a man to Mars and bring him back safely and quickly, my candidate would be H. J. Muller.")[35]

The perfect visual encapsulation of this interrelationship is a birthday card, sent from Carl to Joe: an image of Mars, the best available at the time in 1955. Mars is crossed by a red diagonal thread, above which Sagan has written: "The red thread slowly weaves its way upwards." Muller often referred to the chromosomes as threads, so Sagan's inscription is clearly indebted to his contact with Muller. But why red? And why was this reference so significant to the two of them? As it happens, in his *Out of the Night*, Muller uses a metaphor of a cord or thread to describe the history of "organic evolution on the earth," where each yard stands for 10,000 years. He describes a human generation as occupying less than an eighth of an inch and a cross-section of the cord the size of an aspirin tablet as representing the portion of one generation:

> Now this is just equal to the volume of hereditary material which actually is contained in one generation of mankind, and which is to be passed on to the next generation. . . . Hence our cord now acquire a further symbolic significance, and in that it may be taken as representing in a certain real physical sense the evolving germ plasm of ourselves and our ancestors. . . . Within this cord the fine fibers represent the chromosomes themselves, which are in fact filamentous bodies that intertwine, separate, and reunite in diverse ways as they pass along from generation to generation in the varying combinations resulting from sexual reproduction.
>
> At any given place there is but a single one, out of all the mass of cords, which has led on so as finally to issue in our branch; this may be distinguished, in our figurative representation, by giving it a red color. It is this red cord which may be regarded as the red "thread of destiny," in a rather literal sense. Its free end is even now being spun further, being transfigured by mutation, being twined and interwoven, to give a new sort of living world, dependent on its new properties.

Muller's dreaming of biological possibilities—of life as it could be—not only spanned the astrobiological and the synthetic biological: they were one and the same project. Man could turn life to his own advantage, Muller concluded: "He can take the red cord of life, the thread of destiny, from the hands of Clotho, and spin it for himself"[36] —but not without care for the rest of life:

There are more than a million species on this earth, not counting the races and varieties of each and every one of them is connected in devious and complicated ways with many of the other species, so that the pulling of a thread here may cause unexpected warpings and tearings of the fabric elsewhere. Each type of organism, moreover, is a veritable world within itself.

To talk about the future of life, and life as it could be, is—as Muller already well knew—to talk in different registers about what others have long called fate, or, perhaps even the Fates (and what we might take from their hands). Muller's thoughts about the future of life in astrobiology and synthetic biology were therefore deeply informed by his broad humanistic outlook on the world. Finding and tracing threads of associations across examples and cases was part and parcel of his scientific work, as well as of his communication of that work to broader audiences of undergraduates and the general public. Far from being reduced to downstream implications of scientific work, a humanistic approach provided invaluable upstream roots and contexts for the development of Muller's thought— and for Sagan's own inspiring perspective.

It is no surprise, then, why Sagan would write to Muller in 1958 that "Astrobiology is coming of age." By the next year, Sagan reported to Muller that "Interest in extraterrestrial biology appears to be blossoming at last. . . . Perhaps you were right all along, and my scepticism was unfounded. The prospects are utterly beautiful. Even in places where we had always rejected the possibility of life out of hand we may be mistaken." Sagan described his plans to complete his PhD in the next year, and that his next steps would one way or another enable him to keep his "fingers still firmly in the astrobiological pie." And it is in this letter where the most remarkable passage appears where Sagan writes about the birth of his first son the day before, using Muller's language to describe his own deeply personal experience: "It feels strange adding our fiber to the red thread. I've never before had so strong a feeling of being a transitional creature, at some vague intermediary position between the primeval mud and the stars."[37] Years later, after Muller himself passed away, returning to stardust, Sagan wrote to his widow, Thea, acknowledging Muller's role in fostering his interest in the search life on other planets. Sagan picked up weaving the thread of astrobiology and synthetic biology together still further.

Apart from the 1,705 boxes of Sagan's manuscript material cataloged and available for research at the Library of Congress, some small fraction of Sagan's library is also preserved elsewhere, on a hidden exoshelf in the far reaches of a nearby building. And somewhere on a particular shelf there is a book stamped "Compliments of the Author"—none other than Muller's *Studies in Genetics*. Inside is one of those treasures that only exists on paper, in real life, in a book, somewhere on a shelf on a hidden floor down a long passageway in a grand building of the most august library in the land—an inscription from Muller to Sagan:

> Here is a record of some explorations into the universe inside—explorations of a type now as common as a voyage across the Atlantic. But to us these trips were as wondrous as, in its day, Hanno's circumnavigation of Africa. What a triumph it will be when we can coordinate and combine our outer and inner quests. Christmas and New Years greetings to you and yours for 1962-1963, and may we some day meet "in spirit" on the tundras of Mars.

That one of the greatest visionaries of engineered futures for biology should be the mentor to one of the most famous advocates for looking for life on other worlds is remarkable enough on its own. But it also suggests that not only have astrobiology (in discovering new contexts for life) and synthetic biology (in seeking to engineer new forms of life) shared common threads and not only are both concerned in some general way with "life as it could be"—but that these endeavors are themselves descended from complexly interwoven sets of common ancestors. As we seek to explore the nature and numbers of trees of life on Earth and perhaps elsewhere, and as we consider the many meanings of "life as it could be," it may be worth recalling the ways in which seemingly separate intellectual and scientific traditions might themselves demonstrate an unexpected common genealogical heritage, one that is ours to discover *and* engineer. Genetic sputniks, indeed!

## Notes

1. Regis, *What Is Life: Investigating the Nature of Life in the Age of Synthetic Biology,* Oxford University Press, 2009, 133.
2. John Desmond Bernal, *The World, The Flesh, and the Devil: An Enquiry into the Three Enemies of the Rational Soul,* Kegan Paul, 1929, 46.
3. Julian Huxley, *What Dare I Think?* Chatto & Windus, 1931, 37.
4. Julian Huxley, *Evolution in Action,* Chatto & Windus, 1953, 32, 149, 173.
5. Hans Gaffron, "The Origin of Life," in Sol Tax, ed. *Evolution After Darwin*, volume 1, "The Evolution of Life," University of Chicago Press, 1969, 41.
6. This was Cyril Ponnamperuma's lab. See Steven J. Dick and James E. Strick, *The Living Universe: NASA and the Development of Astrobiology*, Rutgers University Press, 2005, 36.

7. See, for example, Melvin Calvin, "Origin of Life on Earth and Elsewhere," *Proceedings of Lunar and Planetary Exploration Colloquium* 1, no. 6 (April 25, 1959): 8–18.

8. "Biological Future of Man," in G. E. W. Wolstenholme, ed. *Man and His Future*, Churchill, 1963, 263–273, on 271.

9. J. F. Danielli, "The Artificial Synthesis of New Life In Relation to Social and Industrial Evolution," in F. J. Ebling and G. W. Heath, eds., *The Future of Man*, Academic Press for the Institute of Biology, 1972, 96.

10. See Luis Campos, "Strains of Andromeda," forthcoming.

11. *A Review of Space Research, The Report of the Summer Study conducted under the auspices of the SSB of the NAS at the State University of Iowa, Iowa City, IA, June 17–August 10, 1962*, Publication 1079, NAS, 1962, 9–1.

12. In addition, the report concluded, "the economic costs will be amply repaid in the long run by application of space-oriented biotechnology to other fields of biology and medicine."

13. Sagan to Muller, September 15, 1958. Muller mss., Lilly Library, Indiana University, Bloomington.

14. Luis Campos, "The Gene Irradiated," *Radium and the Secret of Life*, University of Chicago Press, 2015.

15. Muller, "Changing Genes," Indiana University convocation, 1947.

16. Muller, "The Immediate Biological Future," *New Frontiers of Knowledge* 1957: 56–59, on 57.

17. H. J. Muller, "Redintegration [sic] of the Symposium on Genetics, Paleontology, and Evolution," in Glenn L. Jepsen, Ernst Mayr, and G. G. Simpson, *Genetics, Paleontology, and Evolution*, Princeton University Press, 1949, 421–445, on 443.

18. Muller, "Genetic Nucleic Acid, The Key Material in the Origin of Life," 69–70, transcript of symposium on "The Origin and Nature of Living Matter," Yeshiva University, June 8–9, 1959.

19. Muller, "Man's Place in Living Nature," *Science Monthly* 89 (1957): 245.

20. Muller, *Out of the Night: A Biologist's View of the Future*, Vanguard Press, 1935, 62.

21. Muller, "Life Forms to Be Expected Elsewhere than on Earth," *The American Biology Teacher* 23.6 (October 1961): 331–346.

22. Muller, "Genetic Nucleic Acid," 70.

23. Muller, *Out of the Night*, 66, 68.

24. Muller, "The World View of Moderns," University of Illinois Graduate College, 1958.

25. Muller, "Man's Place in Living Nature," *The Scientific Monthly*, 84.5 (May 1957): 245–254, on 254.

26. Muller, "The Guidance of Human Evolution," *Perspectives in Biology and Medicine* 3.1 (August 1959): 1–43.

27. Muller, *Out of the Night*, 54.

28. Muller, "The Human Future," in *The Humanist Frame*, London, George Allen & Unwin, 1962, 401–414, on 410, 414.

29. Muller, "What Genetic Course Will Man Steer?" in Elof Carlson, ed. *Man's Future Birthright*, 150.

30. Muller, "Life Forms to Be Expected Elsewhere than on Earth."

31. Muller, "Man's Future Birthright: An Address at the University of New Hampshire," 1957, 23.
32. Muller to Sagan, April 16, 1952. Muller mss., Lilly Library, Indiana University, Bloomington.
33. Muller to Sagan, October 13, 1954.
34. Sagan to Muller, December 12, 1955.
35. James F. Crow and Seymour Abrahamson, "Seventy Years Ago: Mutation Becomes Experimental," *Genetics* 147.4 (December 1, 1997): 1491–1496, on 1496.
36. *Out of the Night*, 126–127.
37. Sagan to Muller, March 18, 1959.

# PART III

# PHILOSOPHICAL ISSUES IN ASTROBIOLOGY

PART III

PHILOSOPHICAL ISSUES
IN ASTROBIOLOGY

# 8

# Do Extraordinary Claims Require Extraordinary Evidence?

## The Proper Role of Sagan's Dictum in Astrobiology

*Sean McMahon*

## Introduction

> Extraordinary claims require extraordinary evidence. . . . For all I know, we may be visited by a different extraterrestrial civilization every second Tuesday, but there's no support for this appealing idea. The extraordinary claims are not supported by extraordinary evidence.
>
> Carl Sagan, *Cosmos: A Personal Voyage*, PBS, December 14, 1980

The dictum "extraordinary claims require extraordinary evidence" was made famous by one of astrobiology's best-known figures, Carl Sagan (1934–1996), who probably borrowed it from his contemporary and fellow sceptic, the American sociologist Marcello Truzzi (1935–2003). The saying remains popularly associated with Sagan's name and has become a favorite of skeptics and debunkers of all stripes. It is also widely cited in scientific journals to rebut unwelcome results, especially in Sagan's own field, astrobiology. This science is concerned with many extraordinary things: the origin of life, its earliest traces in the geological record, its distribution across and between worlds, its fingerprints on the cosmos, and its ultimate future. Sagan's dictum (as I call it) has been deployed against a host of claims in the astrobiological literature, including microfossils in martian meteorites (Kerr, 1996), methane gas in the martian atmosphere (Zahnle et al., 2011), arsenic-based life (Benner et al., 2013), and billion-year-old fossil animal burrows (Brasier, 1998). Sagan's dictum might be considered an essential part of any astrobiologist's conceptual, or at least rhetorical, toolkit. But what does it really mean?

The gist seems to be that one does not have sufficient reason to credit an extraordinary claim unless one also has commensurately extraordinary evidence

Sean McMahon, *Do Extraordinary Claims. Require Extraordinary Evidence?* In: *Social and Conceptual Issues in Astrobiology.* Edited by: Kelly C. Smith and Carlos Mariscal, Oxford University Press (2020). © Oxford University Press.
DOI: 10.1093/oso/9780190915650.001.0008

to support it. This looks reasonable enough at first glance; it has been compared with Hume's common-sense remark that "a wise man . . . proportions his belief to the evidence" (Hume, 1748, p. 110; Pigliucci and Boudry, 2014). But Sagan's dictum goes further: it suggests that some claims are special and that something *extra* is required to justify them, something *more* than the evidence that would *ordinarily* suffice. It is not easy to make sense of this, regardless of whether ordinariness/extraordinariness is treated as a binary or a continuous attribute. In what way (and to what extent) must a claim be "extraordinary" if it is to require extraordinary evidence? In what way (and to what extent) must the evidence be "extraordinary" if it is to satisfy the requirement? And what follows when the requirement is or is not satisfied? Our choice of answers to these questions will determine whether Sagan's dictum is a useful heuristic for astrobiology, an empty platitude, an irrational double standard, a question-begging error, or a fig leaf for the kind of unscientific dogmatism that Truzzi called "pseudo-skepticism" (1987).

Sagan's dictum is, I suggest, "one of those epigrammatic declarations tainted by smartness [to which] suspicion rightly attaches" (Medawar, 1996, p. 207). Its neat formal symmetry lends it the ring of self-evidence, but its vagueness leaves it open to interpretation and some of its interpretations are disastrous. This chapter aims to set out the conditions under which an appeal to Sagan's dictum is justified and those under which it is not, with special reference to existing and anticipated astrobiological debates.

## Sagan's Dictum in Theory

The word "extraordinary" in everyday use has several overlapping meanings: *unexpected, superb, astonishing, shocking, weird, anomalous, important, improbable,* and so on. Some of these are better candidates than others for the kind of extraordinariness that could justify an appeal to Sagan's dictum. The fact that the aphorism is supposed to express an important truth eliminates some interpretations. Extraordinary evidence cannot be defined simply as whatever it takes to confirm an extraordinary claim (or falsify the alternatives); this would reduce Sagan's dictum to a feeble tautology, unfit for service even as a rhetorical device. Similarly, an extraordinary claim cannot be defined simply as a claim for which there is no prior evidence; such claims are usually trivial (it would be eccentric to demand extraordinary evidence from my friend in Brussels when she tells me over the telephone that it is raining there, just because I have no prior evidence for this claim). I therefore consider more promising forms of extraordinariness, keeping in mind that the following options are not mutually exclusive.

## Psychologically Extraordinary Claims Do Not Necessarily Require Extraordinary Evidence

A natural first thought is that an extraordinary claim is one that is sensational, exciting, or provocative: the sort of thing to make one perk up and say, "Wow!." I call this *wow!-extraordinariness*. Any claim to have detected extraterrestrial life would be wow!-extraordinary. There are at least two reasons why we might wish to demand extraordinary evidence for such wow!-extraordinary claims. First, wow!-extraordinary claims may tempt us to accept weaker-than-usual evidence because we *want* to believe them. Second, such claims may have been given undue prominence and uncritical attention by their original proponents and by the media (including the less scrupulous academic journals) because of the interest and attention they attract. These worries point to irrational biases *in favor* of wow!-extraordinary claims, which we should resist and beware. But rationality only requires us to hold such claims to the *same* standards of evidence as we would other claims, not to demand *extraordinary* evidence for them. Indeed, to insist on extraordinary evidence for a claim just because it is wow!-extraordinary would be to adopt an unreasonable double standard and thereby to violate norms of objectivity.

What about a claim that is deeply bizarre, counterintuitive, or strange? I call this *weird-extraordinariness*. Weird-extraordinary claims differ psychologically from wow!-extraordinary claims in that we are typically predisposed *not* to believe them. But this tendency is also to be resisted, not celebrated. If we demand "extraordinary evidence" for a claim merely because it strikes us as bizarre, we overvalue our own intuitions and fail to be objective. Consider, for example, the claim, "every minute, you typically inhale at least one molecule exhaled in Caesar's last breath." Although this claim is completely counterintuitive (weird-extraordinary), it can be convincingly demonstrated without much effort; contrary to Sagan's dictum, only a back-of-the-envelope calculation with ballpark figures and a couple of reasonable assumptions is required (von Baeyer, 1986). The fact that the claim is "extraordinary" reflects the fallibility of our intuitions, not the claim's unlikeliness. The solution to the well-known "Monty Hall" brainteaser is likewise famously counterintuitive but easy to prove (e.g., by constructing the appropriate truth table). Rationally speaking, weird-extraordinary claims do not automatically require extraordinary evidence.

Sagan's dictum faces at least two other objections when applied to claims that are merely psychologically extraordinary. First, it offers no insight into what sort of "extraordinary evidence" might be appropriate to justify an extraordinary claim. It would be irrational to demand weird evidence for a weird claim or astonishing evidence for an astonishing claim; the very claim to have obtained such evidence would then be as vulnerable to Sagan's dictum as the original claim was, leading to an infinite regress. Second, both wow!- and weird-extraordinariness are "in the eye of the beholder," and in practice the claimer is likely to find his or her claim much less weird or astonishing than the skeptic does. This is not to say that no argument can ever settle whether a claim is or is not psychologically extraordinary. It is just that this argument must be settled before Sagan's dictum can be invoked; it has no force against opponents who do not accept that their claim is extraordinary to begin with.

## Probabilistically Extraordinary Claims Do Require Extraordinary Evidence, But Are Hard to Identify

If I claim to have won a lottery, my claim is extraordinary simply because its probability is very low. I call such claims *improbable-extraordinary*. As many people have pointed out, it follows from the axioms of the probability calculus that strong new evidence is required for such claims to have an appreciable probability of being true, in agreement with Sagan's dictum (e.g., Beauregard, 1978; Pigliucci, 2009). This can be seen from Bayes' Theorem:

$$p(\text{claim is true} \mid \text{new evidence}) = p(\text{claim is true}) \times \frac{p(\text{new evidence} \mid \text{claim is true})}{p(\text{new evidence})}$$

The expression on the left-hand side is the "posterior probability," that is, the probability that a claim is true given some particular piece of new evidence. The first expression on the right-hand side represents the "prior probability" that the claim is true, that is, the probability that it is true on the evidence available before the new evidence is introduced. If the claim is improbable-extraordinary, this expression has a very low value.

To render an improbable-extraordinary claim probable, we must maximize the ratio on the right-hand side by obtaining evidence whose probability of being obtained *if* the claim is true—the numerator—is much higher than the denominator, its probability of being obtained whether it is true or not.[1] Thus, the ratio acquires a value much greater than unity. This is a tenable definition of "extraordinary evidence"; it is strictly independent of the extraordinariness of the claim, and it accords well with scientific practice: a result is good evidence for a hypothesis if we would not obtain that result if the hypothesis were false (hence the importance of negative controls and small p-values) and would obtain it if the hypothesis were true (hence positive controls). Such good evidence need not be obtained all at once: it may accumulate stepwise, with each new observation producing a new posterior probability that serves as a prior probability for use "next time." In this way, the balance of evidence can incrementally reduce the extraordinariness of the claim and eventually support it.

A probabilistic reading of "extraordinary claims" (and "extraordinary evidence") renders Sagan's dictum intelligible, nontautologous, and rational according to the axioms of mathematical probability. In practice, however, it may still be very difficult for disputing parties to agree that a claim is improbable-extraordinary, partly because of the well-known "reference class problem." The claim that I possess the sole winning ticket in a million-ticket lottery has an obvious reference class (the million tickets with an equal probability of winning), which supplies the claim with a prior probability of one in a million. But claims of unique discoveries in science may have no reference class that stands out as correct. The first black swan to be discovered was an extraordinary thing set against the reference class of all the swans that had been observed before, since they were all white, but it was an unextraordinary thing set against the thousands of bird species already known to be black. Similarly, the claim that "this bacterium-shaped blob of carbon in an extremely ancient rock specimen is the oldest fossil ever found" seems extraordinary if set against all the equally ancient rocks that *lack* fossils, or even against all the carbonaceous blobs that *are not* shaped like bacteria in the same specimen (e.g., Brasier et al., 2002). But set against the enormous number of bacterium-like structures widely accepted as fossil bacteria in younger rocks, the claim may look probable (e.g., Schopf, 1993). Such disagreements about whether or not a claim is extraordinary are commonplace. They may eventually be settled one way or another—but not, of course, by appealing to Sagan's dictum itself. Once again, Sagan's dictum has no force against claims that are not admitted to be extraordinary in the first place.

## Revolutionary Claims Require Extraordinary Evidence; Some Are Easy to Identify

Sagan's dictum probably originated in the work of the American sociologist Marcello Truzzi, who cofounded the Committee for the Scientific Investigation of Claims of the Paranormal (CSICOP) with Sagan and others in 1976. The same year, Truzzi published an editorial in the first issue of CSICOP's magazine, *The*

*Zetetic*, arguing that "when . . . claims are revolutionary in their implications for established scientific generalizations already accumulated and verified, we must demand extraordinary proof." He elaborated on this view in an excellent 1978 article, "On the Extraordinary: An Attempt at Clarification," which goes into some detail about the rational response to "claims alleging paranormal events." This article emphasizes again that "extraordinariness must be measured against theoretical expectations provided by the general body of scientific knowledge at the time."

I think Truzzi's view can be distilled as follows: claims require extraordinary evidence if they entail the falsehood of established scientific ideas that are themselves well substantiated by evidence and fundamental in their importance. I call such claims *Truzzi-extraordinary*. The claims of astrologers, for example, are Truzzi-extraordinary because if they are correct then our existing understanding of fundamental physics is seriously mistaken (e.g., the diminution of fundamental forces across vast distances of space).

Why should Truzzi-extraordinary claims require extraordinary evidence? One reason given by Truzzi is that extraordinary claims are "revolutionary [in their] effects upon fundamental ideas," such that accepting such a claim would involve one in a far more consequential error than accepting an ordinary claim. A second reason is that the well-established results contradicted by Truzzi-extraordinary claims are supported by evidence "already accumulated and verified." The new evidence must indicate not merely that the new claim is true; it must also either outweigh or somehow account for all the evidence for the traditional view.[2] Of course, as Popper emphasized, a single counterexample (e.g., one black swan) is sufficient to refute any universal generalization ("all swans are white"). But as Bayes' Theorem helps to bring out, the evidence that such a counterexample has genuinely been found needs to be strong enough to overcome the low prior probability that we are justified in assigning to such a claim, given that it flatly contradicts prevailing views in which we have a high degree of prior confidence.

One might think that there is very little scope for the proponents of Truzzi-extraordinary claims to deny that they are extraordinary, since if anything is an "established result," it should be easy to find out that it is one. It has been pointed out, however, that what seems like "common knowledge" differs from person to person, from laboratory to laboratory, and especially from discipline to discipline—a particular problem for the interdisciplinary science of astrobiology (Benner et al., 2013). Benner et al. review the claim made by NASA-funded astrobiologists to have discovered bacteria capable of substituting arsenic for phosphorus in their DNA and other biomolecules (Wolfe-Simon et al., 2010). This claim was comprehensively refuted by later work and provoked a strong backlash in blogs and comment pieces as soon as it appeared.

Sagan's dictum was frequently aired. But Benner et al. point out that: "we cannot understand what a community finds 'extraordinary,' and what that community requires to meet a burden of proof, without understanding the expectations of the community."

These authors argue that arsenic-based DNA is "extraordinary" to an organic chemist because it conflicts with well-established conclusions about arsenate chemistry, but unextraordinary to a geologist, who is familiar with ionic substitutions in mineralogy and does not immediately see why they should not also occur in biochemistry; extraordinary to a biologist for whom the occurrence of a radically different biochemistry in an otherwise unremarkable organism—far from the root of the tree of life—conflicts with basic tenets of natural selection, but unextraordinary to a physicist, who may have essentially no relevant background information against which to assess the claim's extraordinariness.

There is no reason, however, why a geologist or a physicist should not easily understand why arsenic-based DNA is extraordinary once they have absorbed the relevant background information and explanations from chemistry and biology.[3] The lesson for astrobiology is not that we should reject Sagan's dictum for being hopelessly relative—only that we need to talk to each other and explain how the experiences and norms of each field bear upon questions of shared interest. Established scientific conclusions are a common resource upon which, with the right kind of effort and mutual assistance, all of us in our various fields can draw to determine whether or not an astrobiological claim is extraordinary.

Nevertheless, what is "established" in science—and so what is extraordinary in the light of Sagan's dictum—still depends to a large extent on contingent and dynamic social and historical factors. In the 16th century, the heliocentric model of the solar system conflicted outrageously with basic physics and astronomy (the apparently undeviating vertical trajectory of falling bodies; the lack of stellar parallax) and would arguably have fallen prey to Sagan's dictum on Truzzi's terms. But Sagan's dictum does not have to be infallible to be heuristically useful. It must only encapsulate what philosophers call an "epistemic norm": a rule for correct reasoning. Epistemic norms make our beliefs *rational* on the basis of the evidence we have at the time, not necessarily *true*.

It is worth remarking that many psychologically and probabilistically extraordinary claims are not Truzzi-extraordinary. Truzzi (1978) himself pointed out that the existence of fabled beasts like the Loch Ness monster would require only relatively minor, local adjustments in our understanding of evolution and ecology; it would not violate the deep principles of biology as we know it (let alone physics). So it would not be a Truzzi-extraordinary claim.

## Sagan's Dictum in Practice

I next examine two more claims to which astrobiologists have applied Sagan's dictum and one to which they are likely to apply it in future.

## Methane in the Martian Atmosphere

Between 2004 and 2011, four independent research groups using different methods, including both Earth-based telescopes and Mars-orbiting satellites, detected seasonally varying traces of methane gas in the martian atmosphere (Krasnopolsky et al., 2004; Formisano et al., 2004; Mumma et al., 2009; Fonti and Marzo, 2010). Astrobiologists were cautiously excited about these results because the active production and destruction of methane on Mars would imply unexpectedly dynamic geochemical or even biological activity.

Zahnle et al. (2011) made extensive use of Sagan's dictum in a highly cited paper disputing the "extraordinary claim." These authors emphasized not only that the reportedly rapid fluctuations in martian methane concentrations were inconsistent with our understanding of Mars but that all the explanations proposed by other workers violated basic physical or empirical constraints. Zahnle et al. thereby invoked a notion of claim-extraordinariness similar or identical to Truzzi's. Having shown that the claim conflicted deeply with our prior understanding, they then showed why, in their view, the evidence fell short of extraordinariness, arguing that some of the methane signals in the reported spectroscopic data were likely to be false positives (raising p[new evidence]), while other features of the data were actually contradictory to the claimed detection (reducing p[new evidence|claim is true]). In fact, Zahnle et al. went too far: the import of their paper is that the evidence fell short of *ordinary* standards.[4] This makes their appeal to Sagan's dictum apparently redundant.

The same is true, I think, of many other papers citing Sagan's dictum in opposition to astrobiological claims. If we take Sagan's dictum seriously, we can admit that the evidence for an extraordinary claim does meet the standard usually acceptable to the field while *still* rejecting the claim. But I find that scientists are rarely willing to do this,[5] perhaps suggesting that the community is not as committed to Sagan's dictum as it thinks it is. Alternatively, perhaps the dictum is thought to call not for extraordinary evidence but for extraordinary *scrutiny* of the evidence supporting an extraordinary claim, simply to ensure that it really does meet the usual standards of quality.[6] This scrutiny could include the study of "meta-evidence" to test underlying auxiliary hypotheses normally taken for granted, for example, that the relevant instruments and data recorders were functioning adequately when an observation was made. This view, which we might formulate as "extraordinary claims require extraordinary diligence," does not seem to me a literal reading of Sagan's dictum but may be a useful thought in itself.

## Extraterrestrial Visitation

Consider again the quotation from Carl Sagan's seminal TV documentary, *Cosmos*, with which this chapter opens. The claim that we are "visited by a different extraterrestrial civilization every second Tuesday" would certainly be wow!-extraordinary for many people, as hinted at by Sagan's description of it as "appealing." It might also be weird-extraordinary for some. But we have seen that wow!-extraordinary and weird-extraordinary claims do not necessarily require extraordinary evidence. That a *different* civilization should visit us *exactly fortnightly* greatly reduces the prior probability of the claim. But leaving aside the peculiar scheduling, the claim of extraterrestrial visitation per se does not contradict our existing understanding of science (Sagan's "for all I know" seems implicitly to acknowledge this), and so does not qualify as Truzzi-extraordinary. Rationally, it therefore would not require extraordinary evidence.

It might be objected that claims of extraterrestrial visitation are tainted by the late-20th-century craze for UFO sightings, which failed to produce evidence of even ordinary scientific standards. I would agree that claims specifically about flying saucers, little green men, "greys," and so on *can* be held to conflict with our well-substantiated understanding that these are invented tropes and that all previous claims about them were fictional (one could even quantify these false claims to derive a prior probability). Perhaps we need to compensate for being preconditioned by our cultural milieu to the very idea of aliens visiting us in spaceships. But, as I have argued, our obligation in dealing with claims to which we are overly partial should be simply to hold them to *the same* standards of evidence as we would otherwise—not to let down our guard. And this is a distinct epistemic norm, different from anything that can be read in Sagan's dictum. In general, claims of extraterrestrial visits to Earth deserve to be evaluated on their own merits.

Of course I do not believe that extraterrestrials have ever visited Earth—but only because I do not think the "evidence" that has so far been adduced in favor of this claim stands up to basic scrutiny, not because I hold this evidence to an arbitrarily high standard a priori.

## Life on Other Worlds

Any claimed detection of extraterrestrial life, whether in the soils of Mars, the spectroscopic signature of a distant world, or otherwise, will inevitably produce a mass outbreak of Sagan's dictum. And the claimed detection of extraterrestrials would indeed be *extraordinary* in several everyday senses of the word, already considered. But in my view it would not be extraordinary in the right sense for Sagan's dictum to apply legitimately.

The reason for this is that our background knowledge about the distribution of life elsewhere in the universe is practically nonexistent. Results of the Drake equation[7] consistent with current knowledge range from zero to at least hundreds. One of these results is right, *all of them* are wow-extraordinary, but none of them is Truzzi-extraordinary, because none of them conflicts with what we know so far (since what we know so far constrains the very range of values we are talking about).

The epistemic situation is scarcely better with Mars. We have plenty of evidence now that Mars has maintained habitable environments in its geological past and may do so today, but we have no clue whether these habitats have ever actually been inhabited. Fossil, chemical or isotopic evidence could quite easily, I think, tip the balance (at least slightly) in favor of the claim that Mars has supported life. Because it would not conflict with anything else we know so far, such evidence need only meet normal standards of reliability, reproducibility, and so on to justify a proportionate adjustment to our beliefs.

Likewise, it would be enough for anomalies in the atmospheric chemistry of a distant world to agree with model predictions for biotic effects and disagree with model predictions for abiotic effects for the balance of credibility to shift provisionally in favor of the claim that extraterrestrial life has been detected.

## Conclusion

Although it is clearly related to much older epistemological arguments, Sagan's dictum in its modern form originated in work on the skeptical investigation of claims of the paranormal. It expresses one of the many and various "epistemic norms" smuggled into the education of scientists, which routinely guide our evaluation of evidence, hypotheses, and explanations (e.g., Occam's Razor). Although it is not an infallible guide to the truth, I have argued that Sagan's dictum is a justified skeptical response to claims that can be independently evaluated as highly improbable or contrary to well-substantiated prior scientific knowledge, for reasons that Truzzi understood, which are further clarified by Bayesian reasoning. However, it is irrational and contrary to scientific objectivity to demand extraordinary evidence for a claim that is merely amazing or bizarre; Sagan's dictum must be handled with caution.

There are many social factors, not considered here, which may incline a scientist to a posture of hard-headed immovability; it is probably better for one's career to be known as a trenchant skeptic than as a credulous dupe. But the cautious scientific skepticism appropriate to such a young field with such profound subject matter should not be allowed to spill over into reactionary "pseudo-skepticism." We should follow the evidence where it leads (guided by rational modes of inquiry), not impose double standards. In particular, there is no good reason to think that the detection of extraterrestrial life should require extraordinary evidence. It would be a shame if the astrobiological community failed to recognize one of the most "wow!-extraordinary" discoveries in the history of science out of a misguided belief that such claims are somehow automatically unacceptable.

## Acknowledgments

This chapter is based on a talk given at SoCIA 2016, Clemson University, South Carolina. The bibliographic support and critical questions of D. Marosi are gratefully acknowledged, as are helpful comments from C. S. Cockell and three anonymous reviewers.

## Notes

1. Suppose we know that one person in a million has a rare gift: they can predict, 50% of the time, the output of a four-digit random number generator. A certain person, C. Voyant, claims to possess this gift (p[claim is true] = $10^{-6}$); to support her claim, she correctly predicts the next output of the generator. The quality of this evidence can be evaluated as follows: p(new evidence|claim is true) = 0.5; p(new evidence) ~ $10^{-4}$. This yields a posterior probability of about 0.005: the evidence is not yet strong enough to support the extraordinary claim. However, if C. Voyant were able to predict the next output correctly as well, the posterior probability would rise to ~0.96; the claim would be supported by the evidence.
2. As Truzzi was no doubt aware, his line of reasoning recalls Hume's famous argument about miracles, that "no testimony is sufficient to establish a miracle, unless the testimony be of such a kind, that its falsehood would be more miraculous than the fact which it endeavours to establish" (Hume, 1748, p. 115). See Pigliucci (2009) for a Bayesian reading of Hume.
3. Chemists and biologists are likewise fully capable of understanding geology and physics.
4. I make no comment here on whether Zahnle's assessment of the evidence to 2011 was correct but note that NASA's Curiosity Rover subsequently detected atmospheric methane in situ on the martian surface, with seasonal variability comparable to the detections disputed by Zahnle et al. (Webster et al., 2015, 2018).

5. An astrobiological exception may be the rejection by Martin Brasier (1998) of purported animal trace fossils in Precambrian rocks, older by half a billion years than any other fossil evidence for animal life. Identical markings would certainly have been accepted as evidence for animal traces if they had been found in a later part of the rock record. And indeed, Brasier's conclusion was that these rocks probably were younger than they had seemed.

6. Suggested by an anonymous reviewer of this manuscript.

7. Famously, the Drake equation multiplies a string of factors, including the number of solar systems, the number of habitable planets per solar system, the fraction of these planets on which life appears, and the fraction of these that develop observable technological civilizations, in order to estimate the number of detectable civilizations in our galaxy.

# References

Beauregard, L., 1978. Skepticism, science and the paranormal. *Zetetic Scholar*, *1*(1), 3–10.

Benner, S.A., Bains, W. and Seager, S., 2013. Models and standards of proof in cross-disciplinary science: The case of arsenic DNA. *Astrobiology*, *13*(5), 510–513.

Brasier, M., 1998. Animal evolution: from deep time to late arrivals. *Nature*, *395*(6702), 547–548.

Brasier, M.D., Green, O.R., Jephcoat, A.P., Kleppe, A.K., Van Kranendonk, M.J., Lindsay, J.F., Steele, A. and Grassineau, N.V., 2002. Questioning the evidence for Earth's oldest fossils. *Nature*, *416*(6876), 76–81.

Fonti, S. and Marzo, G.A., 2010. Mapping the methane on Mars. *Astronomy & Astrophysics*, *512*, A51.

Formisano, V., Atreya, S., Encrenaz, T., Ignatiev, N. and Giuranna, M., 2004. Detection of methane in the atmosphere of Mars. *Science*, *306*(5702), 1758–1761.

Hume, D. 1748. *An Enquiry Concerning Human Understanding*. Nidditch, P.H. (Ed.) New York: Oxford University Press, 1992.

Kerr, R.A., 1996. Ancient life on Mars? *Science*, *273*(5277), 864–866.

Krasnopolsky, V.A., Maillard, J.P. and Owen, T.C., 2004. Detection of methane in the martian atmosphere: evidence for life?. *Icarus*, *172*(2), 537–547.

Medawar, P. 1996. *The Strange Case of the Spotted Mice and Other Classic Essays on Science*. Oxford, UK: Oxford University Press.

Mumma, M.J., Villanueva, G.L., Novak, R.E., Hewagama, T., Bonev, B.P., DiSanti, M.A., Mandell, A.M. and Smith, M.D., 2009. Strong release of methane on Mars in northern summer 2003. *Science*, *323*(5917), 1041–1045.

Pigliucci, M., 2009. Do extraordinary claims really require extraordinary evidence? In K. Frazier (ed.), *Science Under Siege: Defending Science, Exposing Pseudoscience*. New York: Prometheus.

Pigliucci, M. and Boudry, M., 2014. Prove it! The burden of proof game in science vs. pseudoscience disputes. *Philosophia*, *42*(2), 487–502.

Schopf, J.W., 1993. Microfossils of the Early Archean Apex chert: new evidence of the antiquity of life. *Science*, *260*(5108), 640–646.

Truzzi, M., 1976. Editorial. *The Zetetic*, 1, 4.

Truzzi, M., 1978. On the extraordinary: an attempt at clarification. *Zetetic Scholar*, 1, 11–19.

Truzzi, M., 1987. On pseudo-skepticism, *Zetetic Scholar, 12/13*, 3–4.

von Baeyer, H.C., 1986. Caesar's last breath. *Sciences 26*(6), 2–4.

Webster, C.R., Mahaffy, P.R., Atreya, S.K., Flesch, G.J., Mischna, M.A., Meslin, P.Y., Farley, K.A., Conrad, P.G., Christensen, L.E., Pavlov, A.A. and Martín-Torres, J., 2015. Mars methane detection and variability at Gale crater. *Science, 347*(6220), 415–417.

Webster, C.R., Mahaffy, P.R., Atreya, S.K., Moores, J.E., Flesch, G.J., Malespin, C., McKay, C.P., Martinez, G., Smith, C.L., Martin-Torres, J. and Gomez-Elvira, J., 2018. Background levels of methane in Mars' atmosphere show strong seasonal variations. *Science, 360*(6393), 1093–1096.

Wolfe-Simon, F., Blum, J.S., Kulp, T.R., Gordon, G.W., Hoeft, S.E., Pett-Ridge, J., Stolz, J.F., Webb, S.M., Weber, P.K., Davies, P.C. and Anbar, A.D., 2010. A bacterium that can grow by using arsenic instead of phosphorus. *Science*, 1197258.

Zahnle, K., Freedman, R.S. and Catling, D.C., 2011. Is there methane on Mars?. *Icarus*, *212*(2), 493–503.

# 9

# Rethinking Conceptual Intelligence and the Astrobiology Debate

*Jason J. Howard*

One of the most contentious problems in current debates on astrobiology is the N = 1 problem—can we make warranted scientific inferences about life beyond our planet if the only sample we have is life on our planet?[1] From a philosophical point of view, what makes this question so significant is that it brings to a head assumptions not only about the character of life elsewhere but also the reliability and scope of our own cognitive abilities. My interest is in exploring how we think about the relationship between the evolutionary origin of self-conscious conceptual intelligence on Earth and the likelihood of life, specifically conceptual intelligence, existing elsewhere in the universe. Any thorough reflection on this relationship will inevitably confront questions about the evolutionary contingency of our conceptual capacities. Working through the deeper philosophical and conceptual implications of the N = 1 problem brings us face to face with an unavoidable but extremely complicated question: In what sense do the evolutionary factors that conditioned the emergence of conceptual intelligence on Earth impact or otherwise constrain the validity and epistemological legitimacy of conceptual activities like math, logic, and scientific reasoning?

It is likely that for many people, especially scientists, the evolutionary conditions surrounding the emergence of conceptual intelligence on Earth and the prevalence of this type of intelligence in the larger universe have no bearing on the confidence we should have in our ability to do math, logic, or science. It is a fact that we engage in these types of conceptual activities all the time, and whether other intelligent extraterrestrials would understand these activities, should we ever meet them, is irrelevant. My overall aim in this chapter is to show that such a view is not only shortsighted but likely based on troublesome presumptions that undermine the credibility of scientific reasoning. I demonstrate why these presumptions are so troublesome by way of examining how they shape, often inexplicitly, the N = 1 debate.

One of the things that makes the N = 1 debate so valuable are the different assumptions the debate brings directly into play about conceptual intelligence — by which I mean a being's capacity to explicitly utilize formal conceptual systems

Jason J. Howard, *Rethinking Conceptual Intelligence and the Astrobiology Debate*. In: *Social and Conceptual Issues in Astrobiology*. Edited by: Kelly C. Smith and Carlos Mariscal, Oxford University Press (2020). © Oxford University Press.
DOI: 10.1093/oso/9780190915650.001.0009

like logic and mathematics. My specific interest is unpacking why one might conclude that we will not find evidence of higher-order, conceptual intelligence elsewhere in the universe. Although skepticism about the prevalence of conceptual intelligence rests upon a host of reasons, many legitimate, I want to focus on what implications follow for our understanding of conceptual intelligence if we assume that life, and all its higher conceptual capabilities, is an exceedingly rare if not unique occurrence. As we shall see, skepticism about the likelihood of conceptual intelligence is widespread. The problem with much of this skepticism, however, is that the rarer, more unprecedented and contingent we insist is the phenomenon of life on our planet, and so by implication our conceptual capacities as *Homo sapiens,* the harder it will be to explain our actual conceptual achievements, especially in the domain of scientific understanding. This is a point rarely acknowledged, and so one could describe what I want to do as disclosing the missing premise behind the skepticism that surrounds the N = 1 problem and provide a more compelling alternative.

The structure of my overall argument is organized in the following manner. First, I explain why the consensus of the scientific community on conceptual intelligence holds that it is extremely rare, if not unique. Second, I define the type of formal-logical reasoning indicative of conceptual intelligence and then demonstrate why such reasoning cannot be derived solely from local evolutionary adaptations. Third, I explore different strategies, like that of Conway-Morris, that see conceptual intelligence as a cosmological phenomenon and then sketch an alternate account that avoids the most serious shortcomings with evolutionary explanations of conceptual intelligence. I conclude my analysis with an explanation of why it is important for scientists, especially astrobiologists, to comprehend the deeper conceptual implications of the N = 1 problem. What my overall argument demonstrates is that a thoughtful consideration of the N = 1 problem reveals that *Homo sapiens* do not exhaust the prospects of conceptual intelligence but are rather one instantiation of it. Consequently, this should be one's default starting point on matters of conceptual intelligence if one wants to be consistent and transparent about the knowledge claims made by the sciences, and if one believes our scientific descriptions of the larger universe have some measure of objectivity.

## Setting Up the Problem

Whether or not life is abundant in the universe and, if so, whether it routinely evolves beyond the microbial is very much a live debate. Over the past few decades there has been a plethora of evidence —the existence of extremophiles on Earth, the abundance of exo-planets, the vast cosmological distribution of the chemical elements required for life, research on self-organizing systems— accumulated from biology, cosmology, ecology, astronomy, and numerous other disciplines that have worked to recast the debate on extraterrestrial life; the rise of astrobiology is testament to this mounting cache of evidence.[2]

The N = 1 problem arises from the fact that all of our knowledge about life comes from our understanding of life on Earth, and life on Earth, however diverse and abundant, can be traced back to a common ancestor whose evolutionary development is unique to this planet. The incredible range and precision of our knowledge about biology and evolution should not tempt us into forgetting that such bounty is ultimately derived from a single example. As Carol Cleland warns, even the most fundamental of biological concepts, like replication, metabolism, and other core evolutionary processes, "might rest on mistaken assumptions and we cannot say anything definitive about life until we discover a second genesis" (one independent of conditions here on Earth).[3] This sentiment is anticipated by Francis Crick when he writes: "An honest man, armed with all the knowledge available to us now, could only state that in some sense, the origin of life appears at the moment to be almost a miracle, so many are the conditions which would have had to have been satisfied to get it going."[4]

It is certainly true that belief in the prevalence of life, for example microbial life, and belief in the prevalence of higher-order intelligence are separate claims, for life need not lead to higher-order intelligence (even if life is abundant in the universe).[5] Yet one of the peculiar things about the development of astrobiology over the past fifty years, a point raised up by Lori Marino, is the little attention the field has invested into researching the link between "the evolution of early life" and the development of intelligence.[6] Whatever the link may be, she observes that the default assumption of the scientific community skews toward extreme skepticism on the issue. Marino writes, "over the last several decades many leading scientists have vociferously argued that the emergence of a human-like intelligence is based on a highly improbable set of events (contingencies) that cannot be repeated elsewhere."[7] Although the details of particular positions on this point vary, the common assumption appears to be that the defining characteristics, traits, or capacities of any organism are determined by what adaptations will best ensure the survival of that species in its specific environmental niche. In addition, these adaptations arise as a result of chance mutations whose success is locally determined and so itself based on historical contingencies. In the words of Ornstein:

> The full set of messages . . . tested by selection, from the beginning of life, constitute only a minute and probably unrepresentative sample of different possible messages from which the sample has been "drawn," . . . Therefore, no matter how prevalent *life* might turn out to be, biological evolution on earth can easily

have generated many "inventions," perhaps including intelligence, which are unique in the universe.[8]

Stephen Jay Gould's famous observation that if the tape of life were replayed constantly we would end up with different evolutionary adaptations every time shares Ornstein's sentiment.

What is potentially problematic about claiming intelligence is so rare as to be a fluke in the universe is that it appears to rest on the assumption that *all* of our cognitive capacities should be understood in a similar vein —as the expression of local chance evolutionary adaptations. The problem with this view, which I coin for simplicity's sake evolutionary conceptual reductionism (ECR), is that it seems to suppose all higher-order conceptual activity, including logic and mathematics, has its ultimate origin in biological utility. The gist of the ECR argument runs something like this: the reason why intelligence is so incredibly unlikely, and conceptual intelligence practically unique, is that the number of chance intervening steps life must take for such biological complexity to arise, and consistently sustain itself, is so astronomically unlikely that it should be treated as a unique accident in the universe. In the words of the famous astronomer Fred Hoyle, "The chance that higher life forms might have emerged in this way [by Darwinian evolution] is comparable to the chance that a tornado sweeping through a junkyard might assemble a Boeing 747 from the materials therein."[9]

Such reasoning implicitly suggests that the capacities expressed by evolutionary adaptations, for example, the ability to make valid logical inferences, is itself a chance natural phenomenon, an invention created here on Earth. I say "implicitly" because one need not assume formal logic/math is a purely human invention to consistently embrace evolutionary (naturalistic) explanations of life and mind. The problem is that the scientific consensus on the rarity of intelligence equivocates on this very point, and it is this equivocation I seek to call out. Simply put, should we understand mathematical and logical reasoning as a local adaptation that has been refined over time and that has improved our ability to survive, or is it a discovery about the nature of inferences open to all reasoning beings everywhere?

However commonsensical the view that formal logical reasoning is ultimately a biological capacity rooted in evolutionary complexity may seem for many of us, it is deeply problematic. It is one thing to be cautious when it comes to anthropocentric projections about life elsewhere but another to suppose all reasoning, including logical and mathematical, is ultimately an expression of human biology and by extension human psychology—that all conceptual reasoning is human reasoning *tout court*. I am by no means the first to see the problem. The difficulty of explaining what ultimately grounds logic and math has been a mainstay of philosophical debate since the ancient Greeks, one that has occupied some of the sharpest minds of the Western tradition, and I am certainly not claiming to have solved that problem. My hope, however, is that reframing this age-old epistemological debate in the context of astrobiology will not only revitalize its relevancy but also refine what qualifies as an adequate answer to it.

## Formal Logical Concepts

To help orient us to the gravity of the problem I first indicate what is distinctive about formal logical reasoning, such as that employed in logic and math. Although there are many thinkers one could turn to here, I rely heavily on Husserl's description of logical categorization (*Logical Investigations, Formal and Transcendental Logic*) to help explicate the defining conditions of formal logical reasoning. I privilege Husserl's account for two reasons. First, at the level of descriptive analysis, he focuses on the invariable features of logical inferences, demonstrates the numerous paradoxes of "psychologism" (reducing logical laws to human psychology), and connects these concerns directly to the objectivity of scientific claims. Second, his account of logical reasoning is tied in with his views of time and temporality, which I argue is crucial in finding an alternative evolutionary account of conceptual intelligence. I use Husserl's account, along with Nagel's *Mind and Cosmos,* to explore the irremediable difficulties that arise once we assume formal logical reasoning is completely a product of contingent evolutionary adaptations here on Earth.[10]

To clarify, by formal logical concepts I mean those concepts whose forms of determination are taken to hold for all judgments about any object in every possible circumstance. To take an obvious example, mathematical forms of combination and calculation have the same validity for all who happen to employ them, whomever, whenever, and wherever they happen to be. For instance, $2 + 2 = 4$ regardless of what objects one counts or who does the act of addition. To take a more complicated example, the law of noncontradiction holds that one cannot consistently judge the same proposition to be both true and false at the same time. This does not mean we cannot be mistaken about what we judge but only that when a true judgment is made about some state of affairs, then an opposing claim of false made about the same state of affairs cannot also be correct.[11] Without this and similar logical principles like the law of excluded middle, it would be impossible to consistently and confidently exclude claims since everything could be true and false at the same time, which rules out the project of building a consistent and objective explanation of something.

Another example is logical laws of consequence (*modus ponens* and *modus tollens*), which allow valid chains of inferences to be built. These logical forms do not mean the truth of what we judge about, for example, the speed of light cannot be mistaken or amended but that the reasoning employed to generate a conclusion is valid. Since we know the forms of reasoning employed to investigate the problem are not mistaken, the focus can turn to the evidence (premises) used to establish the soundness of the argument. The key point is that the logical principles alone do not establish the truth of a judgment but make it possible that any judgments can be further determined as true or false.[12] Husserl summarizes the place of these principles in reference to scientific theories as making possible

> the concept of an "Objective" truth—that is to say: an intersubjectively identical truth—extend[ing] to all the propositions that it erects in its theory: its axioms and also its theorems. They all claim, accordingly, to be valid once for all for everyone.[13]

The notion of something "valid once for all" invites misunderstanding, so I want to say a bit more about how this is best understood. It does not imply these formal principles are innate, if by that we mean all humans are born with the ability to engage in logical reasoning at birth, nor does it imply everyone will eventually come to explicitly employ these concepts to construct arguments, nor that these concepts cannot be misunderstood. Rather they exist as invariable forms of conceptual determination accessible to beings of sufficient cognitive complexity—"comprehensive formula[s] covering coexistent and successive connections that are without exception and necessary." These forms concern how certain types of judgments "should proceed, in order that the resultant judgments should be true."[14] And so they imply "normativity" in terms of how they function for us, as beings whose thought requires logical regulation as a condition for making warranted truth claims, but that does not mean they are *"technical rules of a specifically human art of thought."*[15]

Formal conceptual principles are grounded in their limitless range of secure combination. Our confidence in them holds because whoever applies them, at any conceivable time, and whatever object they are applied to, has no impact on their internal justification. These principles may be applied in innumerable ways by countless human beings, but what we discover "in repeated acts" is that these types of judgments "are not merely quite alike or similar but *numerically, identically,* the *same* judgments, arguments, and the like." Although each act of thinking, "as a real . . . process in real human beings . . . in Objective time . . . are individually different and separated. Not so, however, the thoughts that are thought in the thinking."[16] Consequently, these conceptual forms of logical determination can be justly regarded as "laws of thought," provided we do not restrict this notion to laws of human thought.

Husserl calls these laws of logical determination "ideal" (his other term is "irreal"), but this does not mean they actually exist in some "eternal," Platonic realm, waiting to be discovered. It is rather that their necessity is not directly dependent upon contingent natural facts. Husserl explains that to say the laws of logical reasoning are "ideal" means that what structures the validity of a judgment holds

> without regard to time and circumstances, or to individuals and species . . . whoever judges differently, judges quite wrongly, no matter what species of mental creatures he may belong to. A relation to mental creatures plainly puts no restriction upon universality: norms for judgment bind judging beings, not stones.[17]

I realize for some these points are pedestrian and require no exposition, and the idea that logical necessity needs qualification of some sort is wasted effort (my mathematician friends are largely of this opinion). It is obvious that should we encounter alien intelligences, however incredibly unlikely such an event, the only thing we are certain to agree on is the necessity of formal logical concepts to help chart one's way in the universe. But it is precisely on this very point where we need to slow down. For even if we agree with this point, it is what this thesis commits us to, its logical implications, that need clarification, because it excludes *terra firma* from being the sole location of genesis for such concepts. Whatever ontology ends up being correct about the nature of formal logical concepts, if such concepts carry necessity and universality, this would hold for the cognitively conscious of Andromeda just as surely as it does the cognitively conscious of the Milky Way. That certainly does not mean the fact there are cognitively conscious beings on Earth necessitates they be anywhere else in the universe, or that we will ever meet them if they exist, or that we will have discovered the same facts about the universe, but only that if such beings do exist some of our formal concepts will necessarily be the same.

## Formal Logical Reasoning and Biological Utility

What is worrisome about extreme skepticism regarding the prevalence of conceptual intelligence is that it suggests such intelligence is ultimately grounded in biological processes. If one assumes all elements of conceptual intelligence arise through random adaptations under pressure of fitness optimization dictated by the local environment, then the capacity for formal conceptualization (e.g., logic and mathematics) is the result of biological utility. The problem is that if the capacity for formal conceptualization is ultimately reducible to biological utility (as mentioned earlier, I label such positions ECR), then such conceptualization carries no logical necessity or validity. If our formal conceptualizations carry no logical necessity, then the validity and truth of our scientific claims is open to in-eradicable skepticism.[18]

Now it is certainly possible to believe the laws of logic and mathematics are the result of biological utility. Many prominent thinkers, such as E. O. Wilson, appear to take this route, or at the very least equivocate on what grounds higher-order reasoning. Richard Dawkins is a complicated case in point since he seems fairly confident that intelligent life exists elsewhere and assumes we would have a "huge amount to learn from the aliens, especially about physics, which will be the same for them as it is for us. Biology will be very different, though —just how different will be a fascinating question."[19] However, given the strong reductionist tendencies that run throughout Dawkins' thought, my question is what, in the end, does the possibility of such a shared scientific understanding of reality ulti-mately rest upon?

Although it may be difficult to imagine where logical forms of inferential rea-soning derive their validity if not in terms of local evolutionary adaptations, we should resist this move. The most serious problem with reducing logical laws of conceptual determination to biological utility is that the objectivity and in-ferential validity that grounds knowledge claims is completely undermined. In basing logical validity on biological capacities that happen to be replicable and that happen to prove effective for solving problems for certain species on planet Earth, we consign not only our soundest scientific claims to being hopelessly inconclusive, but the very way we think (our capacity to draw inferences) has no guarantee of being correct. In the words of Nagel, such a position "implies that we shouldn't take any of our conclusions seriously," especially scientific conclusions.[20]

Although endorsing ECR will prove problematic for any science that aspires toward objectivity, it is particularly troublesome for those disciplines whose claims explicitly extend beyond planet Earth to the rest of the universe. In refer-ence to the $N = 1$ problem, if ECR is correct there would be no reason to believe any inferences we draw about the larger universe could tell us anything reliable, predictable, or falsifiable about the way things are, since there would be nothing sufficiently dependable about logical reasoning we could appeal to that would verify any of these inferences as correct. Even denying intelligence is an adap-tation that could be instantiated multiple times on multiple worlds requires we have sufficient evidence to generalize our knowledge of life and its conditions beyond Earth. However, if logic is a chance biological adaptation, we have no reason to believe any inferences we make about such complex and distant states of affairs could be true. As we shall see, if ECR is correct, the very premise of as-trobiology as an effort of scientific discovery goes out the window, as does astro-physics, astronomy, cosmology, and mathematics, among many others; indeed, scientific claims in general become largely indistinguishable from any other assertion of common human beliefs.

What ECR can explain is the capacity for empirical generalizations, and this capacity does explain many things we do and believe, but the ability to generalize from everyday experience severely limits the type of knowledge claims one can justifiably make. In addition, the very ability to confirm and build on experience presupposes wide-ranging agreement in certain beliefs, experiences, and perceptions, all of which, for ECR, ultimately rests on the fact we share membership in the same species. Our capacity to form beliefs is created by natural history and is relative to our environment. The key is that we share the same belief-forming mechanisms; what these beliefs are would presumably change if the natural facts of the human condition were to change. The problem is that if $\sqrt{144}$ is 12 but *only* for beings like us, there is no reason to believe this could not have been different in our own natural past or might be different in the future, since the calculation itself carries no necessity other than our own natural inclination toward agreement as a species. But to affirm this is essentially to give up on mathematics as a science. To say $6 \times 6 = 36$ is true so long as the claim is made by *Homo sapiens* but could be false if made by other beings sophisticated enough to engage in numerical reasoning is to jettison the principles of universality and necessity that give mathematics its incredible precision and explanatory scope.

A large part of scientific inquiry is building complicated chains of inferences from which we can draw credible conclusions. If ECR is true, scientific explanation of how things stand can never really get off the ground, because it denies we can appeal to any standard to regulate our beliefs that is not unique to our evolutionary history as *Homo sapiens*. But, as Husserl explains, this is to make truth claims relative to the species that makes the truth claim. For each different species of cognitive consciousness, we would have different truth claims, different laws of inference, from which different laws of nature would be drawn, and so there could be no objective description of one and the same universe.

> To define truth in terms of a community of nature is to abandon its notion. If truth were essentially related to thinking intelligences, their mental functions and modes of change, it would arise and perish with them, with the species at least, if not with the individual . . . the objectivity of being, even the objectivity of subjective being . . . would be gone.[21]

As Nagel argues, however informative the evolutionary account of cognitive consciousness for *Homo sapiens,* at some point one needs to distinguish the accuracy of an organism's perceptual systems and instincts in tracking environments and adjusting behavior from formal conceptual claims this organism makes about the world. The logical distinctions embedded in how we think and speak "enable us to understand reality."[22] This does not imply our understanding of reality ever need be complete or infallible, but the confidence in our ability to unpack the implications of what we perceive could never extend beyond the immediately practical if the conditions that make truth-claims possible are ultimately reducible to biological, or by extension cultural, prerogatives, however this may be understood.

Whatever explanatory variables ECR wishes to employ, pushing back the origin story of cognitive consciousness 200,000 years or 2 million, we are going to run into the same problem, namely, logical principles are grounded in peculiarities of species membership. Now it certainly is the case that many conceptual forms of determination, such as the law of noncontradiction, align with self-preservation as well as facilitate the satisfaction of countless natural desires, but to assume our capacity for logical inferences is inseparable from this is to misunderstand the character of logical principles. In the words of Husserl,

> As soon as the exact character of logical principles is conceded . . . . the possibility of their being changed by changes in the [arrangement] of what actually is, and of consequent transformations of zoological and mental species, is ruled out, and the eternal validity of such principles guaranteed.[23]

It goes without saying that many of our most reliable scientific explanations and models, from the standard model of particle physics to the cosmological account of the Big Bang, describe physical interactions, processes, and natural laws so far removed from the human environment that it boggles the imagination. But as Nagel reminds us, "Is it credible that selection for fitness in the prehistoric past should have fixed capacities that are effective in theoretical pursuits that were unimaginable at the time?"[24] If our scientific explanations of the inner workings of the universe, however incomplete, do track something like what they purport to track, how likely is that going to be if its chance for success is completely random? To suppose that certain contingent adaptations on one particular planet among billions happened to evolve processes capable of allowing one particular species unique access to not only the deepest structures of nature but those furthest removed physically from the organism (by billions of order of magnitude) in space and time strikes me as so implausible it should be taken seriously only after every other plausible explanation has run aground.

In closing this section I want to highlight one last time what is being proposed in defending the irreducible character of logical forms of conceptual determination and what is not. Just because we have access to formal logical principles does not mean we will ever definitively figure out how the universe works. The project of explanation and description is, as Husserl puts it, best construed as an "infinite task" where "*every* truth about reality, whether it be the everyday truth of practical life or the truth of even the most highly developed sciences conceivable, remains involved in *relativities*" in which evidence is collected, discussed, vetted, validated, and, if possible, falsified.[25] The crucial point is that this process is grounded in at least some "regulative ideas" we can access that structure our inferences and that carry universality and necessity; otherwise we are involved in an infinite regress that forecloses any account of nature from being more truthful or accurate than any other. And for concepts like universality and necessity to provide the epistemological scaffolding required to justify scientific claims, they cannot be the chance invention of *Homo sapiens*. In short, if we want to take our scientific conclusions seriously, then this implies other beings of conceptual intelligence would take them seriously as well.

Affirming the validity of logical principles does not mean endorsing the position that science somehow transcends culture, or is not a cultural creation, or cannot be abused, or be awash in the politics of power, or that its language and notions are free of gender bias. The scientific enterprise can and does include all of these things, yet none of these points exclude the possibility of constructing sound theories according to formal logical concepts. We are animals first and foremost, not immaterial souls or thinking machines. That we often make mistakes and let personal bias shape our views does nothing to show we cannot act otherwise.

We now know enough about the human brain and cognition to say with confidence that much of what goes on in our mental life is completely outside our control and that the multiple material configurations that enable perception and large swaths of behavior, as well as many of our decisions, are activated at levels we are neither aware of nor consciously control. If cognitive consciousness is a biological phenomenon, and I believe it is, it stands to reason that much of what conditions and shapes it lies outside of our immediate control and proceeds at a pace of its own accord.[26] Also, mental capacities like "self-awareness," "tool-use," and "numerical ability" are not unique to *Homo sapiens* but shared by other animals here on Earth.[27] None of this, however, proves that how we think is completely determined by random biological processes unique to this particular planet.

## The Link Between Evolution and Formal Logical Reasoning

This section tries to clarify two related claims about formal logical principles; one is the irreducibility of these principles (what Husserl calls their "ideality" or "irreality"), and the other is how we access such formal principles. If it is true that logical principles cannot be reduced to natural facts about species membership without giving up the distinctive epistemological features of these principles, their necessity and universality, then we need to at least be open to exploring what the "ideality" of logical principles commits us to. As Husserl puts it, we need to take seriously

> *the ideality of the formations with which logic is concerned* as the characteristic of a separate, self-contained, "*world*" *of ideal objects* . . . to come face to face with the painful question of how subjectivity can in itself bring forth, purely from sources appertaining to its own spontaneity, formations that can be rightly accounted as ideal *Objects* in an ideal "world."— And then . . the question of how those idealities can take on spatio-temporally restricted *existence,* in the cultural world . . . real existence in the form of historical temporality, as theories and sciences.[28]

My goal here is to suggest a way of understanding the ideality of logical forms and our access to them that avoids relying on teleological notions, such as intelligent design or any form of strong anthropic principles that assumes the universe is specifically fine-tuned for life and intelligence, while also demonstrating the compatibility of such ideality with naturalistic evolutionary accounts. What I offer is a sketch that points out how we should approach the problem once we accept the universality and necessity of logic and mathematics is neither a fiction nor unique invention of a particular subset of primates here on Earth. Admittedly, aspects of this sketch are speculative, but the overarching explanation, I believe, does not require that one accept anything outrageously unrealistic to be plausible while still accounting for the distinguishing characteristics of formal logical concepts.

It is important to note that the ideality of logical forms holds whether or not one agrees with my conjectures on how we access these forms; at least it holds if one believes the sciences provide us with knowledge about reality. Utilizing the $N = 1$ argument as rationale to defend the extreme unlikelihood of higher-order intelligence outside Earth makes an equivocation about the nature of intelligence when it implies *all* forms of higher-order intelligence on this planet are ultimately reducible to biological utility. This equivocation needs to be removed and the irreducible character of logical forms clarified. One of the reasons the equivocation persists is because it is not clear what other explanation of math and logic is feasible besides one that reduces these principles to evolutionary processes here on Earth. Consequently, I think it is important that the irreducible character of logical forms be demystified so people, especially scientists, are more receptive to the idea that logic and math cannot be grounded in species membership. When no plausible alternatives are provided, people are more prone to continue affirming the explanations they are familiar with, even when these fail to make sense of things—hence the need for alternative theories. What follows is an attempt to lay out the groundwork for an alternative that demystifies the "ideality" of logical forms by showing how these forms could relate to evolutionary processes.

The key challenge for any account that accepts the ideality of logic and math is to explain how logical principles relate to evolutionary adaptations without reducing these principles directly to local adaptations. This position implies that logic and math are the kinds of things that could be accessible to *all* beings of higher-order conceptual intelligence, wherever they might be found. I think this implication is unavoidable if we believe in logical necessity, but working out the details of how such principles would most likely become accessible to cognitively conscious extraterrestrial beings is obviously rife with difficulties.

Michael Ruse explores the continuity of intelligence across extraterrestrial species in his article "'Klaatu Barada Nikto'—or, Do They Really Think Like Us?," referencing Richard Dawkins' idea of evolution as a "biological arms race" that encourages complex intelligence as a way of optimizing survival, as well as Simon Conway Morris's notion that certain ecological niches greatly amplify evolutionary complexity; human culture should be seen as one such optimal environmental niche.[29] Ruse does not explicitly side with either approach since he thinks they imply too much evolutionary uniformity among diverse forms of extraterrestrial intelligence, yet he still holds that "alien mathematics and logic" would not be "inherently different from what we have," even if some aspects are likely to be quite different.[30]

Although I completely agree with Ruse's notion that there will be some overlap in formal logical principles, like many that tackle the problem of higher-order intelligence he equivocates on the status of formal logical concepts. The ability to count how many predators have just wandered into a cave (Ruse's example), and the facility for higher-order mathematics, imply more than just a difference in degree of intelligence but also in kind.[31] Mathematics and logic simply cannot retain their necessity if "necessity" is a direct extension of fitness optimization. Saying selection for counting enhances survival need not imply all numerical ability is true because it enable a species to survive, but in glossing over this distinction Ruse leaves us in the dark as to why we should expect any overlap between ourselves and other forms of higher-order extraterrestrial intelligence when it comes to a scientific understanding of the universe.[32]

Although Simon Conway Morris appears to end up in a similar conceptual confusion with his account of higher-order intelligence as the optimal form of evolutionary adaptation, I think his position is highly instructive and provides a plausible way to think about intelligence that is truly universal in scope. His thesis, put simplistically, is that there is much more convergence in evolutionary adaptations than is ordinarily assumed. Over and over again throughout Earth's history we see the same adaptations hit upon by vastly different species—visual, auditory, and olfactory senses; wings; vocalization and communication; socialization—which enable particular species to successfully occupy an environmental niche.[33] The ability to see, hear, move, and communicate all enhance an organism's ability to survive and reproduce. The key point for Conway Morris is that there are a large yet limited number of forms of adaptation that will allow a species to successfully master its environment sufficiently to ensure its survival over time; "baring the physically impossible and adaptationally compromised, it appears that as a general rule all evolutionary possibilities in a given 'space' will inevitably be discovered."[34] He goes on to state that convergence helps to explain "how evolution navigates the combinatorial immensities of biological 'hyperspace,' . . . Convergence occurs because of 'islands' of stability, analogous to 'attractors' in chaos theory."[35]

Intelligence is an adaptation that provides "stability;" the more general, flexible, and efficient the form of intelligence, the better the species will be able to communicate, make predictions, respond to changes in the environment, and so on. For example, the continual convergence of "sensory systems" among diverse species here on Earth provides strong evidence for the "recurrent emergence of such biological properties as intelligence."[36] What makes human intelligence so advantageous is that it creates a "culture"—a symbolic world that is able to engage, predict, and modify the larger natural environment to an unprecedented degree. The real innovation with human intelligence, then, is the creation/invention of culture as a particular type of evolutionary niche, one whose advantages are so multi-faceted that "at least in this biosphere if we had not emerged as the cerebral species than at some point . . . someone else would. In this sense, humans, as a biological property, were inherent from the Cambrian period, if not before."[37]

What is so insightful about Conway Morris's position is his suggestion that the advantages of higher-order intelligence would seem to hold *wherever* life might arise. Any organism anywhere would run into similar types of problems when it comes to replication and survival, not just life on Earth. Finding reliable and stable methods of self-replication would seem to be an extremely difficult problem whenever and wherever life has a chance of occurring. From the perspective of the $N = 1$ problem, Conway Morris's stance is extremely helpful since it does not assume life elsewhere would involve the same biological components as it does here but is a statistical argument based on the likelihood of combining sufficient variables to create and sustain complex biological processes. Conway Morris's position is that we know enough about chemistry, cosmology, and astrophysics to confidently state that whatever the specific variables are, the number of feasible solutions that will allow life to take hold from the immensity of alternatives within biological hyperspace is "strictly limited." As he eloquently puts it, "the vast bulk of any given 'hyperspace' not only never will be visited but it never can be. These are the howling wildernesses of the maladaptive, the 99.9% recurring of biological space where things don't work, the Empty Quarters of biological non-existence."[38]

In contrast to positions like that held by Stephen J. Gould, Conway Morris argues the number of successful adaptations available to any organism is quite narrow, and so, in theory, there are a finite number of "galactic-wide niches" that offer any chance of success. There are literally trillions upon trillions of evolutionary dead ends; evolutionary convergence posits that some adaptations, especially those involving intelligence, optimize the possibility of success, and as a result, life forms will converge toward intelligence.[39] This does not imply those adaptations that facilitate higher, and so more flexible, orders of intelligence will all follow the same exact evolutionary template, but it does imply whatever forms of intelligence do arise will be linked to refinements in biological complexity.[40] Conway Morris summarizes the point in this way:

> convergence is ubiquitous and the constraints of life make the emergence of the various biological properties very probable, if not inevitable. Arguments that the equivalent of *Homo sapiens* cannot appear on some distant planet miss the point: what is at issue is not the precise pathway by which we evolved, but the various and successive likelihoods of the evolutionary steps that culminated in our humanness. To remind ourselves what Robert Bieri noted: "If we ever succeed in communication with conceptualizing beings in outer space, they won't be spheres, pyramids, cubes, or pancakes. In all probability they will look an awful lot like us."[41]

I think Conway Morris's suggestion that higher-order intelligence is so optimal as an adaptation that it represents a cosmic-wide ecological niche is one rich in implications and offers a thoughtful way to approach the larger availability of logic and math. One key factor in the emergence of higher-order intelligence is the refinement of species' sensory systems as the optimal way of solving problems of survival posed by the environment, particularly environments with organisms competing for scarce resources. Both Conway Morris and Bieri suggest such convergence would hold *everywhere,* which limits the variability of shapes that higher-order intelligence can feasibly take. Higher-order intelligence emerges as an optimal problem-solving adaptation, one that effectively cuts down the number of blind alleys any given life-form needs to explore to ensure its survival. To what extent other "conceptualizing beings" might resemble humans thanks to such convergence is a profound question, but my interest is to what extent they might conceptualize reality in similar ways to us. And so even though we would not be so surprised to find aliens who look different from us, we should be extremely surprised to discover they use different kinds of logic.

The accessibility of logical forms of determination like math and logic are dependent on the achievement of higher-order (conceptual) intelligence, and I agree with Conway Morris that such intelligence, given its ability to optimize survival, should be seen as part of a larger galactic niche. In short, the ideality of logical forms become accessible as components of a larger cosmological niche once a certain level of cognitive complexity is achieved. The problem, however, is trying to grasp the nature of this cognitive complexity. On the one hand Conway Morris's position provides an innovative way to approach the issue of intelligence by relocating it in the cosmological landscape of biological hyperspace. Unfortunately, to the extent he construes higher-order intelligence as a "biological property" largely dependent on the convergence of sensory prowess, his position risks backsliding into the same problems we witnessed earlier. No amount of refinements in sensory or neurological complexity alone can account for the ideality of logic and mathematics; this is to reduce such principles of reasoning to particularities of species membership. What we need is a plausible way to think about the notion of evolutionary adaptations that can retain the cosmological scale of Conway Morris but provides sufficient flexibility that it can explicitly avoid reducing the formal necessity of logic and math to biological utility.

## Organisms and Life-Worlds

One of the underlying assumptions that followed the scientific development of Darwinian evolution was that organism and environment referred to radically different elements, with the environment independent and preformed, existing as a problem to be solved through the chance adaptations of the organism.[42] Since the middle of the 20th century, if not earlier, discoveries and developments across the life sciences strongly suggest such dichotomous thinking does not hold up as a characterization of evolutionary change. As Levins and Lewontin persuasively point out, organisms "select their environment," "modify" it in multiple direct ways, "transform structural regularities," "determine what aspects of their environment are relevant," and "respond" to changing conditions. In their words: "Therefore the environment as developmental stimulus helps turn genetic variability into phenotypic variability, which environment as Darwinian filter selects. Much evolutionary theory ignores this double effect of environment." The result, as they aptly put it, is that "the activity of the organism sets the stage for its own evolution."[43] Expanding on this train of thought John Odling-Smee et al. argue that although natural selection is indeed "blind," it inadvertently selects for "*purposive* organisms, namely niche-constructing organisms. This must be true at least insofar as the niche-constructing organisms that are selected by natural selection function *so as to* survive and reproduce."[44]

Since organisms are not just passive placeholders for larger genetic processes but actually shape their environment, thus are active and responsive to the living ecosystems that encompass them, survival and reproduction are core selective pressures among multiple variables rather than the only ones. Consequently, as Francisco Varela puts it, rather than seeing evolution as "forcing a precise trajectory by the requirements of optimal fitness," the key dynamic of the process becomes "how to prune the multiplicity of viable trajectories that exist at any given point." The apt characterization of evolution in this case is not a process of selective optimization but one of "satisficing."[45] What is required are adaptations that are good enough to ensure biological structures of varying complexity subsist over time. In Varela's words, we move "from the idea that what is not allowed is forbidden to the idea that what is not forbidden is allowed."[46] Adaptations still need to be compatible with survival and reproduction, but these conditions can be met in an incredibly diverse range of ways. The most important consequence of this variability, for my account, is that organisms are *undetermined* by the constraints of survival and reproduction."[47] Environments still "trigger" adaptive change but not in a simple linear or one-dimensional route but in terms of "coimplicative relations" that involve "networks" of "self-organizing configurations" in which "organism and medium mutually specify each other."[48] And it is precisely thanks to such "undetermination" that the expanding conditions that gave rise to conceptual intelligence not need be strictly determined by reproductive pressures to gain stability and purchase.

Varela insightfully reminds us that environments do not exist "out there" to serve "as a landing pad for organisms that somehow drop or parachute into the world. Instead, living beings and their environments stand in relation to each other through *mutual specification* and *codetermination*."[49] Organisms and environment exist in a dialectical relationship that enables the co-creation of life-worlds, ecological niches that have the potential to radically restructure the behaviors of a given species. Because the kind of problems the environment may pose are not inert but inherently flexible, some species may find themselves confronted by issues rather far removed from that of biological utility; human culture is an example of one such life-world.

Logical necessity is a conceptual category discovered by *Homo sapiens*; it exists, as do all formal mathematical and logical principles, as ideal objects whose access is dependent just as much on a certain level of neurological complexity as it is a specific type of life-world, one populated by self-conscious beings whose priorities transcend the constraints of survival and reproduction while still "satisficing" them. The forms of inference provided by logic and math assert themselves against the particularities of local conditions and resist the reduction of their intelligibility to empirical considerations of terrestrial origin. That does not mean conceptual reasoning is completely independent of evolutionary processes to scaffold the biological infrastructure required to discover logical principles but that logical reasoning is not reducible to local adaptations.

Conway Morris is correct that we need to see higher-order intelligence as a niche whose appeal is cosmological in extent, but one of the distinguishing characteristics of this niche is how the problem of survival and reproduction would be transformed; it is this transformation that "culture" (or its extraterrestrial equivalent) provides, for what is culture but a symbolic system that gradually emerges through the interactive efforts of self-conscious beings to transform their world? As Nagel emphasizes, any explanation that seeks to explain human intelligence and culture needs to tackle "not only the emergence from a lifeless universe of reproducing organisms and their development by evolution to greater and greater functional complexity; not only the consciousness of some of these organisms and its central role in their lives; but also the development of consciousness into an instrument of transcendence that can grasp objective reality."[50] I think a notion of evolution that encompasses the building of niches that are sufficiently robust to allow at least some organisms to "satisfice" biological utility rather than be wholly conditioned by it is the most feasible way to approach the cosmological significance of logic and math, not to mention the problem of self-consciousness in general.

## Logical Reasoning as a Mode of Temporality

I close out this section by developing what I see as the defining characteristic of those niches, or life-worlds, that enable the ideality of logical principles to become accessible to "conceptualizing beings." For although the notion of niches as "self-organized" environments that "satisfice" biological utility helps us comprehend the ability of "conceptualizing beings" to surpass their basic biological conditions, it still remains unclear how such a niche enables our access to logical necessity. The position I defend follows, in general outline, Husserl's exposition of logical conceptual formations as modes of *"omnitemporality."*[51] In brief, this position states that formal logical principles become accessible once a level of cognitive complexity and intersubjectivity is reached that restructures how a species experiences time and temporality. The discovery of logical necessity is something that cannot occur outside a specific type of life-world; the meaning of the discovery is an intersubjective confirmation and exploration of a given being's conceptual powers. Access to formal logical principles as epistemological norms is dependent on how a species "temporalizes" its experience. Universality and necessity could have no meaning, let alone application as explicit tools of explanation, for an organism that lives perpetually in the present, whose concerns are circumscribed by biological, and by extension, practical needs alone.

What we repeatedly discover in reflecting upon and applying logical and mathematical concepts is that they are completely unaffected by time. As mentioned earlier, although these principles are instantiated by specific people at particular times, factors like duration, location, and circumstance have no relation whatsoever on the sense of the concepts; they are "the same in relation to their repeated productions and . . . the same in relation to the productions of different subjects."[52] Obviously logic and mathematics as disciplines have a history wherein thinkers make discoveries and clarifications, expanding the horizons of these disciplines while teasing out the implications of its propositions, but once discovered, the objects themselves hold as necessary types of conceptual determination, to be taken up and utilized again and again by others. As Husserl puts it, this "timelessness . . . of being 'everywhere and nowhere', *proves . . . to be a privileged form of temporality,* a form which distinguishes these objectivities fundamentally and essentially."[53] And with this clarification we come to the definition of Husserl's rather obscure term of "irreality" and "irreal" objects. Real objects are those conditioned in varying degrees by their involvement in a process of temporal becoming; for example, all entities and objects in the natural world are real objects; these objects are "*individualized by . . . spatiotemporal position,*" whereas irreal objects, although always accessed in determinate places and times, are essentially the same, self-identical object instantiated across all times, worlds, and cognitively self-conscious agents.[54]

Saying this does not make logical and mathematical principles into a creation or invention of *Homo sapiens,* however, since the discovery of these principles is inseparable from the larger mode of temporality from which they become accessible. Discerning formal concepts and applying them to understand reality is to occupy one type of environmental niche. Just as certain species can only perceive specific light waves and hear particular frequencies, so formal concepts of necessity are accessible only to those species that occupy a particular temporal matrix of experience, one that allows a-temporal objects to come into explicit relief. With this discovery, predictions and explanations become available that are completely removed from the confirmatory space of immediate perceptual experience yet nevertheless remain open to validation and confirmation. Following Hans Jonas, insofar as different species have different sensorial capacities, different complexity in brains and neurological systems, we have every reason to believe they process experience differently—that different animal species occupy different time zones in terms of how they organize experience.[55] The upshot of this is that *Homo sapiens* have discovered a *temporal* bandwidth of sorts, one that allows species to move beyond the confines of the present moment to better track the underlying structures of reality.

The key qualification in explaining logical necessity, as Žižek reminds us, is that the discovery of necessity "is a contingent process." The implications of logical necessity are themselves worked out historically; it is a " 'performative' process of constructing (forming) that which is 'discovered.' "[56] The capacity of formal logical principles to refuse reduction to the empirical is something whose implications can only be discovered through the course of experience, but this does not mean what is distinctive about logical principles is simply a generalization from experience. The distinctive sense of formal logical concepts come into relief as principles irreducible to the empirical and its manifold "real" objects, as what resists capture in the immediate present but nevertheless still operates as a necessary form of internal conceptual organization.

Although all living entities share the same universe, the same reality, the horizons of their experience are vastly different. The world of *Homo sapiens* is a historical world saturated in notions of time, of a past removed by billions of years we can still reliably reconstruct, of intervals of duration like nanoseconds that we ourselves cannot consciously experience yet can comprehend, of a future billions of years removed whose general outlines we can nevertheless predict even though there will be nothing alive on our planet to witness it. To discern such a vast scale of time, to transcend the horizon of immediate experience with such a precision that a distinct past and future become possible, requires a deep alignment and congruity between our concepts and the multiplicity of states that different objects can occupy. To experience such an elasticity of temporality, what Husserl calls a plurality of "ideal objectivities" within one all-encompassing horizon of temporal becoming, and confirm it through experimentation, corroboration, and prediction is to occupy a specific cosmological niche that is only available to a range of cognitively conscious entities. We are one specific species on one particular planet that has managed to create a conceptual-symbolic life-world that stretches the horizons of experience sufficiently to read necessity and universality as ties that structure our thoughts about reality. Such conceptual notions may well be available to other cognitively conscious life-forms, and if they are it will be in virtue of their ability to access a horizon of temporality elastic enough it enables them to conceptualize formal patterns and relationships that govern reality in similar ways to us.[57]

Obviously much more needs to be said in explanation and defense of this alternative, which sees a particular form of temporality as the distinguishing characteristic of those cosmological niches that make formal logical concepts available that are inaccessible otherwise. At this juncture, however, a number of things can be provisionally said in defense of it. First, and most importantly, it meshes with how we actually practice science, which is about expanding the accessibility of evidence beyond that available to mundane perception and the present moment, to delve into the distant past and far-flung future, to explore the infinitesimally microscopic and the cosmologically immense, all with the assumption that such horizons of evidence are open to independent corroboration and confirmation by others. Second, it relies directly on a notion of evolution to generate the neurological complexity needed to access the a-temporal concepts of formal logical reasoning without reducing these concepts to peculiarities of species membership. Third, it defends the ideality of formal logical principles but without positing them in some separate world or Platonic realm; these objects come into relief as the experience of temporality is enlarged and do not "exist" in any meaningful sense otherwise. Fourth, it does not require the universe be fine-tuned for life and intelligence but only that a window of temporality can be experienced by some species whose conceptual sophistication opens up a range of experience that makes ideal logical objects possible. Fifth, following Nagel, the explanation is "consistent with our being, among other things, physical organisms," and so eschews reference to any metaphysical posits to explain things.[58]

## Consequences for Astrobiology

Where does this leave us with the N = 1 problem? At a minimum our discussion demonstrates that, in presupposing life, and by extension conceptual intelligence, is so fine-tuned as to be unique to this planet creates an equivocation that opens up our scientific claims to ineradicable skepticism. It is one thing to be doubtful about whether life exists elsewhere, if that life is intelligent, or even if we will ever find it. This is to be expected and healthy, but it is quite another to assume that all expressions of intelligence on this planet are the direct product of natural processes unique to Earth. E. O. Wilson famously claimed that our "genes hold culture on a leash."[59] I do not think it is too much to ask for such reductionist accounts to bite the bullet and follow through on the implications of their own claims, because if it is true that human culture and intelligence remains on a "leash" to the genetic makeup of our species, then the objectivity and validity of scientific claims must also go. If all forms of intelligence are ultimately rooted in local expressions of biological utility, then the necessity and universality that anchors our scientific explanations can be no more than complicated ways of registering our own interspecies agreement on things and thus of no help in critically investigating the larger structure of the natural universe with some measure of objectivity.

*Homo sapiens* do not exhaust the prospects of conceptual intelligence but are rather one instantiation of it. For those interested in exploring the larger universe in a rigorously scientific manner, I think this needs to be the default starting point if one wants to be consistent and transparent about the knowledge claims one aims to make. However, knowing the emergence of conceptual intelligence on Earth is just one instantiation of a larger cosmological niche tells us little about the exact chances that such patterns are instantiated elsewhere.

Even if we find evidence of microbial life on nearby planets, using that alone to fix the likelihood of conceptual consciousness is fraught with difficulty. This may seem to leave us in a similar position to the advocates of the N = 1 problematic, but our caution is of another sort. Clarifying the nature of formal logical principles, their irreducible character, yields no definite predictions on the necessity of conceptual intelligence elsewhere; rather it explains what must be true about these principles if our confidence in the scientific understanding of the universe is to be ultimately defensible. And what must be true is that these principles cannot be the direct result of unique adaptations here on Earth. Instead of leaving the problem there, I tried to tease out the implications of this position and show the plausibility of an extensive horizon of temporality as the defining characteristic of those galactic niches supportive of conceptual intelligence. Whether or not one grants this approach, my analysis of the N = 1 problem and its core equivocation about intelligence demonstrates that even if we never find another instance of "conceiving beings" elsewhere in the universe, we can take some measure of satisfaction in the knowledge that our inability to confirm such beings is not because our conceptual powers, and the sciences that rest upon them, is a fabrication of humans on Earth; we just live in an immense universe.

## Notes

1. For an informative overview of the debate, see Carlos Mariscal, "Universal Biology: Assessing Universality from a Single Example," in *The Impact of Discovering Life Beyond Earth*, ed. Steven J. Dick (Cambridge, UK: Cambridge University Press, 2015), 113–126.

2. There has been an explosion of research in this area over the last two decades; the sources I mention here are ones I have found especially informative. Dirk Schulze-Makuch, "The Landscape of Life," in *The Impact of Discovering Life Beyond Earth*, ed. Steven J. Dick; Caleb Scharf, *The Copernicus Complex: Our Cosmic Significance in a Universe of Planets and Probabilities* (New York: Scientific American/Farrar, Strauss and Giroux, 2014); Dirk Schulze-Makuch and L. N. Irwin, *Life in the Universe: Expectations and Constraints*, 2nd ed. (Berlin: Springer, 2008); M. J. Crowe, *The Extraterrestrial Life Debate: Antiquity to 1900: A Source Book* (Notre Dame, IN: University of Notre Dame Press, 2008); Paul Davies, *The Origin of Life* (London: Penguin Books, 2003); Christian de Duve, *Vital Dust: Life as a Cosmic Imperative* (New York: Basic Books, 1995); Christian de Duve, *Life Evolving: Molecules, Mind and Meaning* (New York: Oxford University Press, 2002); David C. Catling, *Astrobiology: A Very Short Introduction* (Oxford, UK: Oxford University Press, 2013).

3. Cited from Carlos Mariscal, "Universal Biology," 121. See Carol Cleland, "Is a General Theory of Life Possible? Seeking the Nature of Life in the Context of a Single Example," *Biological Theory*, 7 (2012), 368–379.

4. Cited in Simon Conway Morris, *Life's Solution: Inevitable Humans in a Lonely Universe* (New York: Cambridge University Press, 2009), 67. See also F. Crick, *Life Itself: Its Origin and Nature* (London: Macdonald, 1982), 38. It is interesting to note that Crick's extreme skepticism on the likelihood of life arising on Earth led him to posit a version of the panspermia hypothesis, claiming that the materials for life must have been brought here by comets or other extraterrestrial material.

5. This appears to be the position endorsed by Morris in his *Life's Solutions,* although there is some equivocation on the issue in the last two chapters. For similar positions that readily concede that life may be abundant, but not intelligence, see P. D. Ward and D. Brownlee, *Rare Earth: Why Complex Life Is Uncommon in the Universe* (New York: Copernicus, 2000).

6. Lori Marino, "The Landscape of Intelligence," in *The Impact of Discovering Life Beyond Earth* (2015), 95–112, 95.

7. Ibid., 96–97.

8. Cited in Simon Conway Morris, *Life's Solution,* 232. See Leonard Ornstein, *Physics Today,* 35 (1982), 27–31, 29. Conway Morris cites a variety of other thinkers who hold similar views on the prevalence of higher intelligence, such as George Beadle, George Gaylord Simpson, and Irven DeVore.

9. Cited from Gregory Chaitin, *Proving Darwin: Making Biology Mathematical* (New York: Pantheon Books, 2012), vii.

10. Thomas Nagel, *Mind and Cosmos: Why the Materialist Neo-Darwinian Conception of Nature Is Almost Certainly False* (New York: Oxford University Press, 2012).

11. Edmund Husserl, *Formal and Transcendental Logic,* trans. Dorian Cairns (The Hague, The Netherlands: Martinus Nijhoff, 1978), 66.

12. Ibid., 196–197.

13. Ibid., 195.

14. Edmund Husserl, *Logical Investigations,* Vol. 1, trans., J. N. Findley (New York: Routledge, 2001), 43.

15. Ibid., 102–103.

16. Husserl, *Formal and Transcendental Logic,* 154–155.

17. Husserl, *Logical Investigations,* Vol. 1, 93.

18. The issues I reference here are part of a much larger and complicated debate on the evolution of consciousness, of which there is no shortage of perspectives. For those interested in exploring some of the perspectives sympathetic to reductionism, see *The Evolution of Consciousness,* ed. Cecilia Heyes and Ludwig Huber (Cambridge, MA: MIT Press, 2000). Ludwig Huber's article "Psychophylogenesis: Innovations and Limitations in the Evolution of Cognition," is especially instructive in laying out the parameters of the problem (pp. 23–42).

19. Richard Dawkins, "Intelligent Aliens," in *Intelligent Thought: Science versus the Intelligent Design Movement,* ed. John Brockman (New York: Vintage Books, 2006): 92–107, 97.

20. Thomas Nagel, *Mind and Cosmos,* 28.

21. Husserl, *Logical Investigations,* Vol. 1, 87.

22. Nagel, *Mind and Cosmos,* 73.

23. Husserl, *Logical Investigations,* Vol. 1, 98.

24. Nagel, *Mind and Cosmos,* 74.

25. Husserl, *Formal and Transcendental Logic,* 277–278.

26. Francisco Varelas et al.'s work *The Embodied Mind* was one of the first collaborative approaches to tackle the notion of cognition in terms of biological embodiment and remains a seminal text. See Francisco Varela, Evan Thompson, and Eleanor Rosch, *The Embodied Mind: Cognitive Science and Human Experience* (Cambridge, MA: MIT Press, 1993).

27. Lori Marino discusses this point in her article, "The Landscape of Intelligence," esp. 103–107. For another perspective focused just on primates, see Robin I. M. Dunbar's "Causal Reasoning, Mental Rehearsal, and the Evolution of Primate Cognition," in *The Evolution of Cognition,* 205–220.

28. Husserl, *Formal and Transcendental Logic,* 260–261.

29. Michael Ruse, " 'Klaatu Baruda Nikto'—or, Do They Really Think Like Us?' " in *The Impact of Discovering Life Beyond Earth* (2015), 175–188, 177–78.

30. Ibid., 180.

31. Ibid., 179.

32. More recently, Steven J. Dick takes up this problem in his book, *Astrobiology, Discovery, and Societal Impact* (Cambridge, UK: Cambridge University Press, 2018); see chapter 5, "Is Human Knowledge Universal," 141–175.

33. Simon Conway Morris, *Life's Solution: Inevitable Humans in a Lonely Universe* (New York: Cambridge University Press, 2003). Morris's text offers a compendium of data about the most diverse types of species to help make his case for evolutionary convergence.

34. Ibid., 119.

35. Ibid., 127.

36. Ibid., 141.

37. Ibid., 260.

38. Ibid., 309.

39. Ibid., 308–309.

40. Christian de Duve formulates an analogous argument on the similarity between self-replicating organisms throughout the universe, but this similarity is largely founded on the common properties of chemistry. *Life Evolving: Molecules, Mind, and Meaning,* see esp. 95–97.

41. Simon Conway Morris, *Life's Solution,* 283–284. The quote from Robert Bieri comes from his article "Humanoids on Other Planets," *American Scientist,* 52 (1964): 452–458, 457.

42. See Richard Levins and Richard Lewontin, *The Dialectical Biologist* (Cambridge, MA: Harvard University Press, 1985), 52.

43. Ibid., 58. See also Marc Kirschner and John Gerhart, *The Plausibility of Life: Resolving Darwin's Dilemma* (New Haven, CT: Yale University Press, 2005).

44. Originally quoted from Robert N. Bellah, *Religion in Human Evolution* (Cambridge MA: Harvard University Press, 2011), xiii. See F. John Odling-Smee, Keven N. Laland,

and Marcus W. Feldman, *Niche Construction: The Neglected Process in Evolution* (Princeton, NJ: Princeton University Press, 2003), 186.

45. Francisco J. Varela et al., *The Embodied Mind: Cognitive Science and Human Experience,* 196.
46. Ibid., 195.
47. Ibid., 196.
48. Ibid., 196–197.
49. Ibid., 198.
50. Thomas Nagel, *Mind and Cosmos,* 85.
51. Edmund Husserl, *Experience and Judgement,* trans. James Churchill and Karl Ameriks (Evanston, IL: Northwestern University Press, 1973), esp. §64 and 65 (pp. 253–269), 261.
52. Ibid., 260.
53. Ibid., 261.
54. Ibid., 265–266.
55. Hans Jonas, *The Phenomenon of Life: Toward a Philosophical Biology* (Evanston, IL: Northwestern University Press, 2001), esp. 85–87, 99–101. Jonas' concern is not with defending the universality of formal principles but showing the continuity of consciousness throughout life; however, his key point that more sophisticated forms of animal embodiment open up new dimensions of time seems spot on. Jonas is not alone in making this claim, but his work here, originally published in 1966, remains pioneering in this regard.
56. Slavoj Žižek, *Less than Nothing: Hegel and the Shadow of Dialectical Materialism* (New York: Verso Books, 2012), 467. Although Žižek's remarks here are made in reference to Hegel, what he says here is no less true of Husserl.
57. Adam Berg highlights the tension and compatibility between Husserl's notion of time-consciousness and other competing "positivist" or scientific accounts of time in his challenging but insightful book, *Phenomenalism, Phenomenology and the Question of Time: A Comparative Study of the Theories of Mach, Husserl and Boltzman* (Lanham, MD: Lexington Books, 2016). There have been various attempts to "naturalize" Husserl's account of time, the most instructive being the work of Francisco Varela, "The Specious Present: A Neurophenomenology of Time Consciousness," in J. Petitot, F. Varela, B. Pachoud, and J. Roy, eds., *Naturalizing Phenomenology: Issues in Contemporary Phenomenology and Cognitive Science* (Stanford, CA: Stanford University Press, 1999). Although Husserl himself would have resisted such efforts, it seems clear that questions about the origin of internal time-consciousness cannot be evaded indefinitely. Locating the problem to that of a cosmological niche, as I try to do here, only relocates the problem but does not solve it. Ultimately, I think the forms of temporality compatible with logical objects are, borrowing Morris's language "nodes of occupation . . . effectively predetermined from the Big Bang." See Conway Morris, *Life's Solution,* 310.
58. Nagel, *Mind and Cosmos,* 86.
59. E.O. Wilson, *On Human Nature* (Cambridge, MA: Harvard University Press, 1978), 167.

# References

Bellah, Robert. *Religion in Human Evolution*. Cambridge, MA: Harvard University Press, 2011.

Berg, Adam. *Phenomenalism, Phenomenology and the Question of Time: A Comparative Study of the Theories of Mach, Husserl and Boltzman*. Lanham, MD: Lexington Books, 2016.

Bieri, Robert. "Humanoids on Other Planets." *American Scientist*, 52 (1964): 452–458.

Catling, David. *Astrobiology: A Very Short Introduction*. Oxford, UK: Oxford University Press, 2013.

Chaitin, Gregory. *Proving Darwin: Making Biology Mathematical*. New York: Pantheon Books, 2012.

Cleland, Carol. "Is a General Theory of Life Possible? Seeking the Nature of Life in the Context of a Single Example." *Biological Theory*, 7 (2012): 368–379.

Conway Morris, Simon. *Life's Solution: Inevitable Humans in a Lonely Universe*. New York: Cambridge University Press, 2009.

Crick, Francis. *Life Itself: its origin and nature*. London: Macdonald, 1982.

Crowe, M. J. *The Extraterrestrial Life Debate: Antiquity to 1900: A Source Book*. Notre Dame, IN: University of Notre Dame Press, 2008.

Davies, Paul. *The Origin of Life*. London: Penguin Books, 2003.

Dawkins, Richard. "Intelligent Aliens." In *Intelligent Thought: Science Versus the Intelligent Design Movement*. Edited by John Brockman. New York: Vintage Books, 2006: 92–107.

de Duve, Christian. *Vital Dust: Life as a Cosmic Imperative*. New York: Basic Books, 1995.

de Duve, Christian. *Life Evolving: Molecules, Mind and Meaning*. New York: Oxford University Press, 2002.

Dick, Steven. J. *Astrobiology, Discovery, and Societal Impact*. Cambridge, UK: Cambridge University Press, 2018.

Dick, Steven. J. Ed. *The Impact of Discovering Life Beyond Earth*. Cambridge, UK: Cambridge University Press, 2015.

Dunbar, Robin. I. M. "Causal Reasoning, Mental Rehearsal, and the Evolution of Primate Cognition." In *The Evolution of Consciousness*. Edited by Cecilia Heyes and Ludwig Huber. Cambridge, MA: MIT Press, 2000: 205–220.

Heyes, Cecilia, and Ludwig Huber. Eds. *The Evolution of Consciousness*. Cambridge, MA: MIT Press, 2000.

Huber, Ludwig. "Psychophylogenesis: Innovations and Limitations in the Evolution of Cognition. In *The Evolution of Consciousness*. Edited by Cecilia Heyes and Ludwig Huber. Cambridge, MA: MIT Press, 2000: 23–42.

Husserl, Edmund. *Experience and Judgment*. Translated by James S. Churchill and Karl Ameriks. Evanston, IL: Northwestern University Press, 1973.

Husserl, Edmund. *Formal and Transcendental Logic*. Translated by Dorian Cairns. The Hague, The Netherlands: Martinus Nijhoff, 1978.

Husserl, Edmund. *Logical Investigations, Vol. 1*. Translated by J. N Findlay. New York: Routledge, 2001.

Jonas, Hans. *The Phenomenon of Life: Toward a Philosophical Biology*. Evanston, IL: Northwestern University Press, 2001.

Kirschner, Marc, and John Gerhart. *The Plausibility of Life: Resolving Darwin's Dilemma*. New Haven, CT: Yale University Press, 2005.

Lewontin, Richard and Richard Levins. *The Dialectical Biologist*. Cambridge, MA: Harvard University Press, 1985.

Marino, Lori. "The Landscape of Intelligence." In *The Impact of Discovering Life Beyond Earth*. Edited by Steven J. Dick. Cambridge, UK: Cambridge University Press, 2015: 95–112.

Mariscal, Carlos. "Universal Biology: Assessing Universality from a Single Example." In *The Impact of Discovering Life Beyond Earth*. Edited by Steven J. Dick. Cambridge, UK: Cambridge University Press, 2015: 113–126.

Nagel, Thomas. *Mind and Cosmos: Why the Materialist Neo-Darwinian Conception of Nature Is Almost Certainly False*. New York: Oxford University Press, 2012.

Odling-Smee, John. F., Keven N. Laland, and Marcus W. Feldman. *Niche Construction: The Neglected Process in Evolution*. Princeton, NJ: Princeton University Press, 2003.

Ruse, Michael. "'Klaatu Baruda Nikto'—or, Do They Really Think Like Us?'" In *The Impact of Discovering Life Beyond Earth*. Edited by Steven J. Dick. Cambridge, UK: Cambridge University Press, 2015: 175–188.

Scharf, Caleb. *The Copernicus Complex: Our Cosmic Significance in a Universe of Planets and Probabilities*. New York: Scientific American/Farrar, Strauss and Giroux, 2014.

Schulze-Makuch, Dirk, and L. N. Irwin. *Life in the Universe: Expectations and Constraints*. 2nd ed. Berlin: Springer, 2008.

Schulze-Makuch, Dirk. "The Landscape of Life." In *The Impact of Discovering Life Beyond Earth*. Edited by Steven J. Dick. Cambridge, UK: Cambridge University Press, 2015: 81–94.

Varela, Francisco. "The Specious Present: A Neurophenomenology of Time Consciousness." In *Naturalizing Phenomenology: Issues in Contemporary Phenomenology and Cognitive Science*. Edited by J. Petitot, F. Varela, B. Pachoud, and J. Roy. Stanford, CA: Stanford University Press, 1999.

Varela, Francisco, Evan Thompson, and Eleanor Rosch. Eds. *The Embodied Mind: Cognitive Science and Human Experience*. Cambridge, MA: MIT Press, 1993.

Ward, P. D., and D. Brownlee. *Rare Earth: Why Complex Life Is Uncommon in the Universe*. New York: Copernicus, 2000.

Wilson, E. O. *On Human Nature*. Cambridge, MA: Harvard University Press, 1978.

Žižek, Slavoj. *Less Than Nothing: Hegel and the Shadow of Dialectical Materialism*. New York: Verso Books, 2012.

# 10

# What Are Extremophiles?

## A Philosophical Perspective

*Carlos Mariscal and T.D.P. Brunet*

## Introduction

In the 1970s, R. D. MacElroy coined the term "extremophile" to describe microorganisms that thrive under extreme conditions (MacElroy, 1974). This hybrid word transliterates to "love of extremes" and has been studied as a straightforward concept ever since. In this chapter, we discuss several ways the term has been understood in the scientific literature, each of which has different consequences for the distribution and importance of extremophiles. They are, briefly, *human-centric*, at *the edge of* life's habitation of *morphospace*, by appeal to *statistical rarity*, described by *objective limits*, and at the *limits of impossibility* for metabolic processes. Importantly, these concepts have coexisted, unacknowledged and conflated, for decades. Confusion threatens to follow from the wildly varied inclusion or exclusion of organisms as extremophiles depending on the concept used. Under some conceptions, entire kinds of extremophiles become meaningless. Since our understanding of how life works is shaped by what we take to be its extremes, clarifying extremophily is key for many large-scale projects in biology, biotechnology, and astrobiology.

In what follows, we proceed as if a noncontroversial account of life is possible and that it is possible to find complex chemistry in the universe that is similar enough to life on Earth such that both may be considered instances of "life" (but see Mariscal & Doolittle, 2018). We raise, but do not address, the questions of whether the distribution of life on Earth is representative of what we may find elsewhere in the universe, whether the same kinds of extremophiles would exist given a replay of the tape of life. Additionally, each of these concepts assumes life based on some sort of biochemistry in this universe, effectively ruling out claims made by some artificial life proponents that their digital organisms are genuine examples of life (Langton, 1989; Ray, 1995). On the distinction between extremophilic and extremotolerant, we note that all accounts will accept the latter as a broader category than the former, since tolerance of extreme conditions is a prerequisite for extremophily under any conception. Indeed, there will be

Carlos Mariscal and T.D.P. Brunet, *What Are Extremophiles?* In: *Social and Conceptual Issues in Astrobiology.* Edited by: Kelly C. Smith and Carlos Mariscal, Oxford University Press (2020).
© Oxford University Press.
DOI: 10.1093/oso/9780190915650.001.0010

many extreme environments where tolerance is the only option (e.g., *Bacillus marismortui* was extracted and grown from 250-million-year-old salt crystals in the Permian Salado Formation in an inactive yet persistent state; Vreeland et al., 2000). The nature of the environment precluded any organisms *thriving*.

We also note that extremophily, as a functional category, is potentially applicable at many levels of the biological hierarchy. Extremophily at one level does not necessarily extend to higher and lower levels. For instance, a microorganism in isolation might be quite intolerant to certain environmental conditions yet flourish when subjected to the same conditions in the presence of a natural biofilm. Alternatively, a protein molecule might be quite active under certain conditions even if the optimal environment for the organism containing it is far more mesophilic. There is an industry of artificially selecting organisms and proteins to adapt to extreme environments (see van den Burg & Eijsink, 2002), providing some justification to consider "functioning at extremes" as a worthwhile category of investigation.

Finally, we also note certain physico-chemical ranges are rarely considered with respect to extremophily (e.g., time span, size, nutrient availability; Hoehler & Jørgensen 2013), as well as some biological parameters (abundance, isolation, competition, etc.). Perhaps scientific interest must also come into play as to the reason these criteria are not considered relevant. We return to this and other issues later.

In the next section, we give five definitions of extremophily and show their benefits, drawbacks, and unintended consequences. These arguments are summarized in Table 10.1 and represented visually in the three figures. Given research on polyextremophiles, it seems Figure 10.2 is a more plausible representation of the state of current knowledge than the idyllic Figure 10.1 (Harrison et al., 2013). Apparently, life is patchily distributed across various dimensions, which may reflect its contingent history, poor sampling, fundamental limits, or something else. Figure 10.3 shows the conceptual flowchart for all of these views. In the following section, we take a step back to ask whether we should choose between these definitions and how such a judgment could be made. We argue for a limited pluralism, in which some, but not all, of the concepts are acceptable relative to certain practical and theoretical aims.

## Extremophile Concepts

### Human-Centric

**Explanation:** As a first attempt, we might view something as an extremophile if it thrives in the kinds of environments that would be considered extreme for

**Table 10.1.** Summary of the Extremophile Concepts Discussed in the Text, Including Benefits and Drawbacks

| Definitions | Description | Benefits | Drawbacks |
|---|---|---|---|
| Human-Centric | As extremophile is an organism that thrives in environments which would be hazardous to humans or human cells. | Clear, operational, relatively constant | Instrumental, seemingly arbitrary, excludes humans as extreme by definition |
| Edge of Morphospace (Earth) | Extremophiles are known organisms that inhabit the limits of some physical or chemical continuum relative to life as we know it. | Operational, does not require extremophiles to be rare | As research advances, those extremophiles once thought to be at the edge no longer count as such, contingent on the course of evolution on Earth |
| Edge of Morphospace (Universe) | Extremophiles are those organisms in the universe that inhabit the limits of some physical or chemical continuum relative to all other life. | Clear, universal | Empirically intractable |
| Statistical Rarity (Earth) | An extremophile is a known organism that thrives in conditions under which most other organisms do not. | Empirically tractable | Contingent on the course of evolution on Earth, extremophiles can exist by chance, extremophiles may be possible in otherwise average environments |
| Statistical Rarity (Universe) | An extremophile is an organism that exists somewhere in the universe and thrives in conditions under which most other organisms do not. | Clear, universal | Empirically intractable, may imply everything is an extremophile |
| Objective Limits | An extreme is the limit(s) of some physical or chemical phenomenon. Extremophiles are organisms that do well in these environments. | Objective, determinable independent of any examples of life | Appropriate to physics or chemistry, problematic for life. If research thresholds are overly broad, it is unclear what this definition adds that is not better captured by other accounts |
| Near Impossibility | Extremophiles, when they exist, are at the limits of what life's mechanisms can possibly handle. | Useful for very different research questions | Potentially theory-laden, may require an uncontroversial definition of life, may be scientifically impracticable |
| Research Interest | Extremophiles are any organisms, parts, or behaviors of organisms that meet certain research interests much more so than other organisms, parts, or behaviors. | Pragmatic, flexible, compatible with other definitions | Difficult to pin down, potentially unsatisfying |

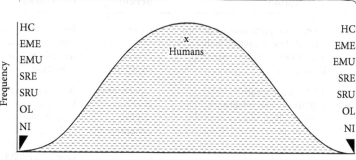

**Figure 10.1.** The easy case of extremophily is when life on Earth is representative of what we will find in the universe. In this case, all our definitions collapse and "extremophile" can proceed as an unanalyzed concept.

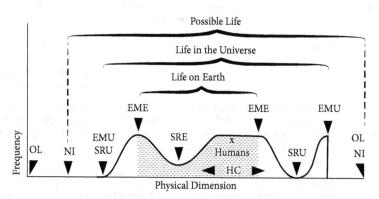

**Figure 10.2.** A situation in which each of the definitions comes apart from each of the others. *Objective limits* (OL) picks out the edges of the physical dimension. Extremophiles under the *near impossibility* (NI) concept may reach the limits of a physical dimension (right dashed line) or fall short (left dashed line). The least populated area in the dimension is picked under one *statistical rarity* (SRU), which may coincide with the actual limits occupied by life in the universe (EMU on left) or not (EMU on right). A similar distinction could be made with respect to life on Earth (SRE and EME in the shaded area). Humans appear somewhere on this space, and extremes may be defined relative to them (HC).

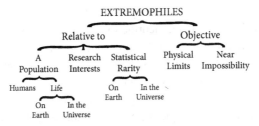

**Figure 10.3.** Extremophile concepts can either be objective or relative. Of the three relative concepts, *population relative, research interest relative,* and *statistical rarity,* the latter seems unmotivated. Abundance itself is neither necessary for being an extremophile, nor sufficient. The *human-centric* approach is likely to only be interesting to a subset of cases. Relative to the edges of what is inhabited by life on Earth or in the universe (*edge of morphospace;* see the text) is particularly interesting for many uses, although the latter less so than the former. *Research interest* definitions of extremophiles are pragmatic in nature and thus not subject to the same theoretical concerns as other approaches. Of two absolute concepts, *objective limits of physical dimensions* and *near the limits of possibility for life,* only the latter is scientifically interesting, although it raises a number of conceptual issues.

humans. This definition[1] is yoked to current human habitability, which allows it to be used for even the study of ancient, distant, or hypothetical samples. We take this definition to be the driving idea behind such claims as:

> Extremophiles are organisms which permanently experience environmental conditions which may be considered as extreme in comparison to the physico-chemical characteristics of the normal environment of human cells: the latter belonging to the mesophile or temperate world. (Gerday, 2002, p. 1)

> Extremophiles survive in environments that would be lethal to humans. (Cohen & Steward, 2001, 1121)

**Benefits:** There is a benefit in such a straightforward, instrumental approach: we are well aware of our tolerance for temperature, pressure, salinity, and so on. Under this definition, we would draw clear boundaries around mesophiles as "organisms that like what we like" and extremophiles as everything else. So this definition is explicit, clear, and relatively constant. Given its clear methodological advantage, it may make sense for many practicing scientists to use this as an operational definition, regardless of whether they ultimately define extremophiles using other criteria in more rigorous settings (see Bich & Green [2018] and chapter 5 in this volume for similar issues with respect to "life").

**Drawbacks:** The most immediate problem for this definition is its arbitrariness. It may strike some as unscientific to have a definition so closely linked to the human condition. It would be akin to using the criterion of "most impressive mountain" to identify Everest rather than "Earth's tallest mountain." Additionally, there are many environments in which humans could not survive, which nevertheless do not seem to be physically, chemically, or biologically extreme. For example, humans could not survive for long at 1 meter under the ocean surface or in the Paleozoic. A variant of the human-centric concept would appeal to the range of survivability conditions of human cells, as in some of the previous quotes. Unfortunately, there are just as many intuitively benign environments that are inhospitable to human cells (e.g., outside of a human body).

**Implications:** One unintended consequence of this definition is that it rules humans as mesophiles by definition. This renders some uses of extremophily as nonsensical. For example, some astrobiologists have claimed "we are extremophiles" to describe the rarity of breathing oxygen (Rothschild & Mancinelli, 2001). According to the human-centric account, humans can never be extremophiles regardless of how rare or unusual they may be, *even* if humans would be considered extreme according to every other definition. Perhaps these implications are unimportant for most purposes. Focusing on humans is, fundamentally, a pragmatic move to highlight important differences. It seems unlikely scientists will keep to this definition if it is inconsistent with the questions they hope to answer.

**Verdict:** The human-centric approach may be useful for many practical purposes, even if it is not the full account of "extremophiles" that scientists should accept.

## Edge of Morphospace

**Explanation:** Biological organisms exist across several physical dimensions, like temperature, pressure, salinity, and so on. Scientists can (and have) mapped those dimensions to multidimensional spaces to show the ranges occupied by life. Such spaces may be called "morphospace," a concept often used in biology to visualize evolution across physical dimensions (Raup, 1966). Similar concepts exist, such as "design space" (Dennett, 1995), "phase space" (Berne & Straub, 1997), and "habitable space" (Harrison et al., 2013). Life has explored a number of physical and chemical limits, although as we will see later, it has reached objective limits only in some of these cases. So perhaps the most natural way to define extremophiles may be with respect to the physical extremes life *has* explored. An extremophile, in this definition, is simply an organism that exists at the edge of the area of morphospace occupied by life on Earth (EME) or in the Universe in general (EMU). The edge, importantly, need not be near an objective limit nor sparsely populated. Extremophiles, in fact, may be the most common creatures with respect to some dimensions. Unlike the human-centric view, this approach would be consistent with humans occupying extremes, or even not existing. Additionally, this concept is not explicitly instrumental. Although this definition is n-dimensional, it may help to picture a three-dimensional space with respect to some variable or another (Harrison et al., 2013). We take such a definition to be the motivation behind such claims:

> Life on Earth is limited by physical and chemical extremes that define the "habitable space" within which it operates. (Harrison et al., 2013, p. 204)

> So the study of terrestrial organisms that can survive on the extreme boundary of these conditions, the so-called extremophiles, greatly informs astrobiology and the search for life beyond Earth. (Dartnell, 2011, 1.25)

Our two formulations, EME and EMU, would be equivalent if life on Earth was the only life in the Universe. EME has the benefit of being empirically tractable, and we might describe some research into extremophily as *extending* the boundaries of morphospace. EMU will never be empirically tractable, but it may be a good conceptual goal for research into extremophily. For EMU, research into extremophily merely *discovers* the boundaries of morphospace.

**Benefits:** EME is empirically tractable and visualizable. Unlike statistical rarity, it is not important for extremophiles to be rare, which is a potentially counterintuitive result. While in many cases, they will be equivalent, edge of morphospace accepts instances in which extreme life is common. In other words, the frequency of life does not decrease as certain dimensions are reached or in which morphospace is occupied evenly across a dimension. Suppose life was evenly distributed across the pH continuum. The edge of morphospace concept would still consider organisms living at pH 0 and pH 14 as acidophiles and alkaliphiles, respectively. Although it is possible for there to be a number of internal boundaries of morphospace in any dimension, we suspect that is rare (see Figure 10.2). For EME or EMU, some examples of life will be extremophilic for every physical and chemical dimension, even if they are not impressive from objectively physical standards. For example, *Deinococcus radiodurans* is an EME extremophile with respect to cold, dehydration, vacuums, and acid, even though it never approaches the objective physical limits of many of these variables.

**Drawbacks:** The edge of morphospace concept has drawbacks, however. First, EME is contingent within the course of evolution (on Earth) and with respect to current scientific sampling. EMU avoids the latter problem at the cost of scientific tractability. Theoretically, EMU may still be contingent within the span of life in the universe if there are possible configurations life may hold but never approximates. This situation may arise, for example, if "islands of stability" exist with respect to evolution. That is, there may be biologically possible life forms that can never arise naturally because the evolutionary path to them is implausible, but that could occur if intelligently designed (i.e., zebras with machine guns). One final curiosity: for any variable in which life exists, there *must* exist an extremophile, as at least some will be closest to the limits within that variable.

**Implications:** This concept seems to underlie two research programs with respect to extremophiles: the synthesis or evolution of extremophily and the seeking out of extreme environments to discover new extreme organisms. Both of these approaches expand the "envelope" of where life is possible. Like several other approaches, this approach requires consideration of life as it actually exists and cannot be determined *a priori*.

**Verdict:** EME is a very useful concept and likely what many researchers intend by the term "extremophile." EMU is less obviously useful, and its theoretical benefits are unclear. We will consider EMU again in the discussion of near impossibility.

## Statistical Rarity

**Explanation:** A broader approach to defining extremophiles might appeal to their abundance. Under this definition, an extremophile would be one that exists in conditions where life is quite rare, either (a) on Earth (SRE) or (b) in general across the universe (SRU). This rarity could be with respect to external, physicochemical properties, such as pressure or temperature ranges, or relational properties, such as isolation or extreme competition. Like the prior approaches, this approach is relative, though it is relative to the broader category of life rather than humans. Something akin to this definition seems to be behind such claims as:

> An extremophile (from Latin extremus meaning "extreme" and Greek philiā (φιλία) meaning "love") is an organism that thrives in and may even require physically or geochemically extreme conditions that are detrimental to the majority of life on Earth. (Gupta et al., 2014, p. 1)

> several organisms are able to thrive in these hostile locations where most life would perish. (Reed et al., 2013, p. 2)

There are two interesting interpretations of the statistical rarity. We call it SRE when we refer to rare organisms which thrive in conditions hostile to the majority of *life as we know it on Earth*. We call it SRU when these rare organisms thrive in conditions hostile to the majority of *all life in the Universe*. SRU includes all life that ever will exist, but not life that is possible but never comes into existence.

Since our only examples of life are those based on Earth, SRE and SRU are equivalent, for practical purposes. But they would also be equivalent in theory if it turned out Earth held the only example of life in the universe. Until a second "example" of life is discovered, SRU is merely a theoretical ideal, albeit one with curious consequences. For example, if life on Earth is significantly different from other life elsewhere (as could happen if Earth is a peculiarly inhospitable environment in the universe), it could be that SRE and SRU pick out entirely different kinds of organisms (i.e., Earth life could be a biased sample of life's extremes; see Figure 10.2).

**Benefits:**   In terms of benefits, it makes sense that extremophiles would be relatively rare organisms. So defining extremophiles based on their rarity is intuitive. With respect to most parameters, it is plausible that life forms a normal distribution. In this normal case, the rarity of organisms will correlate with the extremity of the environment. Additionally, SRE is empirically tractable, although it is subject to change over time as new organisms are discovered. SRU, while not empirically tractable, is appealing in theory. Given many potential instances of life in the universe, the most rare kinds of organisms are also likely to be the most extreme in any number of measures.

**Drawbacks:**   Statistical rarity definitions has unintuitive consequences, however. Consider Figure 10.2, which belies our musings that life on Earth fit a normal distribution across many (or any) physical parameters. There are likely many combinations of pressure, temperature, and so on in which no organisms exist even though they do in nearly identical situations. Statistical rarity is also grounded on contingent natural history. So even if life would thrive in a certain environment, it may never be exposed to such scenarios. If we ran the tape of life again, perhaps the extremes of various metrics might be more inhabited than the more moderate middles (Gould, 1989). Although it is presumably uncommon, it is certainly possible that life might not occupy some environmental variable by chance. For example, the metal iridium is rare on Earth but common in igneous deposits and asteroids (Alvarez et al., 1980). Since iridium is a very rare element, areas once struck by asteroids may have orders of magnitude more iridium than other areas. If areas with moderate amounts of iridium are quite rare, the few organisms that live in these areas would be SRE extremophiles with respect to iridium *even if they have no other extreme properties.* Though critiques from contingency and chance are less effective against SRU, that view carries other, unintuitive consequences. Suppose the vast majority of life worlds in the universe required liquid ammonia. Under SRU, in such a situation, unbeknownst to us, all of life as we know it would be extremophilic with respect to the solvent it uses. Earth might be populated by extremophiles.

**Implications:**   Perhaps extremophily must always be relative. With such a perspective, it would never make sense to describe an organism as simply an extremophile. Instead, extremophiles must always be defined *with respect to* a range of entities. As a result, any time scientists describe a microbe as "an extremophile," they must either be speaking incorrectly or implicitly reference a class of other organisms (e.g., SRE). Consider *Chlamydomonas nivalis*, alternatively described as "cold-tolerant microbes growing on . . . snow fields and glaciers from *many parts of the world*" (Takeuchi et al., 2006; emphasis added), "cryophilic," and "a remarkable extremophile, able to survive and thrive *in an environment that would be fatal to most plants*" (Gorton et al., 2001; emphasis added). If we take relativism seriously, *C. nivalis* might be considered an extremophile with respect to the habitats of most terrestrial plants (SRE), just extremotolerant with respect to snow and glacial environments, and perhaps again extreme with respect to all of the universe (SRU).

**Verdict:** Given the availability of other concepts, we view these critiques as devastating to any formulation of *statistical rarity*.

## Objective Limits

**Explanation:** An alternative to instrumental, relativistic, or contingent criteria may be a mere assessment of objective physical or chemical limits. Certainly, there are some extremes that can be defined in this way: temperature in this universe can range from -273.15°C to at least 200,000°C (Werner et al., 2008). Chemical concentrations (e.g., salinity, oxygen, water) can range from 0% to 100% saturation. Objective limits, then, takes any of these limits and sets some threshold whereby if an organism approaches the threshold, then it may be considered an extremophile with respect to that boundary condition. In this definition, extreme environments are identified first, and extremophiles are defined as creatures that happen to live in those environments. Unlike the previous two definitions, the objective limits account is applicable even if humans, or indeed all life, never existed. We take something like the objective limits view to motivate such claims as:

> Numerous microorganisms are extremophiles, which means they can metabolize and reproduce in extreme conditions of heat, cold, acidity, salinity and other seemingly inhospitable environments. (O'Malley, 2014, p. 5)

> Extremophiles are defined by the environmental conditions in which they grow optimally. (Gupta et al., 2014, 371)

> An organism that thrives in an extreme environment is an extremophile . . . "Extremes" include physical extremes (for example, temperature,

radiation or pressure) and geochemical extremes (for example, desiccation, salinity, pH, oxygen species or redox potential) (Table 1). It could be argued that extremophiles should include organisms thriving in biological extremes (for example, nutritional extremes, and extremes of population density, parasites, prey, and so on). (Rothschild & Mancinelli, 2001, 1092)

**Benefits:** The benefits of setting objective limits is that they can be clearly defined independent of any examples of life. These criteria could be used for as-yet-unknown life and apply universally. To fully flesh out a definition based on objective limits, we would need to specify some threshold or degree of extremophily (i.e. "anything within 10% of the extreme is an extremophile," or "organism X is an extremophile to degree Y"). Some may worry about the arbitrariness and vagueness of a threshold, though these concerns are common in biology.

**Drawbacks:** Problematically, there are certain physical ranges of which life only ever explores a small sampling. For example, although some organisms are intuitively "cryophilic" and "thermophilic," most of these do not come close to the limits of temperature in the universe. Indeed, the closest example at present is perhaps the only extremo*tolerant* tardigrades, or tardigrade eggs, that can be subjected to vacuum and extreme cold conditions without significant damage (Jönsson et al., 2008; Persson et al., 2011). Nor does life thrive at objective extremes of size, pressure, or radiation, among many other parameters. There are no angstrom-sized organisms, no black hole populations, and no species that only thrive in super novae.

**Implications:** There is no guarantee that objective limits be relevant to scientific interests. Life on Earth thrives in the absence of Einsteinium, for example. We are all extremophilic with respect to the absence of Einsteinium. Huzzah. Unless scientific utility comes into play, the objective limits concept would rule every example of life as extremophilic in *some* respect. Admitting scientific utility comes into play with respect to the limits we count as important, interesting, or explanatory, undercuts the objectivity of the definition, which is its primary benefit.

**Verdict:** While objective limits are desirable in the abstract, they are less useful for scientific purposes than the near impossibility concept, discussed next.

## Near Impossibility

**Explanation:** Extremophiles with respect to near impossibility exhibit adaptations to extreme environments that are at the very limit of what is even *possible* for life to tolerate. Most of the preceding definitions dealt with organisms in the real world. In some definitions, notably statistical rarity and edge of morphospace, this led to the unfortunate consequence of extremophily being a contingent concept. Certain versions of these (SRE and EME) also suffered from sampling biases, though this may not concern researchers who are only interested in known extant organisms. One appeal of objective limits was that it avoided both the worries of contingency and sampling biases. Objective limits exist, after all, independent of the existence of life. Problematically, objective limits ruled out many of the paradigmatic examples of extremophily, such as piezophiles, thermophiles, and radioresistant organisms, as the objective limits of those physical dimensions are well beyond what life could tolerate. Unlike objective limits, near impossibility takes living processes into account. Recall that the unabashedly anthropocentric human-centric concept appealed to where humans (or human cells) *could* live. One way to avoid the charge of anthropocentrism would be to take the humans out of the definition and abstract away to the limits of the processes fundamental to life, such as metabolism or evolution. Such an exploration of life's possibilities requires thorough biophysical and biochemical analyses in addition to (or instead of) ecological surveys.

The near impossibility concept, we believe, is a guiding thought in both of the following quotes:

> The limiting temperature above which life cannot flourish is of theoretical and practical importance to many biological and geochemical studies. (Bains et al., 2015, 1055)

> Microbial life exists in all the locations where microbes can survive. (Gold, 1992, 6047)

Many readers may assume that life exists everywhere it possibly can, and so near impossibility may collapse into either statistical rarity or edge of morphospace. But in fact, this is an open question, and there are reasons to assume it is false (Bains, 2004; Schulze-Makuch & Irwin, 2012).

**Benefits:**  Some research in synthetic biology and controlled evolution only makes sense within the context of this definition. Under other approaches, the controlled evolution of radiotolerant bacteria, for example, would be merely the creation of new extremophiles. To make sense of such projects, we must understand them as exploring the theoretical limits of life. Analyses of the possibility conditions for life are important in biology. For example, in 1983, Baross and Deming argued some bacteria grew at temperature ranges of >250°C and 265 atm (Baross & Deming, 1983). In response, many scientists failed to replicate their results, and some argued properties of biomolecules suggested such growth it was "impossible in theory" (Trent et al., 1984; White, 1984).

**Drawbacks:** Although some work has been done to assess the limits of the mechanisms of life (Bains, 2015), these analyses have an inestimable margin of error. It may be the case that life *as we know it* cannot survive above 150°C, for example, but such a claim is dependent on Earth's life resembling all other possible life. We would only expect these analyses to be justified in the cases where we expect all possible examples of life would break down, which is problematic because various definitions of life might set this boundary differently. This analysis could be too narrow, not considering the many ways life could be realized, or it could be too broad, assuming a broader range for living mechanisms than is actually tolerable based on unknown variables. Narrow definitions can be challenged empirically, by attempting to discover or evolve more extreme lineages, but too-broad definitions may be untestable. Because this definition focuses on life that may not exist, it is more subject to theory-laden assumptions than other definitions.¹ One need only take a brief look through the history of biology to see the frequency with which such assumptions and conjectures are overturned by new empirical evidence or new theoretical understanding.

**Implications:** Note that this approach is tantamount to assuming or stipulating a definition of life. As such, near impossibility may come in as many variants as there are definitions of life. They will share the problems of each to boot: Near impossible extremophiles will not settle questions about whether A-Life is alive, for example. Given these worries, near impossibility for life as we know it is not a good enough characterization for this concept. Work in this area ought to highlight the aspects of life in which organisms are bordering on impossibility and why: too hot for proteins, too much sodium for conventional cell membranes, and so on.

**Verdict:** Scientific work in synthetic biology and related areas may rely on a near impossibility characterization of extremophiles.

## Monism, Pluralism, Pragmatism

### Monism

Given the many distinct definitions presented, one might ask whether there is any justified way to decide between the various candidate definitions. Perhaps one could study these definitions, choose the one that most fits one's scientific aims, stomp one's foot, and declare the rest as instances of poor thinking. Alternatively, scientists might opt for a mixed strategy in which we take the best features of each definition and splice them together into an unholy amalgamation. A skeptical reader might conclude these definitions rarely come apart, and so they may insist a precise definition is unnecessary for most scientific use.

Finally, we could stomp our other foot and declare, as in the famous declaration of judicial candor, "I know it when I see it" (Stewart, 1964). Each of these three strategies aims to justify a single definition at the expense of the rest—what philosophers call "monism."

A monist strategy is one that takes a stand on a single, proper understanding of a concept, especially in the face of many proposed alternatives. Monism is conceptually preferable in a scenario in which the object of study is naturally unified. There are also methodological advantages of monism: its delineations are explicit, meaning they can be questioned, tested, and negotiated. Monism is problematic in cases in which the subject matter is not naturally unified. Biology is replete with such examples, and the desire for monism has arguably fueled the interminable debates over the nature of species, fitness, function, and so on.

## Pluralism

Contrasted with monism is pluralism, an approach in which a number of definitions are all entertained simultaneously, sometimes with respect to a particular domain or explanatory issue. The pluralist position is often unappealing to people who desire a single account, for intellectual, personal, or aesthetic reasons. Some versions of pluralism have the methodological disadvantages of being hard to test or falsify, inviting equivocation, and relying on individual researchers to be clear. Nevertheless, extremophily faces similar empirical and theoretical challenges to those that have plagued analyses of other biological concepts (e.g., life, species, and genes). In each of those cases, it seems as if the plurality of natural processes and scientific aims has resisted a single, monist characterization. We hope to take lessons from those debates seriously. Our proposal is pluralist in nature. We conclude there are many aims for research into extremophily: from seeking extreme-tolerant biological products, assessing the abundance, variation, and efficacy of creatures in difficult environments, to inferring the limits of life in the universe.

Before looking into how one might decide on the most appropriate concept of extremophily within a particular research aim, there are overall distinctions to be made between the five concepts of extremophily. We feel the critiques facing the statistical rarity and the objective limits definitions are devastating, and nearly all scientific uses can proceed better with other definitions (see Figure 10.3). While much good science is done with a human-centric definition of extremophiles, it cannot be overlooked that this definition is fully anthropocentric. And epistemic limits preclude the scientific use of those definitions that quantify over undiscovered organisms (SRU and EMU). Thus, most (but not all) astrobiological research into extremophiles is best conducted within the bounds of near impossibility or EME. This timid form of pluralism, we maintain, is a conceptually sound groundwork on which we can vindicate the various roles extremophiles play in our understanding of life in the universe.

## Pragmatism

While a broadly construed and conceptually sound pluralism is required for extremophile-based science in general, each discipline may differ in its preferred definition(s) of extemophily. The research aims of molecular ecologists, in search of extreme habitats, differ from those of biotechnologists seeking new sources of biochemical reagents (c.f. Lentzen & Schwarz, 2006). We might refer to a biotechnological utility (BU) concept to describe the norms of current biotech research. Part of the motivation for this chapter is that such norms are often unclear or inconsistent. The concepts used by most researchers are marshaled without considering other uses within the field, risking serious equivocation. Consider the aims of biotechnological research outlined here:

> As gene sequencing technology becomes more routine, researchers are *determining the sequences of more obscure microorganisms and delving into the diversity of the microbial world with the aim of discovering new products.* It is hoped that genome data on nonpathogenic bacteria will *lead to the discovery of biocatalysts resistant to extremes of pH,* temperature, or solvents; nutritionally beneficial bacteria for probiotics; new types of streptomycete antibiotics; and microorganisms with enhanced capabilities to degrade xenobiotic compounds. (Marshall, 2000, 1026; emphasis added)

> A novel application area for extremophiles is the use of "extremolytes," organic osmolytes from extremophilic microorganisms, to protect biological macromolecules and cells from damage by external stresses . . . A range of new applications, all based on the adaptation to stress conditions conferred by extremolytes, is in development. (Lentzen & Schwarz, 2006, 623)

The act of investigating "obscure" organisms to the end of getting a better picture of the extremes is akin to what we have labeled as SRE, while extremes of pH, temperature, and solvents correspond to something closer to an EME. While we might expect that organisms inhabiting statistically rare environmental conditions would also be those with "biocatalysts resistant to extremes," this is far from certain. Moreover, there seems no a priori reason to suspect that the discovery of new products should be more likely in obscure organisms than in the mesophiles, unless, of course, we had already characterized the more common organisms. While SRE or EME alone are poor proxies for biotechnological applicability, other definitions might be better candidates. A few examples help illustrate this.

The *Taq* polymerase, extracted from the thermophilic archaeon *Thermus aquaticus* (Chien et al., 1976), has had a profound influence on biotechnology since its discovery and eventual use as a reagent in polymerase chain reaction (PCR; Saiki et al., 1988). The utility of *Taq* for PCR amplification derives from its stability at temperatures sufficiently high to denature, or separate, DNA strands in a mixture (Lawyer et al., 1993). At lower temperatures, the DNA strands are bound and thus not available for copying, while at higher temperatures other reaction components begin to degrade. *Taq* is not alone in this capacity, *Pfu* polymerase from *Pyrococcus furiosus* has a similar temperature range, possessing proofreading activity not present in *Taq* and superior to many other thermostable polymereases (Cline et al., 1996; Bargseid, 1991). Indeed, it is probably cases in which many of these concepts converge (SRE, EME, and biotechnological utility [BU]), such as with *Taq* and *Pfu,* that encourage biologists to run these distinctions together.

Biotechnological applications are perhaps one of the best cases for considering extremes as more than just physical and chemical variables and including trophic, ecological, organismal, or population-based extremes. Antibiotic resistance, for example, tends to develop in environments where there is an extremely strong selective pressure caused by antibiotics: conditions that can be anthropogenic or occur in naturally antibiotic rich competitive bacterial habitats. Indeed, Bhullar et al. (2012) have argued nutrient-limited and bacteria (species) rich environments encourage an antibiotic arms-race, suggesting this kind of extremity is of biotechnological utility. As the authors point out, "the diversity in the resistome also suggests that there are a myriad of bioactive molecules with antibiotic properties waiting to be discovered" (Bhullar et al., 2012, e34953). Put another way, we have reason to think that this corner in the intersection of the axes of species richness and nutrient density constitutes a genuine extreme of interest to biotechnologists in the business of antibiotic development. Considerations like these suggest that in context of biotechnological research something like an EME, with special attention given to rare or unexplored habitats, might be most fruitful.

## Conclusion

This chapter explored the multiple, occasionally incompatible characterizations of "extremophiles." We argued that extremophily, far from being a straightforward concept, admits of multiple interpretations, each with extremely different consequences. This concept faces many of the same concerns of vagueness and arbitrariness as other areas of biology, such as defining life itself, species, or genes. Extremophile research is especially prone to these concerns, as it involves basic assumptions about life's nature, limits, and whether we can know either.

We considered a number of possible definitions, including indexing extremophily to the limits of human habitability (human-centric), identifying extreme organisms as those which thrive at the limit of what is inhabited by life (edge of morphospace), as well as those which thrive at the limit of what is inhabitable by life (near impossibility). Each of those definitions had a role to play in the timid pluralism we advocate. Two other definitions, dealing with the rarity of creatures able to survive in one environment (statistical rarity) and one in which extremes were identified by physical boundaries and extremophiles were creatures near those limits (objective limits), had problems that proved devastating to their continued use. One final consideration was the utility of organisms for human purposes (BU), which illustrated how widely divergent research aims may be with respect to extremophiles, although even this view may simply be the pragmatic cousin of the edge of morphospace view. We hope this conceptual exploration of extremophily will guide and buttress further research into this area.

## Note

1. We use the terms "concept," "definition," "account," "approach," and "view" interchangeably, but note that there are scholars with strong opinions on the distinction between these terms.

## References

Alvarez, L. W., Alvarez, W., Asaro, F., & Michel, H. V. (1980). Extraterrestrial cause for the Cretaceous–Tertiary extinction. *Science, 208*(4448), 1095–1108.

Bargseid, M. (1991). A high fidelity thermostable DNA polymerase isolated from Pyrococcus furiosus. *Strategies, 4*, 34.

Baross, J. A., & Deming, J. W. (1983). Growth of "black smoker" bacteria at temperatures of at least 250° C. *Nature, 303*(5916), 423–426.

Berne, B. J., & Straub, J. E. (1997). Novel methods of sampling phase space in the simulation of biological systems. *Current Opinion in Structural Biology, 7*(2), 181–189.

Bhullar, K., Waglechner, N., Pawlowski, A., Koteva, K., Banks, E. D., Johnston, M. D., . . . Wright, G. D. (2012). Antibiotic resistance is prevalent in an isolated cave microbiome. *PLoS One, 7*(4), e34953.

Bich, L., & Green, S. (2018). Is defining life pointless? Operational definitions at the frontiers of biology. *Synthese, 195*, 3919–3946.

Chien, A, Edgar, D. B., & Trela, J. M. (1976). Deoxyribonucleic acid polymerase from the extreme thermophile thermus aquaticus. *Journal of Bacteriology, 127*(3), 1550–1557.

Cline, J., Braman, J. C., & Hogrefe, H. H. (1996). PCR fidelity of Pfu DNA polymerase and other thermostable DNA polymerases. *Nucleic Acids Research, 24*(18), 3546–3551.

Cohen, J., & Stewart, I. (2001). Where are the dolphins? *Nature, 409*(6823), 1119.

Dartnell, L. (2011). Biological constraints on habitability. *Astronomy & Geophysics, 52*(1), 1–25.

Gerday, C. (2002). Extremophiles: Basic concepts. In K. Matsuura (Ed.), *Knowledge for sustainable development: An insight into the encyclopedia of life support systems* (pp. 573–598). Paris, France: UNESCO Publishing.

Gold, T. (1992). The deep, hot, biosphere. *Proceedings of the National Academy of Sciences of the United States of America, 89*, 6045.

Gorton, H. L., Williams, W. E., & Vogelmann, T. C. (2001). The light environment and cellular optics of the snow alga *Chlamydomonas nivalis* (Bauer) Wille. *Photochemistry and Photobiology, 73*(6), 611–620.

Gupta, G. N., Srivastava, S., Khare, S. K., & Prakash, V. (2014). Extremophiles: An overview of microorganism from extreme environment. *International Journal of Agriculture, Environment and Biotechnology, 7*(2), 371.

Harrison, J. P., Gheeraert, N., Tsigelnitskiy, D., & Cockell, C. S. (2013). The limits for life under multiple extremes. *Trends in Microbiology, 21*(4), 204–212.

Hoehler, T. M., & Jørgensen, B. B. (2013). Microbial life under extreme energy limitation. *Nature Reviews Microbiology, 11*(2), 83–94.

Jönsson, K. I., Rabbow, E., Schill, R. O., Harms-Ringdahl, M., & Rettberg, P. (2008). Tardigrades survive exposure to space in low Earth orbit. *Current Biology, 18*(17), R729–R731.

Langton, C. G. (1989). Artificial life. In C. G. Langton (Ed.)., *Artificial life* (pp. 1–47). Santa Fe Institute Studies in the Sciences of Complexity, Proceedings Vol. IV. Redwood City, CA: Addison-Wesley.

Lawyer, F. C., Stoffel, S., Saiki, R. K., Chang, S. Y., Landre, P. A., Abramson, R. D., & Gelfand, D. H. (1993). High-level expression, purification, and enzymatic characterization of full-length Thermus aquaticus DNA polymerase and a truncated form deficient in 5'to 3'exonuclease activity. *Genome Research, 2*(4), 275–287.

Lentzen, G., & Schwarz, T. (2006). Extremolytes: Natural compounds from extremophiles for versatile applications. *Applied Microbiology and Biotechnology, 72*(4), 623–634.

MacElroy, R. D. (1974). Some comments on the evolution of extremophiles. *Biosystems, 6*(1), 74–75.

Mariscal, C., & Doolittle, W. F. (2018). Life and life only: A radical alternative to life definitionism. *Synthese.* [Advance online publication]

Marshall, A. (2000). Review: This month in biotechnology. *Nature Biotechnology, 18*, 1026. doi:10.1038/80174

Persson, D., Halberg, K. A., Jørgensen, A., Ricci, C., Møbjerg, N., & Kristensen, R. M. (2011). Extreme stress tolerance in tardigrades: Surviving space conditions in low earth orbit. *Journal of Zoological Systematics and Evolutionary Research, 49*(Suppl. 1), 90–97.

Raup, D. M. (1966). Geometric analysis of shell coiling: General problems. *Journal of Paleontology 40*(5), 1178–1190.

Ray, T. (1995). An evolutionary approach to synthetic biology: Zen and the art of creating life. In C. G. Langton (Ed.), *Artificial life: An overview* (pp. 179–210). Cambridge, MA: MIT Press.

Reed, C. J., Lewis, H., Trejo, E., Winston, V., & Evilia, C. (2013). Protein adaptations in archaeal extremophiles. *Archaea, 2013*, 373275.

Rothschild, L. J., & Mancinelli, R. L. (2001). Life in extreme environments. *Nature, 409*(6823), 1092.

Saiki, R. K., Gelfand, D. H., Stoffel, S., Scharf, S. J., Higuchi, R., Horn, G. T., . . . Erlich, H. A. (1988). Primer-directed enzymatic amplification of DNA with a thermostable DNA polymerase. *Science, 239*(4839), 487–491.

Takeuchi, N., Dial, R., Kohshima, S., Segawa, T., & Uetake, J. (2006). Spatial distribution and abundance of red snow algae on the Harding Icefield, Alaska derived from a satellite image. *Geophysical Research Letters, 33*(21).

Trent, J. D., Chastain, R. A., & Yayanos, A. A. (1984). Possible artefactual basis for apparent bacterial growth at 250 degrees C. *Nature, 307*(5953), 737–740.

van den Burg, B., & Eijsink, V. G. (2002). Selection of mutations for increased protein stability. *Current Opinion in Biotechnology, 13*(4), 333–337.

Vreeland, R. H., Rosenzweig, W. D., & Powers, D. W. (2000). Isolation of a 250 million-year-old halotolerant bacterium from a primary salt crystal. *Nature, 407*(6806), 897–900.

Werner, K., Rauch, T., & Kruk, J. W. (2008). Discovery of photospheric Ca X emission lines in the far-UV spectrum of the hottest known white dwarf (KPD 0005+ 5106). *Astronomy & Astrophysics, 492*(3), L43–L47.

White, R. H. (1984). Hydrolytic stability of biomolecules at high temperatures and its implication for life at 250 degrees C. *Nature, 310*(5976), 430–432.

# PART IV

# ETHICAL ISSUES IN ASTROBIOLOGY

# 11

# Convergences in the Ethics
# of Space Exploration

*Brian Patrick Green*

## Introduction

With all of the problems here on Earth, we might do well to ask ourselves, "Why should we care about ethics in space?" I think there are three reasons: first, because as we go into space we already are and certainly will be making ethical decisions (which may be of historic-level importance to future humanity; second, because if we meet extraterrestrial intelligences (ETIs) they also will be making ethical decisions (which may also be of historic-level importance to future humanity), and it would be good for us to anticipate what their ethics might be like so that we can interact in a mutually beneficial way; and third, because knowing how ethics works should help in both of the previous issues.

Gladly, scholars have been doing significant work in space ethics for decades, applying the principles of ethical thinking to questions of space. In this chapter I examine some of these perspectives and argue that there has been a convergence in the ethics of space exploration and use: several authors have, in varying ways, advocated similar projects. I do this in six parts. (a) I examine three articles by Lupisella and Logsdon, Randolph and McKay, and Smith. (b) I look to the works of literary theorist Rene Girard and businessman Peter Thiel, who add some practical context to the contemporary ethics of space exploration and use. (c) I look to the traditional resonances I think all of these ethics share, and the general rules that all may be orbiting—a form of natural law ethics based on promoting the fulfillment of nonconflicting purposes. (d) I gather my case by making explicit the connections and convergences between these sources. (e) I consider the significance of these convergences. (f) I conclude.

Brian Patrick Green, *Convergences in the Ethics of Space Exploration*. In: *Social and Conceptual Issues in Astrobiology*. Edited by: Kelly C. Smith and Carlos Mariscal, Oxford University Press (2020). © Oxford University Press.
DOI: 10.1093/oso/9780190915650.001.0011

## Space Ethics

### Lupisella and Logsdon

Over twenty years ago, Mark Lupisella and John Logsdon offered the idea of a "cosmocentric ethic" to the space community.[1] In it, they examine many possible ethical approaches to space exploration and use, and offer their cosmocentric ethic as the best alternative. It "establishes the universe as the priority in a value system,"[2] and acknowledges both the instrumental and intrinsic value of life,[3] yet makes perhaps its strongest argument through a comparison of two stories. In the first, humans kill inferior life forms and are judged negatively by superior ones.

> stark and silent . . . were the Martians—dead!—slain by the Humans against which their systems were unprepared . . . slain, after all the Human's devices had worked, by the blind foreigners that had landed upon their world. Yet across the gulfs of space, minds that were to Humans as Humans were to the Martians that perished, intellects vast and cool and unsympathetic, regarded this earth with contempt, and slowly and surely drew their plans against us—we who had killed another.[4]

In another story, humans respect inferior life forms and are judged positively by superior ones.

> silent . . . were the Martians — silent, yet alive! — preserved by the Humans against which their systems were otherwise unprepared . . . alive, after all the Human's devices had worked, alive from the care shown by those who had landed upon their world. And so, across the gulfs of space, minds that were to Humans as Humans were to the Martians, intellects vast and cool and sympathetic, regarded this earth with admiration, and slowly and surely drew their plans to welcome us to the cosmic neighborhood—we who had evolved beyond our selfish genes—we who had chosen respect.[5]

Through this ethical narrative they clearly lend their weight toward generally protecting life and implicitly advocate for a sort of Golden Rule space ethic, where life is respected in the hope that our lives will be respected in turn. Lupisella has since advanced his theory in much greater detail, but the general sense remains: we ought to protect living things.[6]

Of note in Lupisella and Logsdon's narratives are that higher-alien ETIs become god-like judges of human actions, and the fundamental principle of judgment for these ETIs is a moral test: Can humanity show care for extraterrestrial life life-forms, especially when we do not benefit from (or are even harmed by) showing that care? Interestingly, whether this god-like ETI test is altruistic or egoistic on the part of the ETIs is unclear. Do they wish to destroy or preserve us because of our (potential) threat status toward them (a selfish motive), or because of our threat toward others (an altruistic motive)? Or is there something else?

The moral of the stories is, in any case, that we ought to respect and protect other living organisms. It can, in fact, be interpreted as, basically, the Golden Rule. However, the Golden Rule comes in two forms: the negative form ("Do not do unto others as you would not have them to do unto you)" and the positive form ("Do unto others as you would have them do unto you"). But the rule could actually be more the rule of deterrence and fear, in contrast to the Golden Rule, what I call (recalling the Cold War and nuclear deterrence) the "Uranium Rule": "Do unto others as they do unto you (or *could* do unto you)," but which has many names, being also known as "tit for tat," "eye for an eye," "retaliation in kind," "*lex talionis*," and so on.

This ethical theme—the rule of reciprocity—runs through all of the ethical systems examined in this chapter and thus in itself represents a major convergence in the ethics of space exploration.

## Randolph and McKay

More recently, Richard Randolph and Christopher McKay have proposed their own ethic to guide astrobiology, suggesting that the paradigmatic framework ought to be one that centers on "protecting and expanding the richness and diversity of life."[7] Following Kenneth Goodpaster and J. Baird Callicott[8] and with additional arguments from Cockell and McKay,[9] they argue that this ethic provides a good framework for space exploration and use.

Their framework is founded upon three axioms. First is the universal value of all life: all life has both intrinsic and instrumental value—this gives a bare minimum of consideration to all living things, while still allowing for human interests to be considered.[10] Second, the precautionary principle gives proper time for allowing research and reasonable decision-making.[11] Third is the "cosmic Golden Rule," in its Confucian negative formulation, which states "Never impose on others what you would not choose for yourself."[12] Together, these axioms yield an ethic that Randolph and McKay believe should garner widespread and pluralistic support.

Of note in Randolph and McKay's system is that the formulation of the Golden Rule is a negative "do not do," but the prime moral principle is a positive formulation to "protect and expand." In other words, there is a demand not just to let things be, but to actively promote life, to bring it to dead places. Why would this be?

Valuing something, for Randolph and McKay, does not mean just leaving it alone; it means actively helping it, perhaps analogously to how a parent would help an infant. Infants cannot be left to fend for themselves—they must be actively cared for. But in this case the Golden Rule ought to be in a positive formulation; just as, if we were infants (or alien microbes) we would want to be cared for, so too should we care for other infants (or alien microbes). This expansion of perspective from self to community is, of course, a primary effect of religion, hence the Golden Rule is found in nearly every world religion[13]—which raises to scrutiny another assumption.

This negative/positive formulation issue is a tension that hides a deeper, prior axiom underlying the ethic. This is a belief, a metaphysical assumption, that actively helping nature is good, because life is doing something good and we ought to help it achieve its goal. There are two assumptions here, one of teleology and one of friendliness. The assumption of teleology has long been subject of debate in academic circles, but in Randolph and McKay's estimation the purpose of the universe is to enhance the richness and diversity of life. The assumption of the fundamental "friendliness" of the universe—or that it ought to be friendly—is another question, however. After all, the universe's goal could be quite horrible, and if so we ought to resist it, not help it. Both are most definitely not empirical questions but ones of first principles.

The moral of Randolph and McKay's ethic is that the universe has a teleology that involves the evolution of life toward something: a pursuit of an end. That which impedes this pursuit is bad and that which helps it is good. Therefore, to be good, we ought to help it. Thus humans should become cultivators of life in the universe.

## Smith

Kelly Smith has advocated for a cosmic ethics based on the value of living complexity, especially valuing the triadic concatenation of sociality, reason, and culture (SRC), or the SRC triad (SRCT), which makes ethics possible.[14] In this way he echoes the philosopher Hans Jonas, who similarly advocated protecting the underlying traits that make ethics possible (hence his "imperative of responsibility"—that humankind should exist).[15]

Interestingly, as with Randolph and McKay, Smith's "manifest complexity" assumes there is a teleology to the universe; that is, the purpose of life in the universe is to oppose entropy and become increasingly complex. And a moral cultural group ought to match their own teleology to the teleology of the universe (i.e., assisting the growth of manifest complexity).

Of note in Smith's framework is that the SRCT places "first order moral value [as] a property of groups and only derivatively of individuals."[16] Groups are where moral culture resides; individuals only form particular instantiations of a group's moral culture. Importantly, because of this capacity, universal to all species is a "subjective objectivity" that permits them to at least pursue the possibility of a universal moral framework.[17] Lastly, following Aristotle's notion that a species' particular trait should be its *telos* in life, Smith asks what life's particular trait might be and determines that manifest complexity seems to be the universal teleology toward which life is aiming.[18]

Based on this, the more a creature approaches being an SRCT creature, the more moral value such a creature will have. Ultimately, for Smith, his framework argues that SRCT creatures have full intrinsic value, partially SRCT creatures have intrinsic and instrumental value, and non-SRCT creatures have only instrumental value.[19] This allows humans to explore and use space with some concern for the well-being of extraterrestrial life forms that we may encounter, but the measure of concern that we show extraterrestrial life-forms will be proportionate to their level of SRC.

## The Ethic of Wanting Something New

While these authors have contributed to some of the "traditional" corpus of space ethics, French literary theorist René Girard and his protégé, the billionaire Peter Thiel, have not. However, because Thiel is influencing (partly through his colleague Elon Musk, founder of Space X corporation) the *practical* or *applied* ethics of space exploration (and technological development more broadly), it is worth examining and making explicit, at least in part, what that unspoken ethic and associated motivation might be.

## Girard

René Girard and his theories of mimesis and scapegoating posit that because humans imitate each other in everything, including our desires, we soon come into conflict and competition over those imitated desires because resources are finite.[20] Girard found prominent examples of this imitation of desire in Shakespeare and other authors,[21] but these examples are not only in fictional literature: they are found in reality as well. Economic bubbles and manias are a perfect example of people all wanting the same thing, with disastrous consequences. This same phenomenon can also be found in competition over resources and even animal (and human) competition for mates. In all cases, people can come to want something just because other people want it. Girard powerfully asserts that religions, ranging from the Biblical religions to sacrificial cults, also display this phenomenon, though he takes care to point out that not all religions deal with the phenomenon in the same way.[22]

The standard unethical response to this situation of mimetic desire is to resolve the conflict by blaming and scapegoating an individual or a group for it: hence the near-universality of human sacrifice, witchhunts, warfare, persecution of minorities, and other sorry deeds of human history.

Another response of dubious ethicality is to rigidly structure society in such a way that no one ever considered that they could want the same things as too many other people, given the resources, such as through as strict caste or class system. Peasants could not become kings or ever have the same things as kings, so this was never even contemplated or wanted. Social status was hereditary; social strata were rigid and social mobility nonexistent; one was stuck where one was born and "the American Dream" and the "rags to riches" stories of Horatio Alger were centuries or millennia in the future. However, this social stratification began to fall apart with the appearance and growth of the "middle class" bourgeoisie in the Middle Ages of Europe. Because the middle class made social mobility and *nouveau riche* status possible, the old social structures and impediments to mimetic conflict began to degrade, leaving yet another solution to be found. In many cases, this involved scapegoating wealthy minorities, as pogroms, Communism, and Nazism all did. This can be seen even recently in populist movements such as Occupy Wall Street's ("the 99%") cry against "the 1%."

One ethical response to mimetic conflict could be simply to want something else, so as to avoid conflict. Whenever a desire-generated conflict begins to appear, if enough people would simply decide not to pursue it, then the conflict would be defused. Unfortunately, though it is possible (some religions emphasizing it; e.g., Buddhism's rejection of *taṇhā* [desire] and Judaism's and Christianity's commandment "do not covet"), this solution is very contrary to human inclination and therefore difficult.

Yet another ethical response would be to somehow multiply the resource so that it could be more widely shared, thus "enlarging the pie," so to speak, and allowing for more "win-win" situations. Remarkably, technology allows this "miracle." This non-zero-sum approach leads us to Peter Thiel's technological solution to mimetic conflict.[23]

## Thiel

Though it might seem unusual to look for relevance to space ethics in the realms of technology and business, technology entrepreneur and student of Girard, Peter Thiel, has summed up the two ethical solutions to mimetic conflict in his recent book on technology startups titled *Zero to One: Notes on Startups, or How to Build the Future*. First, in business, one should assiduously avoid competition, or seek to be the only business in one's field.[24] Second, in business, one should create abundance, not scarcity, by pursuing technological innovation that expands resources and that achieves completely new products at least ten times better than the currently existing alternatives.[25]

To expound upon the first solution, Thiel gives two pieces of advice. First, do not do what others are already doing; do something else. Want something else that no one else wants; then no one will compete with you for it. Second, when (or if) you do start to succeed, you will have a monopoly—which is good for business. Thiel provocatively advocates for monopoly, not competition.[26] However, it is important to understand that he is not talking about "government favorites" or monopolies that rely on artificial scarcity. He is not advocating for a monopoly of supply but rather one of demand—one where consumers freely choose the superior product and avoid others. This is a monopoly caused because one has a product so superior to all the alternatives that the alternatives are simply no longer reasonable options.

This brings us to the second piece of advice: leaps forward must be by a factor of ten, or else one cannot escape competition. Thiel argues that this type of technological advance is necessary; it is not optional, because we are consuming non-renewable resources and are headed for disaster. Technological stagnation will get us all killed. Advance is necessary just to stay alive, much less to thrive in a future of abundance.[27]

Without this innovation we are not only on track to stagnate and die from lack of resources, but we are also doomed to political infighting, as scarcity gradually eats away at standards of living. Thiel describes the importance of technological advance in a capitalist democracy thus: "The give-and-take of Western democracies depends on the idea that we can craft political solutions that enable most people to win most of the time. But in a world without growth, we can expect a loser for every winner."[28] This makes it difficult for democracy to function because good political deals can no longer be had, thus leading to partisan rancor and gridlock.

Thiel's technologically driven, growth-minded ethic is deeply relevant to "the Silicon Valley values" of live and let live (and often live and let die) libertarianism. Let others freely pursue their dreams: if they are worthwhile, (naturally) all will benefit. This is particularly relevant for software because software copies freely—there is no competition for a scarce resource.

However, software is not enough. Thiel and his friends like Elon Musk are now pioneering the movement from bits to atoms, with corporations like Musk's Tesla, the Boring Company, NeuraLink, and various hyperloop companies. Most relevant to space ethics, SpaceX, with its reusable rockets, has made spaceflight ten times better by making it potentially ten times cheaper and ten times more frequent. Additionally, space exploration allows, quite dramatically, for "wanting something different" from everyone else. This is one reason why Musk's SpaceX is pursuing the settlement of Mars—no one else is doing it. They have the corner on the market, though the market is nearly nonexistent. What market there is now effectively looking to SpaceX.

It might make sense now to return to space ethics, especially given some of the outrageous remarks Musk has made about using thermonuclear weapons to warm Mars and so on.[29] But before returning to space we must first turn for a moment to tradition, because I argue that all of these ethical systems are converging on one ethic, with deep roots.

## Traditional Resonances

Here I want to try to find the roots of the ethics being proposed. I have previously argued that Immanuel Kant's deontology and particularly John Stuart Mill's utilitarianism have limitations that prevent their broad applicability to space ethics, while Aristotle's theories, carefully sifted and reframed, actually suit space ethics quite well.[30] Following in this vein, therefore, this is my assertion awaiting evidence: underlying all of the previously mentioned ethical systems is a particular kind of Aristotelianism,[31] a natural law virtue ethics, with an unavoidably Thomistic squint.[32]

What is natural law ethics? The natural law is the ethical "law" built in to an organism by its very nature. According to Aristotelianism and Thomism, all living organisms have an entelechy: a built-in purposeful teleology. Achieving natural *telei* is good for that organism, and refining organism-characteristic habits into excellence is virtue. Individual natural laws participate in the overall natural law of the universe and the universe's ultimate end: the universal teleology (for Aristotle and Aquinas, with some differences, divine contemplation).

There is obviously quite a bit of controversy in philosophy and science as to the nature, ontological level (social, psychological, biological, metaphysical, etc.), and even existence of teleology. That debate can go on, but for the sake of a discussion of ethics, teleology must be assumed; otherwise intelligent actions have no purpose and no "ought" to them. As Alasdair MacIntyre has argued, ethics is a bridge from nature to destiny, and losing either the starting or the ending point corrupts ethics beyond repair.[33] As long as anyone wants to discuss ethics, at least some teleology must be assumed, even if it is a purely human-posited one.

For example, in each of the aforementioned ethical systems, we can either identify or surmise a goal. In Logsdon and Lupisella the goal is simple survival—in the first scenario the humans misbehave, are judged, and are then targeted for destruction, while in the second scenario humans make the right choice and are instead welcomed into the cosmic commons. In Randolph and McKay's paper, there is a different goal, which is the enhancement of the "richness and diversity of life" in the universe. In Smith's paper, the goal is "manifest complexity." In Girard's ethic, the goal is to escape from the cycle of mimetic desire and violence and instead achieve a culture where conflict is reduced. And Thiel's ethic is similar: do not compete with others, find something new that will increase abundance, and thereby reduce squabbling over resources.

Some might object that natural law ethics is a clear violation of various philosophical proscriptions such as Hume's law that "one cannot derive an 'ought' from an 'is,'"[34] Kant's separation of the noumenal and phenomenal,[35] G. E. Moore's "naturalistic fallacy,"[36] and Carnap's fact/value split,[37] but these dicta are remarkably weak. Hans Jonas dispatches Hume by noting that his logic is invalid: if one assumes there is no entelechy (that "is" contains no "ought"), then one concludes that there is no entelechy.[38] But one could always assume something else.

Others have criticized Hume's is-ought problem as well, including such varied figures as Alasdair MacIntyre,[39] Daniel Dennett,[40] Patricia Churchland,[41] Sam Harris,[42] Joan Roughgarden,[43] and so on. G. E. Moore and Carnap have been dealt with by such thinkers as William Frankena,[44] Bernard Williams,[45] Mary Midgley,[46] and Hilary Putnam.[47] Kant, of course, is still extremely influential in philosophy and ethics, but the metaphysics of his separation of ethics from nature is not one of his ideas that has withstood the test of time. While the works of Hume, Kant, Carnap, and Moore dammed the flow of naturalistic ethics for two centuries, the dam has now collapsed and ethics has returned to nature.[48]

Of note in connecting the dots between nature, ethics, and goals is that if there is an ethic (Point B) and a goal (Point C), then there is also an assumed human nature (Point A) from which to draw the path from origin to destiny. This assumed human nature—and every ethic does assume one, even if it is as lax as "human nature is to posit our own ethics" or "human nature is to have no nature"—is sometimes explicit and sometimes implicit. But the fact that it is there is always enough for a natural law claim to be made. Any system with an entelechy and an ethical code to guide from origin to destiny is a natural law ethic. Every ethic has these three components and therefore is in this sense a natural law ethic. Even an ethic that is positivist, or nihilist, or purely metaphysical, or theological, still has an implicit understanding of human nature (i.e., that our nature is to make up ethics, or have no ethics, or to follow the ethics of an external force or deity).

Returning again to the ethical systems discussed earlier, in Logsdon and Lupisella, for example, there is an assumed human nature, which is that we would like to survive. If, oppositely, we had a death wish and knew aliens would kill us for violations of interstellar law, then violating that law could become our purpose, and the means to violate it would become our moral "oughts."

Likewise, if the goal of Randolph and McKay's ethic is to "protect and expand the richness and diversity of life" in the universe, and humans are to be the ones to carry out this task, then it assumes that human nature is to care for other living things and help them achieve their own *telei*.

Smith's goal of "manifest complexity" is sought by an ethic, which respects creatures in accordance to their capacity to express the SRCT. In this ethic, human nature is to express this triad itself, and further, to evaluate its presence in other creatures. In that way, our nature is to be social, rational, and cultural, and to go forth and search for creatures like ourselves and, in the process, finding creatures of varying SRCT levels, to judge then our ethical obligations toward them.

Girard's goal is to reduce mimetic violence and thereby promote peace at all levels. Girard's ethic, then, is to reveal this natural human proclivity toward mimetic desire, understand its method of action, then show us the path to avoid it by using our better sensibilities to override this destructive tendency. Importantly, Girard stands out on this point because while mimetic desire is naturally good (allowing us to quickly learn from each other), it also presents a natural tendency toward conflict that needs controlling via other natural tendencies.

Thiel's goal is to promote human survival, which he sees as threatened by scarce resources and mimetic violence. His ethic, then, is to avoid competition and increase abundance through technological research, primarily and most effectively performed by startups of like-minded individuals.

Each thinker has an anthropology, an ethic, and a teleology: the three key parts of a natural law ethic.

In other places I have developed a natural law ethic explicitly based on the metaphysical assumptions underlying the scientific method.[49] I argued that morally-capable organisms should strive to have nonconflicting *telei* and avoid cross-purposes, thus maximizing their collective potential to flourish. Organisms should pursue this if they have a choice (as humans do, both with ourselves and with navigating the *poiesis* of new ecosystems). When *telei* do necessarily conflict (as with parasites, pathogens, and predators), purposes can be prioritized on the basis of capacity for excellence (e.g., humans have more capacity for excellence than mosquitoes and mosquitoes more capacity for excellence than a virus). The ethical principles can be summarized thus:

1. Respect *telei*, both human and non-human, promote their fulfillment and minimize their thwarting, and if possible prevent cross-purposes. Discriminate between cross-purposes via capacity for virtue.
2. Respect life itself, as the ground of all *telei*.
3. Humans (and other intelligent and moral creatures), as the only creatures capable of respecting *telei*, ought to exist.

While the progression of thought runs from first to third, the order of moral priority is third to first.[50]

This is a synthetic ethic in the neo-Aristotelian tradition which additionally has several resonances with Girard's and Thiel's ethical ideas. It offers additional theoretical context for the three systems of space ethics discussed previously, as well as highlighting some of their convergences.

## The Connections

Each of our authors in question meets the same criteria in their work. All assume (a) an anthropology/human nature—what organisms are and what the universe is; (b) a teleology—the point of it all, whether for an organism or the universe; and (c) an ethics—the path from anthropology/nature (an origin) to teleology (a destiny). In this way all five are built on the same axioms as neo-Aristotelian/neo-Thomistic natural law virtue ethics. This means they have expected similarities (i.e., the structure just described), but they also have deeper similarities that might not at first be apparent

One of the first deep similarities is that, to a certain extent, living things should just be left to be—that is, not meddled with, allowed to pursue their own ends. This is a negative formulation of the Golden Rule ("Do not do unto others what you would not have them do unto you"), which respects the life and freedom of organisms with inbuilt entelechies. Because life forms naturally have their own goals to pursue, our interference may well harm their ability to pursue their ends; therefore the first rule is to "do no harm." And the easiest way to do no harm is to do nothing.

However, sometimes doing nothing is not the right solution to a problem. The second deep similarity between these ethics is that sometimes we ought to help things grow. This is a positive formulation of the Golden Rule ("Do unto others as you would have them do unto you") and involves the cultivation of natural tendencies, or providing resources to creatures in distress. Life forms are naturally pursuing goals, but for various reasons many will not be able to fulfill them. Under certain circumstances, then, it may become reasonable to help these creatures. For example, it may be reasonable to defrost a freeze-dried Mars to help its native life. If there is no native life to an extraterrestrial object, this rule can justify extending life there from other places.

As a third deep similarity, all of these ethics acknowledge that problems appear when one organism's natural goals conflict with those of another organism. This, then, promotes the idea that we should avoid conflicts that harm other living things, especially lose-lose situations. Lose-lose situations might seem easy to avoid, but, unfortunately, as Girard describes, they can come about relatively easily because, to a certain extent, we might be naturally inclined toward them. Win-lose situations are more common in nature, and understandable: predation, pathogens, and parasitism all make sense in a way, with one organism exploiting another for its own gain. Yet we might ask, is there a way to get beyond the win-lose paradigm and instead into the win-win paradigm? Thiel's ideal for the use of technology may allow this, if we can use technology to keep expanding our resource base faster than we can use it and explore lifeless places in space and bring life there, to flourish beyond the heightened competition of Earth.

As a fourth similarity, all of these ethics must deal with the idea that some goals are better than others, thus allowing a ranking of goals (e.g., ones that protect life, avoid conflict, increase resources, promote complexity, help organisms pursue their good ends, fulfill the universal telos, etc., in no particular order). While, for example, Smith might quibble with Randolph and McKay over whether "manifest complexity" is more important than "richness and diversity," both should mutually acknowledge that either's goals are superior than their opposites (destroying manifest complexity or richness and diversity) and that other goals of particular groups or individuals would lie on a spectrum of priority in between the extremes.

This leads to the fifth similarity, which is that intelligence is the key trait that allows for discrimination between good ends. Intelligence is vital in every one of the ethical systems. For example, in Lupisella and Logsdon, intelligence allows us to choose better ends and avoid cosmic punishment. In Randolph and McKay, intelligence is what allows us to make the best choices on behalf of all organisms within the scope of our power. In Smith, intelligence (construed as the SRCT) is not only what gives an organism value but allows us to know that they have value and how we ought to react to them. In Girard, it is intelligence that allows us to outsmart our own destructive tendency toward mimetic rivalry. In Thiel, intelligence allows us to choose different goals and develop miraculous new technologies. And in neo-Aristotelian and neo-Thomistic natural law ethics, it is our intelligence that sets us apart from the other animals as creatures with a particular role in the universe: to act to discover and fulfill our own nature.[51]

It is worth noting here that each ethic—to a greater or lesser degree—assigns to humanity, in our intelligence, a role in facilitating the purpose of the universe, as it assumes that purpose to be. The overall purpose of the universe appears to be something along the lines of "stay alive and reproduce, and help other things stay alive and reproduce." This is an ethical understanding in consonance with evolution and one that gives humankind a special role in the universe as the only creature (that we know of) capable of fulfilling this quest.

The sixth and deepest similarity between these ethics is that all of them, like all ethics in general, assume an ontology of violence, not an ontology of peace.[52] The universe, at its core, seems to be one of struggle, and, therefore, to perform our own tasks within that universe is also a struggle. But perhaps we are misled—perhaps we assume violence when we ought not to. Perhaps if we changed our perspective to one which assumed peace we might either (a) come to accept that we are already living in a peaceful existence (as Buddhism might say) or that (b) it is only humanity that violates peace, and the rest of nature is peaceful (as some theistic traditions might argue). Ethics—as an entire endeavor—is necessary to us because conflict is assumed. If we had no desires, we would not compete or fight. Perhaps not all species in the universe are like we are. But what we do know is that we can reduce conflict by reducing our wants and expanding our abundance.

## Significance

Connecting this back to space exploration and use, large areas of space are relatively unwanted (there is little human motivation to go there) yet are free of competition from any potential native life (e.g., lifeless asteroids and moons) and thus provide a perfect environment in which to expand Earth life, particularly human life, in ways that do not impinge on the well-being of other creatures. This shows remarkable consonance with Randolph and McKay's ethic for "protecting and expanding the richness and diversity of life," as well as Lupisella and Logsdon, Smith, Girard, Thiel, and natural law ethics.

Space, then, gives humans and Earth life the "space" we need to reduce competition for resources here on Earth and, perhaps, in the words of the activist Peter Maurin, "make the kind of society where people find it easier to be good."[53] Technology and space exploration in combination could help in that regard because resource constraints are slackened—though of course mere abundance can be an impediment to the good in ways that scarcity is not (such as temptation toward gluttony and greed).

This also provides a broadly applicable framework for ethical guidance when dealing with extraterrestrial life. In parts of space where life may exist, this ethic likewise protects that native life in proportion to its own capacity for excellence (including unique excellences).

One truism of ethics is that any ethic for humans ought to be possible for humans to fulfill, as in the saying: "ought implies can." But it also ought not to be so easy for us that we can lazily do whatever crosses our minds, no matter how evil. Ethics ought to stretch our capacities toward our better selves, to become, in Aristotle's words, morally speaking, more "divine" and like "the activity of God."[54] (Compare this to Lupisella and Logsdon's alien "gods"—we should, apparently, want to be like them as well).

Selfishness is equivalent to no ethic at all; it merely follows the natural flow of baser human nature and extols us to float downstream. A better ethic is one that teaches us to swim and points us toward the wonders we might find farther upstream, closer to the source. This is an ethic that exercises our character and helps to develop us into the best people we can be. All the ethical systems described here give an ideal of who we are and what we ought to do in the context of the goal we are pursuing—one in accord with the universe.

Lastly, this convergence is significant because it means that these seemingly disparate ethical systems might not actually need to compete with each other. In fact they are, at their cores, quite similar, while perhaps emphasizing slightly different points. From a Girardian perspective their similarity might spark mimetic rivalry, but if we step above this competitive paradigm we might instead see that all are working toward the same end: the fulfillment of a perceived universal entelechy consisting of the harmony of living creatures.

## Conclusion

In conclusion, there is a convergence toward neo-Aristotelian/neo-Thomistic natural law ethics among not only practitioners of space ethics but also such divergent fields as literary criticism and business. Why would that be?

That all of these diverse fields have converged on a similar answer, which is also a very old one, could indicate several things:

1. There is a cognitive/psychological bias (or biases) influencing our thinking.
2. There is a cultural constraint (or constraints) influencing our thinking.
3. There is a biological "current" (or more than one) influencing our thinking.
4. There is a metaphysical "grain" to the universe (or several) influencing our thinking.

Whichever of these it may be (and it may be all of them), it is of profound importance not only to our own ability to think and understand the universe but also to our thinking about our own origin, identity, role, and destiny in the universe.

## Acknowledgments

First presented as Brian Patrick Green, "Convergences in the Ethics of Space Exploration," SoCIA 2016 Conference, Clemson University, Clemson, SC, September 24, 2016. I would like to thank the SoCIA 2016 conference organizers, conference participants, and the editors and reviewers of this volume for their insightful questions, hard work, and excellent comments.

## Notes

1. Mark Lupisella and John Logsdon, "Do We Need a Cosmocentric Ethic?" Paper IAA-97-IAA.9.2.09, presented at the International Astronautical Federation Congress. American Institute of Aeronautics and Astronautics, Turin, Italy, 1997, pp. 1–9.
2. Ibid., 1.
3. Ibid., 8–9.
4. Ibid., 1.
5. Ibid., 9.
6. E.g., Mark Lupisella, "Cosmocultural Evolution: The Coevolution of Culture and Cosmos and the Creation of Cosmic Value," in *Cosmos and Culture: Cultural Evolution in a Cosmic Context*, ed. Steven J. Dick and Mark L. Lupisella (NASA Publications, SP-4802, 2008), 321–359.
7. Richard O. Randolph and Christopher P. McKay, "Protecting and Expanding the Richness and Diversity of Life: An Ethics for Astrobiology Research and Space Exploration," *International Journal of Astrobiology* 13 (2014): 28–34.
8. Ibid., 29–30, citing J. Baird Callicott, "On the Intrinsic Value of Non-Human Species," in *The Preservation of Species*, ed. Bryan G. Norton (Princeton, NJ: Princeton University Press, 1986), 138–172,; and Kenneth E. Goodpaster. "On Being Morally Considerable," *Journal of Philosophy* 75 (1978): 545–551.
9. Ibid., 29–30, citing, e.g., Charles S. Cockell, "The Value of Microorganisms," *Environmental Ethics* 27 (2005): 375–390; and Christopher P. McKay, "Does Mars Have Rights? An Approach to the Environmental Ethics of Planetary Engineering," in *Moral Expertise*, ed. Don MacNiven (New York: Routledge, 1990), 184–197. Additionally, by utilizing Cockell's "biorespect," developed as a correction or expansion of Albert Schweitzer's "reverence for life," Randolph and McKay continue in the Schweitzerian tradition.
10. Ibid., 30.
11. Ibid., 31.
12. Ibid., 32.
13. See, for example, "Golden Rule," *Wikipedia* (2019). Available at: https://en.wikipedia.org/wiki/Golden_Rule.
14. Kelly C. Smith, "Manifest Complexity: A Foundational Ethics for Astrobiology?" *Space Policy* 30 (2014): 209–214.

15. Hans Jonas, *The Imperative of Responsibility* (Chicago: University of Chicago Press, 1984), 10–11, 37, 43, 118, 129–30, 136, 139.
16. Smith, "Manifest Complexity," 209.
17. Ibid., 209, 211.
18. Ibid., 211–213.
19. Ibid., 214.
20. René Girard, "Ch. 1, Mimesis and Violence," in *The Girard Reader,* ed. James G. Williams (New York: Crossroad, 1996), 9–19. Citing René Girard, "Mimesis and Violence: Perspectives in Cultural Criticism," *Berkshire Review* 14 (1979): 9–19.
21. See, for example, René Girard, *A Theater of Envy: William Shakespeare* (New York: Oxford University Press, 1991).
22. For example: René Girard, *I See Satan Fall Like Lightning* (New York: Orbis, 2001).
23. This approach is also considered, though in a different way, by the journalist and author Robert Wright in his book *Nonzero: The Logic of Human Destiny* (New York: Vintage Books, 2000).
24. Peter Thiel with Blake Masters, *Zero to One: Notes on Startups, or How to Build the Future* (New York: Crown Business, 2014), 1–43.
25. Ibid., 48–49.
26. Ibid., 23–43.
27. Ibid., 2, 9, 194.
28. Peter Thiel, "The End of the Future," *National Review*, October 3, 2011, available at: http://www.nationalreview.com/article/278758/end-future-peter-thiel
29. Todd Leopold, "Elon Musk's New Idea: Nuke Mars," CNN, September 11, 2015, available at: http://www.cnn.com/2015/09/11/us/elon-musk-mars-nuclear-bomb-colbert-feat/index.html.
30. Brian Patrick Green, "Ethical Approaches to Astrobiology and Space Exploration: Comparing Kant, Mill, and Aristotle," in Special Issue: "Space Exploration and ET: Who Goes There?" *Ethics: Contemporary Issues* 2, no. 1 (2014): 29–44.
31. Aristotle, *Nicomachean Ethics,* translated by W.D. Ross, revised by J.O. Urmson, in *The Complete Works of Aristotle*, Volume II, ed. Jonathan Barnes (Princeton, NJ: Princeton University Press, 1984), 1729–1867.
32. Thomas Aquinas, *Summa Theologiae,* translated by the Fathers of the English Dominican Province, 1st complete American ed., 3 vols. (New York: Benziger Bros., 1948).
33. Alasdair MacIntyre, *After Virtue*, 2nd ed. (Notre Dame, IN: University of Notre Dame Press, 1984), 54–55.
34. The phrase "Hume's Law" originates in R. M. Hare, "Universalisability," *Proceedings of the Aristotelian Society* 55 (1954–1955): 303, and has gained widespread use, see, for example, Hakan Salwen, *Hume's Law: An Essay on Moral Reasoning* (Stockholm: Almqvist & Wiksell International, 2003). The concept of Hume's Law originates in David Hume, *Treatise on Human Nature*, edited by L.A. Selby-Riggs (Oxford, 1888; 1978 reprint), 469 (Bk. III, part I, sec. I).

35. Immanuel Kant, *Foundations of the Metaphysics of Morals* and *What Is Enlightenment?*, trans. Lewis White Beck, 2nd ed., Library of Liberal Arts (Upper Saddle River, NJ: Prentice Hall, Inc., 1997), 5 (389).

36. G. E. Moore, *Principia Ethica* (Cambridge: Cambridge University Press, 1959), 10–14, 21, 38–40, 48–54, 58–62.

37. Rudolf Carnap, *The Unity of Science* (London: Kegan Paul, Trench, Hubner, 1934), 22. Cited in Hilary Putnam, *The Collapse of the Fact/Value Dichotomy and Other Essays* (Cambridge, MA: Harvard University Press, 2002), 18.

38. Jonas, *The Imperative of Responsibility*, 44.

39. Alasdair MacIntyre, "Hume on 'Is' and 'Ought,'" *The Philosophical Review* 68 (1959): 453.

40. Daniel Dennett, *Darwin's Dangerous Idea: Evolution and the Meanings of Life* (New York: Touchstone, 1995), 467.

41. Patricia S. Churchland, *Braintrust: What Neuroscience Tells Us about Morality* (Princeton, NJ: Princeton University Press, 2011), 4–7.

42. Sam Harris, *The Moral Landscape: How Science Can Determine Human Values* (New York: The Free Press, 2010), 10, 38, 42.

43. Joan Roughgarden, *Evolution's Rainbow: Diversity, Gender, and Sexuality in Nature and People* (Berkeley: University of California Press, 2004).

44. William Frankena, "The Naturalistic Fallacy," in *Perspectives on Morality: Essays of William Frankena*, ed. Kenneth E. Goodpaster (Notre Dame, IN: University of Notre Dame Press, 1976), 4–7.

45. Bernard Williams, *Ethics and the Limits of Philosophy* (Cambridge, MA: Harvard University Press, 1985), 121.

46. Mary Midgley, "The Withdrawal of Moral Philosophy," in *The Essential Mary Midgley*, ed. David Midgley (London: Routledge, 2005), 176.

47. Hilary Putnam, "Objectivity and the Science-Ethics Distinction," in *The Quality of Life*, ed. Martha Nussbaum and Amartya Sen (Oxford: Clarendon Press, 1993), 143–145; and Putnam, *The Collapse of the Fact/Value Dichotomy*, 19–30.

48. For a fuller analysis of this topic see Brian Patrick Green, *The Is-Ought Problem and Catholic Natural Law*, Dissertation, Berkeley, CA: Graduate Theological Union (ProQuest), 2013.

49. Ibid.

50. Ibid., 204.

51. Of note is that these five similarities are also similar in some ways to "Principlism" in bioethics, where the four principles of non-maleficence, beneficence, justice, and autonomy, roughly align, respectively, with the negative formulation of the Golden Rule, the positive formulation of the Golden Rule, avoiding conflict and justly choosing among competing ends, and allowing for intelligent freedom within bounds. Tom L. Beauchamp and James F. Childress. *Principles of Biomedical Ethics*, 5th ed. (Oxford: Oxford University Press, 2001), and the *Belmont Report: Ethical Principles and Guidelines for the Protection of Human Subjects of Research*, Report of the National Commission for the Protection of Human Subjects of Biomedical and

Behavioral Research, Department of Health, Education and Welfare (Washington, DC: U.S. Government Printing Office, September 30, 1978).

52. John Milbank, *Theology and Social Theory: Beyond Secular Reason*, 2nd ed. (Malden, MA: Blackwell, 2006).

53. Dorothy Day, "Letter to Our Readers at the Beginning of Our Fifteenth Year," *The Catholic Worker*, May 1947, 1, 3, available at: http://www.catholicworker.org/dorothyday/articles/155.html. Also "to make that kind of society where it is easier for men to be good." Dorothy Day, "Peter's Program," *The Catholic Worker*, May 1955, 2, available at: http://www.catholicworker.org/dorothyday/articles/176.html.

54. Aristotle, *Nicomachean Ethics*, 1861–1863 (X.7, 1177b30–1178a1 and X.8, 1178b8–23).

# 12

# Ethics and Extraterrestrials

## What Obligations Will We Have to Extraterrestrial Life? What Obligations Will It Have to Us?

*Adam Potthast*

## Introduction

This chapter explores what ethical obligations we will have toward extraterrestrial life that almost surely exists elsewhere in the galaxy. I use "life" in the most liberal fashion possible, to include everything from advanced civilizations to silicon-based bacteria. I argue that we are unlikely to have basic ethical obligations to extraterrestrial life because it is very unlikely that extraterrestrial life could have moral status, by the lights of our leading moral theories. I make the same argument regarding whether extraterrestrial life would have any ethical obligations to us. Furthermore, even if extraterrestrial life did have moral status, consistency with moral practices on Earth would suggest that we have very few basic ethical obligations to this life. However, it does not follow that we could treat extraterrestrial life in any matter human beings see fit, because extraterrestrial life may require protection as we follow through on basic ethical obligations to ourselves.

Throughout this chapter, I use the term *basic ethical obligation* to characterize my position. By this term, I simply mean a direct obligation to another being or group of beings and their interests the way we have ethical obligations directly to other human beings or animals on Earth. Later, I contrast a basic ethical obligation to a *derivative ethical obligation*, which would be a duty to another being or group of beings that one has as a consequence of some other, basic ethical obligation. For instance, we do not generally believe we have basic ethical obligations to a common house plant, but I might have a derived or derivative obligation to the plant if I agree to take care of it for its owner while the owner is away on vacation. My basic obligation is to the plant's owner, but my derivative obligation could be said to be to the plant itself.[1]

Adam Potthast, *Ethics and Extraterrestrials*. In: *Social and Conceptual Issues in Astrobiology*. Edited by: Kelly C. Smith and Carlos Mariscal, Oxford University Press (2020). © Oxford University Press. DOI: 10.1093/oso/9780190915650.001.0012

# Simple Answers

I believe there is a great deal of confusion at work in ordinary discourse about ethical duties to extraterrestrial life. The influence of contemporary science fiction literature, movies, and television shows has created a prototype concept of extraterrestrial life in much of the Western world.[2] The extraterrestrial life depicted in science fiction tends to consist of conscious, rational, communicative beings that can experience happiness, pleasure, and pain. It is rare to see nonrational extraterrestrial life depicted in any kind of detail, especially plant and/or microbial life.

Since this prototype concept likely generates some of the interest around the question of our duties to extraterrestrial life, I think it is best to declare from the outset that yes, we would in fact have basic ethical duties to conscious, rational, communicative beings that can experience happiness, pleasure, and pain. Since the basis of ethical duties to humans and animals on Earth rest on these concepts,[3] it would strain any argument to make an exception to such beings just because they originate from another world. Grounds for excluding such extraterrestrial life from the sphere of ethical obligation would have to be racist or speciesist (terrestrialist?). Even if the beings were a direct threat to us, we would have to consider their rationality or sentience in whether to attempt to harm them, exactly as these ethical obligations work on Earth.

Similarly, conscious, rational, communicative extraterrestrial beings that can experience happiness, pleasure, and pain would have ethical duties to us. If they could consider courses of action and had basic control over their responses, we could invoke any one of a number of arguments that ethical behavior is rational and such beings would be under the authority of the categorical imperative or a utilitarian ethical obligation. To deny this would require denying the same of any conscious, rational, communicative terrestrial beings that can experience happiness, pleasure, and pain.[4]

# Will Extraterrestrial Life Be Rational and/or Sentient?

If we do know that we will have ethical duties to beings that are rational or sentient, the next important question is whether extraterrestrial life is at all likely to be rational or sentient. Many believe that the answer to this question is yes. Immanuel Kant believed that his ethical theories applied to all rational beings, implying that other rational beings could exist.[5] Despite objections from Descartes, it is now widely accepted that many animals other than human beings are sentient. It cannot be denied that it is possible for extraterrestrial beings to be rational or sentient. However, I believe we should resist the inference from possibility to probability in this case. There are factors that make the existence of extraterrestrial rational and/or sentient beings extremely unlikely.

I start from the premise that our models of what it means to be rational or sentient come from our experience of life on Earth—human and animal life in particular. While there are some who claim evidence of sentience in plants,[6] all noncontroversial examples of sentience and rationality occur in kingdom Animalia and possess highly specialized neuronal cells (in the case of sentience) or even more highly specialized organization of neuronal cells into brains (in the case of rationality). Neurons and brains are the result of specific evolutionary pathways that exist because of a history of certain selection pressures and other evolutionary factors.[7] These pressures resulted in the existence of more organisms with the capacity to produce the proteins necessary for formation of neurons and brains. It is difficult to say how common such selection pressures might be on worlds other than our own, but as we discover more exoplanets with similar profiles to our own, some would argue that our confidence that such selection pressures would produce brain-like (or neuron-like) structures should increase.

However, it is worth questioning whether exoplanets—even in the large numbers that exist in our galaxy—exist in numbers sufficient to make an evolutionary history similar to our own likely. Since we can only hope to interact with a thin sliver of the exoplanets in our own galaxy (if we are lucky), the likelihood of finding a planet with a similar evolutionary history is further diminished. Even if initial conditions and selection pressures were similar, there would be an enormous amount of evolutionary history, full of contingencies, that would have to play out just right before anything like brains evolved. According to current understandings, 800 million years elapsed before life itself even took hold on Earth (and we can imagine that many other unstable forms of organization like life came and went). And it took 3.2 billion years after that for vertebrate life to evolve out of the chaos of early life. It would be another 590 million years before human brains came to be in something like their current form. As with other forms of life in general, we can imagine that for the one time neuronal cells evolved or human brains evolved, there were many other patterns of organization that failed. The particular set of selection pressures, resources, and luck that resulted in neurons and brains seems to have only occurred once on Earth, after billions of years of stops and starts. There do not appear to be alternative, nonneuronal forms of sentience or non-brain-based forms of rationality that have coevolved with the forms we see now on Earth.

It is true that there is no way of knowing whether Earth is representative of what happens in general on exoplanets (though it does appear to be unique among planets in our solar system). We have no way of knowing whether 3.8 billion years is a (comparatively) long time for rational, sentient life to have evolved or a short time. In addition to evolutionary factors on Earth, there are also nonevolutionary factors, such as meteors, comets, toxic gas explosions, weather conditions, the stability of other geological systems, and other forms of cosmic luck to contend with. We have no way of knowing whether we have been lucky or unlucky when dealing with nonevolutionary factors. But we do know that sentience appears to have survived and thrived in only one form (neuron-based sentience) and rationality appears to have survived and thrived in only one form of sentience (brain-based rationality present in humans[8]). It seems like a staggering number of factors have to be just right for life in general to take hold, let alone sentience and rationality. Furthermore, if we assume we have been lucky, it still took roughly half of the time planets have existed in the history of the universe[9] for sentient, rational life to evolve here.

Finally, and perhaps most importantly, our experience of sentient, rational life on Earth is based on a highly iterative evolutionary history where each tiny step forward relies on the millions or trillions of steps that came before it. The existence of DNA that codes for neurons or brains relies on millions of pre-existing genes, adaptations to particular evolutionary niches, epigenetic factors, and random mutations to say nothing of the fact that it relies on the proto-evolutionary process which produced deoxyribonucleic acid as a method of gene inheritance in the first place.[10] If we assume there is no divine hand guiding these processes, the process would be akin to hitting a thousand-to-one jackpot billions of times in a row.[11]

To summarize my argument, then, it is certainly possible that rational, sentient life could evolve elsewhere in the universe. However, our view of the odds that it could occur on a planet we may one day encounter (or may encounter us) should be guided by a view of the length of the evolutionary process needed to produce rationality or sentience on Earth, the enormous number of contingencies present in the process on Earth (including the contingencies of original conditions on Earth), and the lack of any knowledge about the representativeness of Earth's example. The universe is a big place, and the numbers lead some people to increase their confidence in the odds. But the portion of the universe we could possibly interact with is much smaller should be an important correction to that confidence.

In short, the actual existence of sentient, rational life on Earth shows that it is possible elsewhere but not that it is probable elsewhere. To make the inference from the fact that we do not know if our experience of the evolution of sentience and rationality on Earth is typical to the claim that it must be typical is too much of an appeal to ignorance. It may sound like the opposing argument—that it would be atypical—would be an appeal to ignorance as well. But the uniqueness of sentience and rationality on Earth, coupled with an intuitive sense of the sheer number of things that could block budding life, sentience, or rationality, and the iterative nature of evolutionary change does give us reason to dramatically lower the probability of sentience or rationality evolving elsewhere in the universe (and certainly within the proximity we would need it to evolve within in order to have potential ethical relationships to it).

## Quasi-Sentience, Quasi-Rationality, and other Possible Ethical Values

One objection to this general line of argument is that sentience and rationality are special features of human beings and animal life on Earth, but life on other worlds might have *other* ways of being special that give them moral status. This might take the form of something like sentience (quasi-sentience) or something like rationality (quasi-rationality) or it might be something entirely new. Even if it is unlikely that sentience or rationality evolved on other worlds, surely quasi-rationality or quasi-sentience would generate some ethical obligation on our part, right?

I think the point is well taken here if quasi-rationality or quasi-sentience are descriptors of functional states that overlap with what we commonly refer to as rationality or sentience. Rationality or sentience could easily be instantiated in non-carbon-based life-forms or even life-forms that exist as information. We are probably even closer to creating such beings on Earth than we are to meeting such beings from other worlds. Such beings would certainly generate ethical obligations on standard ethical theories. However, as I have argued, our only experience with rationality or sentience comes from beings with neurons or brains on Earth and other evolutionary stories on other worlds are unlikely to come to the same ends. One of the main reasons we can so easily imagine extraterrestrial life as rational and sentient is the same reason we are likely to create artificial rationality or sentience: it is rational, sentient beings that do the imagining and creating. Such beings will not be led through a process of evolution from unicellular organisms which themselves emerged from pre-existing protein structures. They will be created fully formed. Essentially, if we create artificial rationality and sentience, we will smuggle our entire evolutionary experience onto new hardware.

What then of other ways of being special? Here I think there are strong reasons to take care with ethical language. When asking about ethical obligations, we have to pay special attention to the meaning of the word "ethics." While common usage of the word may be broad enough to encompass slightly more, when ethicists use the word "ethics" in an evaluative fashion, they almost always adopt a pre-existing theoretical perspective. In this chapter, I have been using a perspective broad enough to include both utilitarian and deontological perspectives on ethics (more on virtue ethics later).[12] If it is unethical to cause harm, the harm is unethical because of the pain caused or the threat to someone's human dignity and responsibility to make their own choices.

It is tempting to ask, "what could harm be?" or "what would dignity look like?" in beings without brains. With beings specifically imagined or created to instantiate the functional states of sentience or rationality without brains, these are fruitful questions to ask. However with extraterrestrial beings whose evolutionary history is in response to very different selection pressures—perhaps even built on top of different protein structures—states which approximate the functional states of rationality or sentience may be a strained metaphorical connection if they are found at all. The functions present in extraterrestrial life will be a direct result of the functions the beings needed to instantiate to survive and reproduce—if reproduction is even one of these functions. And we do not even need to contemplate extraterrestrial beings to follow through with this line of argument. If we had never discovered bacteria or viral life on Earth and suddenly stumbled across it, few would argue that we have ethical duties to maintain it (at least in areas where it was not actively helping us out, as in our gut biomes). We know this because such discoveries have occurred often in human history and our ethical concepts did not widen to encompass it. No one seems to have ethical qualms about our immune systems fighting off previously unknown viruses when they threaten people we care for and love.

While the possible evolution of structures that support functional states like rationality or sentience on other worlds seems unlikely to me given the complexity of the evolutionary story that has preceded them here on Earth, I do think it is slightly more likely that we could find some form of life that resembles microbial life or basic plant life here on Earth. Coming earlier on in response to more primordial selection pressures seems to make them more probable as starting points—even if their physical forms may be quite different. Some of these structures may even appear quite beautiful to us. Should we strain our ethical concepts to include such beings in our basic ethical obligations?

With one exception that I note in the section that follows, I think the answer is still "no." For such a disposition to count as an ethical obligation, it must be in some way related to the structure of ethical obligations and the history of the concept of ethics that we have constructed here on Earth. Ethics cannot, after all, mean everything. Someone who believes we have a basic ethical obligation to protect all plant life on an exoplanet while eating salads on Earth is either exhibiting inconsistency or uses the word "ethical" in a way that is different than the way the word has been used to this point in Western philosophical traditions. Since the word "ethical" has a meaning, this chapter attempts to explore the implications of its existing meaning. If one chooses to use the word in new ways to mark out new features for protection, one might be entertaining new concepts entirely and owes an account of why these new features still count as ethics. So it is not a naiveté of foresight that certain entities (even living entities) have been excluded from the sphere of ethical consideration to this point. The boundaries of our concept of ethics are always deliberate not just in their conception of what to include but also what to exclude. With the same exception I noted earlier (and explore in the next section), there are not large movements to include microbes, plants, and fungi within the sphere of ethical consideration. It would strain consistency to say that beings on other worlds that resemble plants should generate ethical obligations to us if plants on this world do not engender the same obligations. Since, by the lights of our best ethicists and common practice, these obligations consist in protecting, promoting, and cherishing the interests of rational and/or sentient beings, to move to protect beings that lack these states would break (or at least dramatically strain) the concept of ethical obligation.

We include many beings in our sphere of ethical consideration. But so far it has been a minority position to extend consideration to life itself[13]: causal structures that simply take in fuel, trap energy in a self-sustaining pattern, and/or reproduce other structures that do the same.[14] In fact, we have made deliberate choices to exclude mere life from our sphere of ethical obligations. Of the many novel and beautiful things we are likely to encounter when humanity starts expanding into the universe, I have argued that it is not likely these things will be rational and/or sentient. Since the concept of ethical obligation has come about on Earth to map on to rationality and sentience and cannot be easily extended, I conclude that it is doubtful we will have basic ethical obligations to extraterrestrial life.

## Our Potential Ethical Obligations to Extraterrestrial Life

So far I have argued that while we would have basic ethical obligations to extraterrestrial life if it were rational and/or sentient, extraterrestrial life is not likely to be rational and/or sentient because these states are particularly improbable results of our evolutionary heritage here on Earth. I have also argued that it will not be possible to enlarge the concept of ethics to encompass plant-like or microbial life on other planets. I believe many would fear taking this line of argument seriously because they believe it would lead to a repugnant conclusion: that humanity is ethically permitted to do whatever it likes to the life it may find on other worlds or elsewhere in the cosmos.

I do not believe this conclusion follows directly, though, for an important reason: while we may not have basic ethical obligation to nonrational, nonsentient extraterrestrial life, we may have nonbasic, derivative ethical obligations to such life—namely ethical obligations that derive from ethical obligations we have to ourselves.

All three major theoretical approaches to ethics (utilitarianism, Kantianism, and virtue ethics[15]) have an important place for duties to self. For virtue ethics and Kantian ethics, the duty is basic: cultivating a virtuous character and not treating oneself merely as a means to an end constitute core responsibilities to both theories. Utilitarian theories also make one's interests part of any utility calculation (though the choice of optimal utility may not always promote maximum personal utility). Correspondingly, we can see that multiplying these effects means that human choices about what to do with nonsentient, nonrational extraterrestrial life (and even inorganic matter) have ethical consequences for individual humans. These choices will matter to individual or communal virtue, our abilities to make rational decisions, and the overall health and well-being of the human race.

Here are just a few of the ways that protecting extraterrestrial life has a direct effect on human virtue, dignity, and well-being:

- **Protecting humanity:** There are many clear and present dangers to human life on Earth itself, and these dangers are likely to multiple as human beings reach out into the cosmos. Preserving extraterrestrial life in order to study it scientifically, learn from its survival strategies, and even harness its usefulness in defending ourselves from terrestrial or extraterrestrial dangers would be infinitely more useful than exploiting or depleting it. All three ethical traditions support the importance of protecting human life, and extraterrestrial life could be the most important tool ever discovered to accomplish this mandate.
- **Practicing stewardship and conservation:** Our abilities to practice stewardship and conservation of resources on Earth is highly limited by the history and regimen of property rights on Earth. New worlds with new life and resources would provide the opportunity to build truly sustainable practices that could be transplanted back to Earth or used as models for further cultivation of resources as more new worlds are discovered.
- **Protecting beauty:** If we find extraterrestrial life, it is as likely to inspire new art as it is new science. Generations will gaze upon or listen to or otherwise sense the contours of beings which may very well be judged as amazing and beautiful. Appreciating and contemplating the beauty of alien worlds and lifeforms can inspire virtue and contribute to happiness, and Kant (2000) argues that appreciating natural beauty is an indicator of a "mental attunement favorable to moral feeling."
- **Practicing reverence:** Similar to stewardship and conservation, reverence is an important feature of human flourishing that can be hard to find an excuse for in modern living. By seeing extraterrestrial life as something other than a resource to be depleted, we have the ability to practice this virtue

and extend our human sensibilities. To simply destroy extraterrestrial life for cheap energy would rob humanity of the opportunity to cultivate this virtue.

- **Promoting wonder and joy:** The emotions of joy and wonder are integral in the creative process and have themselves been instrumental in igniting human achievement throughout human history. Few discoveries on Earth would bring about the joy and wonder as the discovery of extraterrestrial life would. Refusing to protect extraterrestrial life would deprive humanity of this source of wonder and joy, thus depriving it of the insight and achievement that goes along with it, to say nothing of the naked utility of joy and wonder in the first place. While wonder and joy are of obvious value, the destruction of opportunities for wonder and joy are particularly devastating, and one cannot discount the suffering that could be created by failing to protect extraterrestrial life.

- **Practicing restraint:** The utility of restraint in our own lives and the use of resources as a species as a whole is very high. Yet the scarcity of resources on Earth has not led to the ability to practice restraint. A new source of wonder and resources in extraterrestrial life would provide another opportunity to practice the kind of restraint that leads to virtue, maximal utility, and the ability to pursue ethical projects and goals.

- **Increasing resources:** While the other reasons concentrate on the value of preserving extraterrestrial life in a more or less pure state, there could also be a tremendous benefit to humanity in the careful use of resources provided by extraterrestrial life and the resources underlying extraterrestrial habitats. Given the scarcity of resources on Earth, other sources could, if wisely used, promote human happiness, support projects to aid human beings in poverty, and lead to opportunities to cultivate virtues like courage as well. While this reason does not support quite as much protection as the other reasons, it still requires a careful cost-benefit analysis, as do questions about ethical usage of resources on Earth. We should not count on the existence of resources in another part of the cosmos to save us from the tough work of ethical thinking.

Thus, while we do not have basic ethical obligations to nonrational, nonsentient extraterrestrial life, we do have basic ethical obligations to ourselves and these obligations will often entail that we cherish, promote, and protect extraterrestrial life, perhaps even for reasons that do not hold for organisms with similar capacities on Earth. These arguments, I believe, are the proper grounds on which to have ethical arguments about what to do with extraterrestrial life, and I believe such arguments lead to a strong presumption in favor of a policy regime that would promote strong protection of extraterrestrial life in whatever form we find it.

## Conclusion

It can be somewhat jarring to learn that we do not have basic ethical obligations to extraterrestrial life as such, but I believe that the unlikelihood of the existence of rational, sentient life that we are likely to encounter, together with the concept of what it means to be an ethical obligation here on Earth, leads to just that conclusion. At very least the debate over such a topic needs to be rescued from simplistic notions of functional states present in contemporary science fiction. However, as I have argued in this chapter, it does not follow that we have no ethical obligations at all to extraterrestrial life. Rather, by paying careful attention to ethical theories and human interests, I believe that there are several strong reasons to prefer policies that protect extraterrestrial life and whatever habitat that life occupies.

## Notes

1. I use this terminology at least in part to avoid debates about intrinsic and extrinsic value; however, I believe that the same argument could be made in those terms.
2. Interestingly, the search for extraterrestrial life has not captured the imagination of non-Western peoples, until recently (Schneider, 2013).
3. Here, I consider just Western, Euro-American ethical traditions. My limited familiarity with Eastern traditions suggests the arguments I present could be extended to Eastern traditions, but making that argument is beyond my scope of expertise.
4. While it is not an ideal term, for the rest of this chapter I will use the word "sentience" to refer to the capacity for conscious experience, pleasure, and pain. I use the word "rational" in the Kantian sense, for the deliberative capacity to give oneself ends to follow rather than operating automatically based on causal stimuli.
5. Kant (2012) even considers possible inspiration we could draw from inhabitants of other worlds who are "sensible," though they may be composed of "lighter" and "more volatile" matters than humans on Earth.
6. Brenner et al. (2006).
7. The model I assume for this chapter is the "traditional view" of evolutionary theory put forward by Stephens (2004).
8. Famously, Kant (2011) even argues that we cannot be certain that we are rational, but exploring this angle would take us far from the central topic of the chapter.
9. See Shelton (2015).
10. This is, I think, a point that is masked in one of the most popular arguments for the existence of intelligent life elsewhere in the cosmos: some weightings of the Drake

equation. The equation is typically used to show that the number of worlds (the numerator of the equation) is so high that even if one claims intelligent life is rare the number of worlds with intelligent life will be high. One way of making the point I am arguing for in this chapter is to suggest that the denominator of the equation is dramatically higher than almost anyone estimates, due to the contingency of evolutionary steps necessary to get intelligent or rational life.

11. To hit even a five-to-one jackpot billions of times in a row would be an incredible achievement.

12. I believe that the three Western theoretical traditions I mention here are the current leading contenders in Euro-American versions of ethical theory; however, I believe the arguments within this section could be adapted to approaches that stem from divine command theories or natural law theories.

13. I do not mean to minimize the efforts of environmental ethicists here who extend some kind of ethical consideration to life itself. Callicott (1999) and Katz (1997) both argue for intrinsic value for plants and other natural objects independent of sentience and rationality. Their positions, however, have not motivated large-scale change in ethical thinking yet, possibly because their positions clash with the anthropocentricity of the concept of ethics.

14. As a further illustration of the boundaries of our ethical concepts, stars and other nuclear reactions meet these criteria (with the exception of reproduction). Yet I think that—our ability to have an influence on them aside—few would say that it makes sense to say we have ethical obligations to stars, let alone sustaining nuclear reactions on this world. Though perhaps we could imagine immensely powerful rational beings that could snuff out stars the way we cut down a tree, I think it would still strain the concept of ethics to think of these beings as having ethical obligations to stars.

15. See Singer (1979), Timmerman (2006), and Foot (2009).

# References

Brenner, E., Stahlberg, R., Mancuso, S., Vivanco, J., Baluska, F., & Van Volkenburgh, E. (2006). Plant neurobiology: an integrated view of plant signaling. *Trends in Plant Science, 11*, 413–419. 10.1016/j.tplants.2006.06.009

Callicott, B. (1999). *Beyond the Land Ethic: More Essays in Environmental Philosophy*, Albany: SUNY Press.

Foot, P. (2009). *Virtues and vices and Other Essays in Moral Philosophy*. Oxford, U.K.: Clarendon Press.

Kant, I. (2000). *Critique of the Power of Judgment* (Cambridge Edition of the Works of Immanuel Kant) (P. Guyer, Ed.; E. Matthews, Trans.). Cambridge: Cambridge University Press. doi:10.1017/CBO9780511804656

Kant, I. (2011). *Immanuel Kant: Groundwork of the Metaphysics of Morals: A German–English edition* (Cambridge Kant German-English Edition) (M. Gregor & J. Timmermann, Eds.). Cambridge: Cambridge University Press. doi:10.1017/CBO9780511973741

Kant, I. (2012). Universal natural history and theory of the heavens or essay on the constitution and the mechanical origin of the whole universe according to Newtonian principles (1755). In E. Watkins (Ed.), *Kant: Natural Science* (Cambridge Edition of the Works of Immanuel Kant, pp. 182–308). Cambridge: Cambridge University Press. doi:10.1017/CBO9781139014380.007

Katz, E. (1997). *Nature as Subject*. New York: Rowman & Littlefield.

Schneider, J. (2013). Philosophical issues in the search for extraterrestrial life and intelligence. *International Journal of Astrobiology*, *12*(3), 259–262. doi:10.1017/S1473550413000025

Shelton, J. (2015, January 27). Astronomers discover a replica solar system. Retrieved from https://news.yale.edu/2015/01/27/astronomers-discover-replica-solar-system

Singer, P. (1979). *Practical ethics*. Cambridge: Cambridge University Press.

Stephens, C. (2004). Selection, drift, and the "forces" of evolution. *Philosophy of Science*, *71*(4), 550–570. doi:10.1086/423751

Timmermann, J. (2006). Kantian duties to the self, explained and defended. *Philosophy*, *81*(317), 505–530. Retrieved from http://www.jstor.org/stable/4127406

# 13

# METI or REGRETTI

## Ethics, Risk, and Alien Contact

*Kelly C. Smith*

## Introduction

As Space Guy learns the hard way, there are risks posed by actively *messaging* extraterrestrial intelligence (METI). METI is an offshoot of the *search* for extraterrestrial intelligence (SETI), though most members of the SETI community are careful to distance themselves from their more ambitious cousins. Most of the discussion within the space community concerning METI has been about the level of risk it poses. Advocates of METI usually argue that the risk is so low it is not worth worrying about (Shostak, 2015; Vakoch, 2016), while opponents typically counter that we cannot assess the risks accurately enough to make a well-informed decision (Brin et al., 2014; Peters, 2017).[1] In other words, the debate has been mostly over the kind of empirical question that physical scientists feel comfortable dealing with. My goal is to cast the problem in a new, more

Kelly C. Smith, *METI or REGRETTI*. In: *Social and Conceptual Issues in Astrobiology*. Edited by: Kelly C. Smith and Carlos Mariscal, Oxford University Press (2020). © Oxford University Press. DOI: 10.1093/oso/9780190915650.001.0013 philosophical, light in the hope that this will reveal some critical, overlooked features of the problem. As much as possible, I steer clear of the empirical details that have consumed so much attention, partly because I lack the relevant expertise to enter deeply into these disputes but more importantly because entanglement in the details often distracts us from seeing the big picture clearly.[2]

Assessing the empirical dimensions of METI risk is a useful exercise, to be sure, but it is often unappreciated that these details just do not resolve key questions. In particular, if we look at METI through an *ethical* lens, the central question is not what the *level* of risk is but whether those who are exposed to that risk (in this case, all of humanity) *agree* to it. This sort of consideration is nothing new to social scientists (see Denning, 2010), who are used to dealing with human subjects in their research, but it is terra incognita for most space scientists.

I admit to a certain ambivalence concerning METI. I would *personally* love to push the transmit button and send a signal to every nearby system. If I had to bet, I would say the odds very much favor either a good outcome or no outcome at all, and thus I would gladly take the risk for myself. But that is an answer to the wrong question. The right question is whether I am willing to *unilaterally impose* such a risk *on everyone else*. I simply cannot justify that, as I clearly lack the requisite moral authority, no matter how confident I am in my assessment of the potential risks and rewards. Admittedly, the cognitive dissonance creating by supporting METI in one sense and opposing it in another is uncomfortable. But, as I say to my students: "If ethical principles were always convenient, it wouldn't be so hard to uphold them."

## Background on METI

Since ancient times, people have speculated that the heavens might contain other forms of life, even other civilizations (Dick, 1982; Crowe, 2008). But it was not until 1960 that the astronomer Frank Drake formulated the famous Drake equation in an attempt to bring speculation about aliens closer to the realm of science (see Tipler [1981] for an excellent historical survey of the Drake equation). The equation is designed to calculate the number of alien civilizations we might, in principle, be able to contact:

$$N = R * f_p * n_e * f_l * f_i * f_c * L$$

Where the number of contactable civilizations (N) is the product of:

- the rate of star formation (R)

- the fraction of stars having planets ($f_p$)
- the fraction of planets that can support life ($n_e$)
- the fraction of habitable planets where life actually evolves ($f_l$)
- the fraction of planets with evolved life where intelligence arises ($f_i$)
- the fraction of planets with intelligent life where a civilization capable of interstellar communications develops ($f_c$)
- the period of time during which such civilizations attempt communication

The extent to which the Drake equation is actually *science* is debatable of course, since we do not yet have enough information to fill in most of these variables with confidence. However, we have made enormous progress in recent years on the first few. For example, the deep field survey pointed the Hubble space telescope at a tiny piece of sky that seemed especially empty, collecting every photon it could for the equivalent of twenty-two days. The result is the famous picture revealing no less than 10,000 *galaxies*. Hubble analysis of similar fields in the Ursa Major and Fornax regions revealed 3,000 and 10,000 galaxies, respectively. This kind of research allows scientists to estimate that there are approximately 10,000,000,000,000 (10 trillion) galaxies out there. If we use a conservative figure for the number of stars in the average galaxy (100 billion), we can then estimate the number of stars in the visible universe as about 1,000,000,000,0 00,000,000,000,000, or $10^{24}$ (Howell 2017).

Further, in the twenty-five years since the discovery of the first planet beyond our solar system, we have compiled a catalog of almost 4,000 of them in 3,000 different systems (numbers which will increase dramatically in the next few years when the Transiting Exoplanet Survey Satellite and the James Web Space Telescope come online). Planets seem to be far more common than we had thought, with at least one detectable planet per star on average. Rocky and water worlds like Earth are also much more common than we assumed, which means there are many more opportunities for life to evolve in ways similar to what occurred on primordial Earth (Hawkes, 2017; Goldschmidt Conference, 2018). Indeed, a recent article by Frank and Sullivan (2016) calculated that the only way Earth life could be unique in the universe would be if the evolution of life, given suitable conditions, were *virtually impossible*—with odds of less than 1 in 100,000,000,000,000,000,000,000, or $10^{-23}$. So we have excellent reasons to think there is indeed life out there.[3]

There have been at least thirty-one intentional attempts to message alien worlds since 1974 (Zeitsev, 2006; Quast, 2018). Most of these do not represent major risks as they involve one-time, relatively low-power transmissions. Several are just publicity stunts, as when Pepsi hired one of the EISCAT radar arrays in 2008 to share the good news concerning their breakthrough tortilla chip technology with any aliens living in the Ursa Majoris constellation

(Barras, 2008). However, plans are being made by several players to do this in a much more serious and sustained way—for example, by using powerful lasers aimed directly at nearby star systems with known habitable planets. The goals and motives of these new METI projects range widely. On the one hand, METI International has assembled a team of scientists and other experts to preparation for the kind of METI effort one would expect serious scientists to undertake. On the other hand, the amusement park millionaire William Kitchens has created the Interstellar Beacon Project, which plans to "backup humanity" by beaming the entire contents of Wikipedia to the stars.[4] As a result of these developments, the debate concerning whether this is a good idea has blossomed into something of a cottage industry (Duner et al., 2013; Vakoch, 2014; Brin, 2014; Baum, 2016; Johnson, 2017; Peters, 2017; Peters, 2019).

Before addressing the risks, it is important to be clear concerning two aspects of messaging that might not be obvious to those outside the field. First, if we do make contact, it will almost certainly be with a civilization that is far, far older than ours (and thus, presumably, technologically far superior to our own). This is a simple matter of probability based on the relevant ages:

- The universe began 13,700,000,000 years ago
- Life on Earth began 4,500,000,000 years ago
- Multicellular life began 1,500,000,000 years ago
- Humans began 200,000 years ago
- Human science began 10,000 years ago
- The human Industrial Age began 250 years ago

It seems the conditions were right for life to have evolved at any point in the last 12 billion years or so, but terrestrial life did not appear until 8 billion years of that window had passed by, so it is extremely unlikely that we were the first life in the universe. Further, *humans* have only been around for the blink of an eye in cosmic terms—just the last 0.0002 billion of the 4.5 billion years life has been on Earth (0.00016% of the tenure of terrestrial life)—so even in the unlikely event that *life* on Earth evolved before life elsewhere, it still seems likely that alien life evolved into *intelligent forms* before we did. And it is worth noting that it does not take much of a head start to make a major difference: to paraphrase Arthur C. Clarke, alien technology would likely be indistinguishable from *magic*,[5] even to beings a mere 10,000 years less developed.

Second, debate about precisely what *content* should go into a message is largely beside the point as far as risk is concerned, since *any* message will reveal both the location of Earth and the relative state of our technology. While we might not care if aliens were to learn of our delight in salty snacks, most people would be more concerned about their knowing the precise location of our home world. Furthermore, our electromagnetic transmissions would likely be a very primitive technology compared to neutrino beams or whatever the next generation(s) of communication might look like. Thus *any* signal we send will contain at least two vitally important pieces of information: (a) these guys are technologically backward, and (b) they live right *there*. That seems clear prima facie grounds for concern.

## An Ethical Framework

Asking whether we *should* be attempting METI is ultimately an ethical question, though the scientists and engineers who discuss it do not always realize the implications of this simple fact. Therefore, I want to begin by examining an explicitly ethical argument based on an analogy with medical ethics that will set the tone for the discussion that follows.

In modern medicine, it is not considered ethically appropriate to treat a patient without his or her consent (Beauchamp, 2012; Hall et al., 2012; American Medical Association, 2016). Why? The basic argument is based on the moral concept of *autonomy*: since the risks of whatever procedure one performs are risks *to the patient*, it should be up to her to decide whether or not she is willing to accept those risks.[6] In other words, it is her life, so it should be her decision. Of course, there are exceptions to this principle, but they are fairly narrow. Thus a doctor can treat a patient without her consent if any of the following conditions are met:

1. The patient is *not competent* to decide. There are cases where the patient is simply unable to understand the situation, as with a small child or Alzheimer's patient. But even in these cases, we must try to find a surrogate decision-maker so that someone—in particular, someone with the *patient's* best interests at heart—decides whether the risk is acceptable.
2. There is a life-threatening *emergency*. If a patient shows up in the emergency room bleeding to death, there may not be time to get consent. Yet we still allow treatment on the reasonable assumption that patients typically want to live and thus *would* consent to being saved if they could speak for themselves.
3. *Other* people would be harmed. A patient with a dangerous communicable disease (e.g., Ebola) might well refuse to be treated. But if the risks to others are sufficiently serious, forcing her to undergo treatment is usually preferable to allowing harm to others.

These are all relatively straightforward. However, physicians are often tempted to allow other exceptions that *seem* to pass ethical muster but are in fact extremely problematic. For example, suppose the doctor sincerely believes both that:

1. The treatment is truly in the best interest of the patient. In particular, there may be no other option that will save the patient—if the patient refuses the treatment he will surely die, perhaps in a horrible way.
2. The reasons the patient gives for refusing treatment are silly—for example, the patient might believe he is not actually sick or that his illness is caused by a magic spell instead of an infection.

These seem intuitively plausible exceptions because the intent of the physician is both benevolent and a relevant expert well positioned to conclude the treatment is medically indicated. So why aren't they permissible? One way to put the question is this: Why shouldn't we just let the physician—a well-intentioned expert—decide?

For one thing, we have to be careful assuming that the physician is only motivated by benevolence—in particular, he might have a *conflict of interest*. This does not necessarily mean his judgement is misguided, of course, but only that he has an interest in the decision being made in a particular way, so we have to be on guard for bias (conscious or otherwise). For example, the treatment might be part of a research study the physician wants to see through to completion or he might stand to gain monetarily. Even the most cursory scan of the history of scientific research can leave no doubt that highly educated, well-intentioned people regularly do things under these circumstances that are truly horrible (Institute of Medicine, 2009; Elliot and Stern, 2011). It is thus simply naïve to assume that every physician is motivated solely by beneficence.

For another, it is not clear the physician is expert *in the right way*. While it is reasonable to expect physicians to be familiar with the standard rules concerning ethical *practice*, this does not insure that they are especially good at ethical *reasoning*. And ethical reasoning is the *relevant* expertise when it comes to navigating complex tradeoffs that cannot be resolved with standard rules of practice. After all, the most difficult ethical dilemmas pose a choice, not between good options and bad ones but between competing good options (or competing bad ones).

It is certainly good, all else being equal, to save the patient, but it is also good to give the patient freedom of choice. Any decision will thus reflect *which* ethical principles we take to be most important and, unfortunately, there is often no *objective* standard for "good." Thus, all too often there is just no objectively defensible way to argue that one good course of action is inherently more moral that another,[7] and the physician's expertise in medical science does not give him any privileged insight into these complex moral choices. He may be convinced that saving his patient's life is for the best, but the patient may be just as convinced that enduring heroic measures to extend a life of suffering is a bad bargain. Whenever there are zero-sum choices and what counts as the best course of action is unclear, it seems best to leave the decision to those most directly affected. The patient may be wrong, but she should have the right to be wrong, since she is the one who must shoulder the consequences of the decision.

What is the connection to METI? If METI poses a risk, then it is a risk affecting *all of humanity*—the public is in the position of the patient. And with METI, none of the standard exceptions apply: people can understand the choice being offered, at least in lay terms; there is no emergency, and no harm can reasonably be expected to others from a failure to act. The precise reasons the public might give as to why they consider the risk unacceptable are beside the point—especially since many reasons they could provide are perfectly reasonable.[8] Moreover, space scientists are no more expert in ethical judgment than physicians and may even have their own potential conflicts of interest.

It is thus *immoral* to pursue METI under the present circumstances.[9] In particular, no unrepresentative group (whether scientists, politicians, or eccentric millionaires) should be allowed to proceed until, at the very least, we better understand public attitudes about the risks. But METI proponents will no doubt respond that there is more to be said about the possible exceptions to such a blanket injunction, so I now turn to an examination of these in more detail.

## Competence

### Scientific Paternalism

It might not be admitted openly, but it seems that METI proponents are motivated in part by a significant skepticism concerning the competence of a scientifically illiterate public.[10] Certainly, anyone (including myself) who has spent time in the trenches of public education, defending science against a proudly ignorant faction of the public, will be sympathetic to this attitude. But we should call it what it is so we can see the implications clearly: this is *scientific paternalism*. Paternalism refers to any system that infringes on the personal freedom and autonomy of a person (or class of people) with benevolent intent. And in modern medical ethics, paternalism is considered inappropriate unless there is no other viable option.

Of course, every parent has occasionally demonstrated their paternalistic inclinations by declaring, "It's for your own good." —that is why it is called *paternalism*, after all. But with parents, we can generally say both that the child is not competent to decide for themselves and that the parent has the child's best interests at heart—at the very least, the burden of proof is on someone claiming this is not the case. We also have to keep in mind that, for both medical and METI contexts, the basic standard of competence only requires that the one impacted understand *in lay terms* what the experts tell them. This is a low bar that the vast majority of the adult population can clear, even if they have unscientific ideas about the efficacy of vaccines or the dietary habits of aliens.

So what does this tell us about the advisability of scientists making decisions for a relatively uninformed public? When it comes to METI, the public could be said to be much like a young adult, clearly passing the basic standard of competence, though perhaps lacking an appreciation of the finer points that come with age, wisdom, and specialized training. Such individuals are often well advised to listen to their elders, but it certainly does not follow that we should act without consulting them, much less that we should *force* them to take risks they are unwilling to accept.[11] The relevant question is whether the public is able to understand the nature of their decision and its possible consequences in lay terms, since we cannot possibly require the public to be as expert as space scientists, just as we cannot require a medical patient to understand his health issues in the same nuanced way his physician does—if we did, there would be no room at all for personal choice. And, when it comes to METI, there is simply no good evidence that the public fails to meet this standard.

## Space Scientists as Public Policy Experts

Another problem involves the competence *of those advocating METI*, since part of what makes their arguments intuitively appealing is the assumption that their expertise gives them unique insight into what is best. But, as with physicians, there is no reason to think that space scientists are better positioned to make *ethical* judgments than the average person. Moreover, scientific paternalism delves into not just ethics but public policy as well. Does knowledge of astronomy convey a superior ability to make good public policy? That seems doubtful.

In fact, there are good reasons to suspect that scientists may be *worse* at public policy than a random sample of the public. After all, scientists (and academics in general) are hardly known for their high social IQs—indeed, it seems likey that many self-select into such careers precisely because they are relatively free from political entanglements.[12] Yet by definition, public policy occurs in a social context, which means it must be tailored to appeal (in some sense, anyway) to *everyone*, whatever their level of understanding. And even if they had the temperament, space scientists represent the extreme tail end of the bell curve in terms of what they value, since even a member of the general public who watches the NASA channel regularly would likely never consider dedicating thirty years of her career to a single space mission. Thus, a policy that passes the strictest *scientific* scrutiny, particularly when that scrutiny is entirely within a particular scientific discipline, can be (and often is) ill-advised. For pragmatic reasons if nothing else, we should be wary of any public policy that rides roughshod over the opinions of non-scientists, since beginning one's sales pitch by telling those who disagree that they are ignorant children whose opinions should be ignored is not an effective way to build consensus.[13] In a democratic society like ours, where science is funded in large part from the public coffers, such an approach can easily backfire. Indeed, Michaud (2007) notes that in modern times the public is *especially* prone to demand accountability concerning the use of new technologies for controversial purposes, and METI certainly fits this bill. So, while it is certainly true that public policy would be greatly improved if it paid more attention to science, it simply does not follow that it would be wise to turn over political decision-making to scientists.

Finally, experts in general tend to discount the ability of non-experts to solve complex problems, and METI advocates are no exception. But there is growing evidence concerning the "wisdom of crowds" that suggests groups of non-experts are often *better* than experts at solving complex problems (Surowiecki, 2005). Of course, this does not mean the public will always choose correctly. Proponents of METI rightly bemoan the unconscious biases that infect opposition to the project, such as excessive loss aversion (Korbitz, 2014; Shostak, 2015; Vakoch, 2016). But it is not as if scientists are free from bias themselves—just like physicians, they are usually well intentioned and well informed, but they are also human. When it comes to assessing risk in particular, they have been known to *systematically underestimate* the dangers of low probability events, as famously occurred with the Challenger space shuttle disaster (Presidential Commission, 1986). As Arthur C. Clarke observed: "If an elderly but distinguished scientist says that something is possible, he is almost certainly right; but if he says that it is impossible, he is very probably wrong."

## Consent Versus Consensus

When I present the ethical argument proposed here at professional conferences, the inevitable response is that it is simply impossible to secure anything like "consent" from all of humanity: "What would you have us do, arrange for 7.5 billion people to vote on this?" This is a legitimate point as the model of consent we use in medicine quite clearly will not work—getting every human being to sign a legal document is just not possible. But that is also quite clearly a straw man, since a looser but still important type of *consensus* is not only possible but often achieved—as has been exhibited many times through collective action by the United Nations, governing scientific bodies, and so on. Consensus is more easily obtained than consent—in particular, it is not an all or nothing phenomenon. A signed consent document is thus at one end of the continuum, while letting anyone with access to a radio telescope do whatever they wish is at the other. It is a false dichotomy to force a choice between these two options, as there are plenty of alternatives between them that reasonable people can (and should) pursue.

One difficulty is that these other options are not always apparent to those who do not work in public policy. For example, the average scientist immediately thinks that imposing restrictions on METI must involve developing international *laws* along these lines. This would be, to put it mildly, extremely difficult. Fortunately, public policy often relies instead on *soft laws*—that is, rules and policies that, while neither legally binding nor formally enforceable, nevertheless set clear best practices and expectations (Christians, 2007; Guzman and Meyer, 2010). Such agreements can be quite effective and are commonplace in other cutting-edge fields like nanotechnology and genetic engineering.[14] To be sure, it would still not be *easy* to put soft laws in place, especially if we lack the foresight to recruit the relevant expertise to our cause.[15] But difficulty is not impossibility. Is it truly *impossible* to forge an agreement among those operating radio telescopes and powerful lasers (most of whom are dependent on public funding) to the effect that they will not send messages without *some* kind of review? Could we not construct a system of review that is more representative than an informal agreement among a handful of space scientists? Given that every single research university and most hospitals in the United States have Institutional Review Boards that do precisely this, the answer seems to be a resounding "no."

Of course, public consensus might *actually* be impossible if the public is adamantly opposed to METI. But that is a very different matter, since in that case we are talking not about the practical difficulties of securing *evidence* of consensus among a willing (or at least indifferent) public but about our paternalistic desire to force the opinion of a small group of experts down the throats of an unwilling public. To be sure, that is not the way the METI advocates see their actions. They would no doubt argue that they are trying to do what is best, and it is obvious to them what that is—so obvious, in fact, that they have little interest in considering non-expert opinion. This is actually quite a common point of view in scientific research and one that regularly causes problems. For example, when I served on a Human Subjects Review Board at my university, I would often hear researchers complain, "If I tell research participants about every potential risk in the informed consent document, they will freak out and refuse to participate!" I would have to respond that their objection makes my point: although it is inconvenient *to the researcher*, theirs is not the only point of view, and certainly not the most important one from an ethical perspective (Smith, 2017).

## A Slippery Slope?

A radical twist on our supposed inability to reach consensus often lurks just below the surface of these discussions. For example, some METI advocates oppose any broader discussion of these issues on the grounds that, since the public stubbornly refuses to *ever* accept *any* level of risk, requiring public consensus will inevitably result in a *moratorium* (Gilster, 2017). Moreover, they often muse that, once such a moratorium is in place, it will likely stay in place *permanently*. Shostak (2017) thus concludes that "Limiting strong transmissions skyward will straitjacket our descendants, not just in their efforts to do active SETI but for many other projects," while Nielsen (2013) takes the angst to cosmic levels, observing that:

> If our civilization determines that METI is too great an existential risk to bear, then existential risk perception begets risk aversion and possibly culminates in permanent stagnation . . . an entire galaxy (or more) might be plunged into permanent stagnation, flawed realization, or subsequent ruination as a consequence of this perceived existential risk.

This kind of thinking may play well when one's audience is composed entirely of scientists skeptical about public competence, but it clearly goes too far.

First, while it is certainly true that the public does not approach risk in the same way scientists do, and that this can be problematic, it is hardly the case that they are *never* willing to accept risk. For example, despite extensive public debate concerning the potential dangers of genetically modified organisms, most Americans seem not to mind the fact that the vast majority of their grain-based food products contain them. Second, this argument shifts the goal posts by replacing the discussion about whether we should be attempting METI *at present* to one about whether we should *ever* attempt it. This paints those advocating any restrictions at all in an unfair light, since a permanent moratorium on METI is a far more radical solution than they propose.

Second, this argument is a textbook example of a *slippery slope*: if we do X, it will inevitably lead to Y and thus, since Y is completely unacceptable, we should not do X. But slippery slope arguments are typically classified as logical *fallacies*. The problem is not that slippery slopes are never of concern but rather that the slipperiness of the slope is typically *asserted* rather than established. So if we want to pursue such an argument, we must ask: "Are there compelling reasons to think that, if we impose regulation (even a moratorium) on METI *today*, we will *never* be able to change our minds?" Once the question is posed explicitly, it seems clear that there are no good reasons to think this. Worse, it is hard to see even how we *could* have such reasons. That is because in the case of METI we have hundreds or even thousands of years to act. Who can predict with any confidence what public opinion will be that far into the future—or for that matter what kinds of social and political decision-making frameworks will even exist? The truth is that we do not have a good handle on what the public thinks about this *right now*, since there is almost no good data available.[16] As scientists, should we not agree that we must better understand the present state of affairs before attempting to extrapolate unsupported intuitions into the distant future?

## Emergencies and Conflicts of Interest

Emergencies are a clear exception to the need for consent in medicine, so is there any sense in which our current lack of significant METI efforts constitutes an emergency requiring immediate action? No, there is not. METI is a *very* long-term project: even if there is an advanced civilization in the closest system to ours (a mere 4 light years away); and they were to detect our signal and respond immediately; at least eight years would elapse before we even knew they were there. Moreover, as the difficulties of establishing meaningful communication with an alien species over such distances are truly enormous, it would likely be several more decades before any substantive information could be transferred. Of course, the alien civilization would probably be far more distant than this, in which case communication could not possibly occur in the lifetime of anyone currently alive on Earth. So what's the rush?

There is, however, one subset of humanity who would suffer from delay: current METI proponents. These scientists would likely lose any chance to be involved in what would doubtless be, if successful, one of the most important projects in human history. This fact is ethically significant, since it represents a clear conflict of interest. Just as in medical ethics, we must be sensitive to situations where those advocating a course of action stand to gain from what they advocate. Just as in medical ethics, this does not necessarily mean that the conflicted person is actually acting in bad faith; only that we have a responsibility to weigh all the possible influences on his thinking in order to assess its objectivity. Perhaps few METI proponents are influenced at all by such considerations, though it is important to note that such influence can be subtle, even unconscious.[17] After all, we know that even highly educated experts can be influenced in problematic ways without realizing it. Consider the case of drug reps in medicine: for many years, it was standard practice for them to provide small incentives to physicians to get them to listen to their marketing pitches. Physicians typically argued that this was harmless, since there was no way their prescription practices would be unduly influenced by a free lunch or a baseball cap with the company logo. But when social scientists actually looked, the data revealed what drug companies had long known: this is precisely what happens (Rose et al., 2017). So not only is there no emergency; we actually have reason to worry about the objectivity of METI proponents who urge immediate action.

## Risks to Others

### Possible Benefits of METI

I do think an underappreciated argument in favor of METI involves the possible benefits of contact. If we were to develop a good relationship with an advanced, benevolent alien civilization, they could certainly help us in innumerable ways: cure cancer, solve global warming, even advise us on how to secure world peace. And there could be indirect but critically important benefits as well—perhaps just *knowing* that we are not alone in the universe would change our culture of self-absorption in beneficial ways—for example, by encouraging us to put aside our internecine squabbling and act together for the common good. Thus, one could argue that *failing* to engage in METI imposes a significant opportunity cost on humanity. The problem here is that it also seems possible that contact with aliens could be *detrimental* to humanity, and thus a rational choice would require a careful assessment of the relative likelihood and magnitude of costs versus benefits. At this point, I have to dip a cautious toe into the debate about the level of risk posed by METI. If, as many METI advocates maintain, there is simply *no* risk a reasonable person should worry about, then perhaps the possible benefits tip the balance in favor of METI. But is that really the case—are all the worries about aliens truly without merit?

### Silly Risks and Speculative Disagreements

As METI proponents are quick to note, many in the general public focus on silly risks. Thus, the average person worries about Hollywood visions of evil aliens coming to Earth, motivated by avarice and aggression—fictional aliens tend to reflect deep-seated evolutionary worries about others coming to take our stuff, whether that stuff is human flesh or lebensraum. It is perfectly legitimate to point out that such concerns are vastly overblown. Even an extremely advanced alien civilization might lack the capacity to travel the vast distances between stars in order to pose a *direct* threat to Earth, since as far as we know the speed of light is a universal speed limit and the distances in question are truly vast. And if aliens could make physical contact, this would imply both star faring technology and a drive to explore, in which case (arguably anyway) they would likely have established themselves in multiple star systems before meeting us. It is difficult to see how a multisystem alien civilization would covet the comparatively meager resources of Earth. Of course, it is hard to *completely* discount these concerns—perhaps Earthly resources are unique in some way we cannot appreciate with our current level of technology, for example, but it is fair to point out that this kind of threat seems quite unlikely at present.

Many will note at this point that the discussion is becoming highly speculative—and that is the point. Since we know nothing about aliens, parties on *both* sides of the debate are forced to rely on speculation of one form or another. Purely speculative battles are rarely conclusive, since both sides can deploy intuitively plausible arguments that cannot be definitively countered. For all we know, aliens might have motives we have never encountered before and thus we completely overlook in our deliberations. For example, an alien race with a very long time horizon might adopt a policy of exterminating any young race that could pose a competitive threat—not right now, but 100,000 or even 1,000,000 years in the future (Soter, 2005; Liu, 2016b). Nobody can really say how likely this particular scenario is, but it at least seems far more plausible than fevered dreams of rapacious alien invasion fleets. And there is certainly no dearth of other plausibly worrisome scenarios in science fiction to choose from. The bottom line is that it is hard to see how *either* set of warring speculations could ever decisively carry the day, and certainly neither side can legitimately claim the imprimatur of *science*.

Yet the argument many METI advocates seem to be making is that *their* speculations are much more reasonable than those of their opponents. Even if we do not concern ourselves at all with *public* opinion and only focus on what experts say, this strains credulity. The fact of the matter is that many respected scientists have publicly expressed concern, even alarm, about the possible consequences of METI. Indeed, a recent petition ("Regarding Messaging," 2015) calling for a moratorium on METI transmissions pending "a worldwide scientific, political and humanitarian discussion" was signed by thirty-five distinguished space scientists. Michaud (2018) recently summarized the worries of an even broader group of scientists, including the physicist Stephen Hawking; Nobel laureate George Wald; biologist Michael Archer; physicist George Baldwin; and astronomers Robert Jastrow, Eric Chaisson, and Zdenek Kopal, among others. He concludes that the best-case scenario for METI relies on personal belief and preference, not proven facts. He is right. In addition, continuing to assert that the pro-METI speculations are somehow inherently superior, despite widespread *expert* dissent, seems like an act of faith, unmitigated hubris, or both.

## The Appeal to Nice Aliens

There is one twist on this line of reasoning that might initially seem more defensible—an argument I label "the appeal to nice aliens." The argument postulates that any aliens we encounter would be much more "advanced" and that this surely applies to their *social* development as well as their technology (Harrison, 2000). Lemarchand (2000) puts this in an especially stark fashion when he says, "it is impossible to have superior science and technology, and inferior morals." The basic idea is that any civilization that has survived for thousands of years, much less millions, must have overcome its aggressive tendencies long ago or they would have destroyed themselves. Thus, we can safely assume that they would exhibit a benevolent tendency toward other beings like themselves, including humans. While I think this argument may have potential, it often takes on the (undeserved) character of an obvious truth despite a number of serious problems.

For one thing, there are difficulties with using a term like "advanced" in such a blanket fashion, as it conflates a number of different, and quite controversial, claims. It is one thing to extrapolate *technological* development from humans to aliens, since we have good reason to think that the basic facts of science are the same for everyone (e.g., there is only one periodic table of elements and only so many ways to refine metals). But it is another thing entirely to postulate *ethical* convergence arising out of social and political dynamics we do not even understand well *in humans*. For example, the very idea that social development is *linear* is rejected by most social scientists (Traphagan, 2017). But even if we accept that social development has a natural endpoint, what evidence do we have that benevolence is it? Our limited data shows that civilizations on Earth can persist, even thrive, for very long periods of time despite extremely aggressive tendencies. They can often mitigate the *internal* negative effects of aggression on their society by drawing a psychological distinction between "us" and "them," allowing them to persecute outsiders while acting benevolently toward those in their own society. Indeed, this is a distressingly easy feat in humans. Phillip Zimbardo, the investigator behind the infamous Stanford Prison Study, argues that this ability is *the* central feature of all groups that have committed systematic atrocities like the Holocaust (Zimbardo 2007). It thus seems perfectly possible, given the data we have, for an alien civilization to feel no moral obligation whatsoever toward "others." Of course, there could still be *external* effects of aggression a nasty alien civilization would have to contend with, most notably causing one's neighbors to retaliate in kind. But these might be negligible, for example if the civilization were so technologically superior to those neighbors that retaliation from these "primitives" was not a real concern. After all, this exact scenario has played out many times in our own history.

Thus, it is far from clear that *only* societies that are peaceful and benevolent would be sustainable, even over cosmic time scales. And even if the vast majority of alien civilizations are benevolent, it only takes *one* to pose an existential threat to humanity. This is another feature of the speculative battle that seems under-appreciated. Even if we grant that the METI advocates' speculations offer vastly superior predictions of alien psychology and behavior in general, they are still *probabilistic.* There will almost certainly be exceptions to the rule, and thus we still should worry that perhaps humanity had the hard luck to be born in a bad cosmic neighborhood.

The bottom line is that, while this kind of argument may initially look less speculative because it draws on inferences from human history, that impres-sion fades once we realize that those inferences are highly suspect. Any intel-lectually honest scientist must admit that the error bars are simply too high to make predictions here with any confidence. Or, as Freeman Dyson (1964) put it:

> Our business as scientists is to search the universe and find out what is there. What is there may conform to our moral sense or it may not. . . . It is just as un-scientific to impute to remote intelligences wisdom and serenity as it is to im-pute to them irrational and murderous tendencies.

## Other Harms

For the sake of argument, though, let us set thoughts of belligerent aliens aside and assume that advanced aliens are invariably benevolent and not one of them would ever dream of harming humans. Can we stop worrying? Unfortunately, no. For one thing, benevolent aliens could still wreak *inadvertent* havoc on humanity. It is not difficult to find situations on Earth where contact with "ad-vanced" societies has ended badly for the "primitives," even when there is no nefarious intent or overt aggression (Traphagan, 2018). It seems plausible, even likely, that the mere *awareness* of another, incomprehensibly advanced civi-lization would result in existential panic. After all, what would be the point of human striving when it has all been done already, and far better than we puny humans ever could?[18] How many scientists would welcome demotion from in-trepid investigators pushing the boundaries of knowledge to remedial students attempting to fathom advanced alien technology? All in all, it seems safe to as-sume that any robust communication with aliens would lead to the collapse of *some* important aspects of Earth culture, though in ways and with impacts we can only dimly foresee. That alone suggests we should be very careful what we wish for.

There is also a huge blind spot in virtually every discussion of METI: they always focus on risks *to humans*. But couldn't contact with an alien society have a deleterious impact *on them*? Michaud (2013) makes what should be an obvious point, but one that is all too often overlooked: aliens might well look upon us as a deeply mysterious, and possibly threatening, surprise.[19] Traphagan (2017) argues our blind spot here is due to several unconscious but highly suspect assumptions about what an advanced society must be like. For example, we tend to assume that any advanced society would have to be (a) so stable that nothing we do could have much of an impact at all; (b) politically unified, such that one faction could not use contact to further its own internal agenda; and (c) receptive to contact and the revelations that come with it. But it is just these kinds of assumptions that cause social scientists who study contact situations on Earth to recoil in horror. They have learned humility the hard way, with a long history of well-intentioned sociologists and anthropologists inadvertently harming the societies they seek to study, usually because they failed to check their assumptions at the door. And Traphagan goes on to illustrate specifically just how contact might wreak havoc on an alien world with an advanced, but very different, society. Of course, maybe the social scientists are worrying too much. Maybe. But since they *are* the relevant experts here, perhaps we should at least include them in our discussions.

In the final analysis, all this speculation concerning the risks of contact is, well, speculation. We simply are not in a position at present to assess with any confidence either the relevant probabilities of the different outcomes or the nature and magnitude of their impact. A reasonable person has to admit the possibility of both risks and rewards, but it is entirely unclear what to do with such unspecific information. All we can know for certain is that contact with a civilization much more advanced than ours would be momentous, probably for both civilizations, which is sufficient reason to think carefully before we act. And, if it is morally problematic for a small group of METI advocates to make decisions on behalf of humanity, how much worse is it for them to make decisions on behalf of an entire alien civilization about which they know nothing?

## Risk Per Se Versus Acceptable Risk

If we have an ethical duty to consult those impacted by any decision that carries real risk, then we have to consult the public concerning METI and actually *listen* to what they say. Experts can and should educate the public about risks they do not understand, but we should not *impose* expert opinion on them as long as they meet a basic standard of competence. The question is whether, having been informed of the relevant risks and rewards, including an honest assessment of any scientific uncertainty, they are willing to *accept* those risks. This brings us to a distinction that is all too often glossed over in METI discussions: *objective* risk versus *acceptable* risk.[20] Even with a risk that is objectively quite low, it might be perfectly rational to refuse to accept it, especially if the negative consequences are sufficiently dire.

Allow me to indulge in a philosopher's thought experiment to make this point clear. Suppose, through some bizarre chain of events, you are presented with a button. You have all the information you need to objectively assess the possible outcomes of pressing the button as follows:

1. Nothing happens (98.9% objective probability).
2. $1,000 is added to your bank account (1% objective probability).
3. Everyone you have ever loved dies a painful death (0.1% objective probability).

I suspect few would press the button under these conditions and almost no one would press it repeatedly. Some would surely refuse to press it no matter how we altered the magnitude of reward or the relative probabilities of the outcomes, as long as the dire outcome were still on the table. Of course, we could debate the various permutations endlessly, but this only underscores my point: calculations concerning what risks we find acceptable are *much* more complex than calculations of objective risk. It is hard to see how we could come up with a scientifically respectable theory to determine what people *should* decide, since this would require (at the very least) a theory of cultural and psychological dynamics that is far beyond our present capabilities.[21] The bottom line for our present purposes is that it does not seem *irrational* for someone to refuse, in principle, to *ever* risk such a truly dire outcome[22].

When it comes to METI, we simply cannnot rule out a true existential risk to humanity—indeed, for all life on Earth. In other words, there is clearly *some* chance METI could result in what would literally be, from a terrestrial perspective anyway, the *worst conceivable outcome*. This puts the disagreement between experts advocating METI and a putatively reluctant public in a very different light. Rather than the public simply being ignorant about objective risks, they may just value the relevant risks and rewards differently. And, since there is no way to specify in a scientifically respectable fashion how they *should* value these things, we should allow them the freedom to choose for themselves.

It might be argued that even this existential risk is really not so different from others humanity has taken in the past.[23] For example, there were worries that the detonation of the first atomic bomb might cause a chain reaction, burning away Earth's atmosphere (Buck, 1959). More recently, there were fears that operation of the Large Hadron Collider could create a black hole that would consume the planet (Siegel, 2016). But these cases are quite different from METI in two critical ways. First, there was a much clearer benefit to be gained from carrying out the experiments. Second, the debate about the risks was necessarily a *scientific* one, since whether the risks were even *real* could only be answered by theoretical physicists. The general public thus could not meet even a minimal standard of competence. But neither of these considerations apply to METI.

## The Problem of Future Generations

There is one final consideration I wish to consider briefly: the problem of future generations. Although I am not aware of anyone discussing this in detail in the context of METI (though see Brin [2013] and Goetz [2016]), it is a nexus of lively debate in the environmental ethics community (Lawrence, 2014; Nolt, 2017). Simply put, the question is: How we should value the desires and welfare of humans who will exist in the far future? Our intuitions do not speak with one voice here. On the one hand, future humans will presumably be just like us in every morally relevant way. Thus, we must admit in the abstract that they deserve the same ethical consideration we give humans who happen to be alive at present. On the other hand, if we equate the interests of future generations with those of present-day humans, it becomes extremely difficult to justify a whole host of actions that are important to us right now. In particular, every decision to utilize a nonrenewable resource risks robbing future generations of whatever opportunities that resource might have afforded them. In addition, since there will likely be far more humans in the future than at present, assigning their interests equal value could easily swamp out every short-term benefit present-day humans seek to attain. For practical reasons if nothing else, it seems we have no choice but to *discount* the interests of future generations relative to present-day humans.[24]

When it comes to METI, such discounting creates an odd sort of conflict of interest. It is not that the *benefits* of METI unduly influence our decision-making, since given the time frame involved there is almost certainly no benefit to present-day humans (other than METI advocates). Instead, here the *risks* of METI will be born solely by future generations. Suppose we admit the following possibility: 1,000 years from now, an alien civilization detects our METI signals and responds by sending a huge rock hurtling toward Earth in order to wipe out the competition. How much should we really be concerned about this, given that it will not occur for at least 2,000 years? Psychologically, our worry seems to decrease with the time interval—people would worry far more if the rock were due in 50 years and more still if it were due next week. It is even possible, given the very, very long time scales involved, to make an argument to the effect that we should allow METI advocates to exercise *their* freedom of choice and do whatever they like, since no harm will come to those alive at present.

This is not just a theoretical dispute about the relative moral value of present versus future humans, as it also involves complex empirical questions that we are not in a position to answer with any confidence. After all, it is clearly possible that what seems like a massive problem today could be easily resolved with future technology. If so, imposing morally significant costs on present-day humans in order to spare future generations might turn out to have been both unnecessary and immoral, at least in hindsight. But future technological progress cuts both ways, as we could also argue that it makes no sense to stick our grandchildren with a fait accompli in the form of a METI signal that cannot be recalled, since they will likely be much better positioned to decide whether and how to attempt contact.

Yet again, one's position on this depends on subjective perspectives such as the level of technological optimism one is willing to entertain. But the mere *possibility* of avoiding risks does not seem to be sufficient to *ignore* them. For example, an oil industry executive might argue that we should not impose stringent controls on carbon emissions now, since it is quite likely we will find a much simpler and cheaper fix for global warming eventually.

Unfortunately, I cannot even begin to suggest a solution to this conundrum. I will say, however, that it is critically important for us to be *consciously aware* of it. We should not allow a tacit lack of concern for future generations to bias our thinking in favor of METI (or any other technology). Perhaps this is a legitimate reason to favor METI, perhaps not. But we will never know unless we *explicitly* consider the arguments and subject them to careful critical scrutiny.

## Conclusion

In this chapter, I have tried to avoid discussion of very specific empirical arguments requiring complex technical discussions. Instead, I propose a general ethical framework that emphasizes the need for consensus. The public is sufficiently qualified to have opinions worthy of consideration, since there are perfectly legitimate grounds for concern and how one values acceptable risk seems inherently subjective. So even if we had scientific consensus as to how we should act—and we certainly do not—that alone is not sufficient reason to impose a risk on the public they may be unwilling to take.

My argument is not that METI is a bad idea in general—just that we have not yet thought it through carefully and, until we do, we should err on the side of caution. It seems undeniable that METI poses *some* risk even if we all agree that we lack the information needed to characterize it in a scientifically respectable way. Speculation is a temptation for both sides, but it ultimately only identifies raw possibilities without shedding much light on how to make a truly informed decision. We are all of course free to have personal opinions about what aliens are like and how to approach METI in consequence, but we must keep firmly in mind that they are just that—*personal opinions*. In particular, when a scientist advocates METI, she is speaking not with the authority of science but as just one human being among many.

Fortunately, there are a number of perfectly reasonable steps we can take now to insure we do not act in haste. Rather than simply allowing anyone with access to the necessary resources do whatever they wish, we need to involve public policy, social science, humanities, and other fields of expertise to develop explicit best practices and then work to enshrine these in soft law. Exactly what such agreements should look like is a debate for another time, but some obvious first steps have been floated repeatedly by many experts in the field (e.g., Billingham and Benford, 2014; Brin, 2014; Gertz, 2016), though with surprisingly little uptake so far. These include:

1. Permanently archiving basic information concerning messages that have been sent in a publicly accessible format. It is hard to make good policy for the future when it is unclear precisely what has already happened.
2. Classifying the different kinds of transmissions in terms of their risk profile so that we can tailor regulation appropriately. Transmissions that seem to pose extremely low or extremely long-term risk, such as those directed at very distant stars or that do not exceed the power of unintentional transmissions (EM leakage), might even be judged harmless—but only *after* these facts are established.
3. Requiring that any new METI transmission be explicitly proposed and subject to some level of review. Such review should be conducted by a group that includes voices outside the space science community and could operate something like existing Institutional Review Boards.
4. Collecting better information about public attitudes towards METI. If it turns out that the public is actually *in favor* of METI, or even indifferent, then many of the concerns expressed here become moot. Since this is an empirically tractable question, there is no justification for continuing to either assume or ignore public opinion.

Unfortunately, there is no simple solution to the METI question—as I often tell my students, "The most important thing to know about ethical issues is that they are always more complicated than you initially think." The appropriate response is to make sure we take the time to think carefully *before* we act—especially if there is a possibility of a truly disastrous consequence. Until we can thoroughly evaluate our options, it is probably best to err on the side of caution.

## Acknowledgements

This chapter has benefitted enormously from the critical feedback of many experts who were very generous with their time and ideas—in particular: James Benford, David Brin, Michael Michaud, Paul Quast, and John Traphagan. Any inadequacies that remain despite their best efforts are entirely mine.

## Notes

1. For a critical review of many of the common METI arguments, see Gertz (2016).
2. To take just one example, METI advocates often argue that the risk is very low because we have surely already been detected. They then engage in a complex technical argument to support this claim while failing to grasp that any victory they secure through these means will surely be pyrrhic: what is the point of taking *any* risk if nothing new will be accomplished (Smith, 2018)?
3. I have always suspected that people, even scientists who make the case for METI, have a very difficult time thinking and acting as if there are *truly* aliens out there. These debates thus have a certain unreal, even humorous, quality to them that is hard to dispel through purely rational analysis.
4. It should be noted that I serve on both METI International's Advisory Council and the Interstellar Beacon's Advisory Board, so both organizations are open, at least in theory, to dissenting opinion.
5. Michael Michaud suggests a corollary to Clarke's observation, to the effect that such technology might even be indistinguishable from *nature*. For example, Liu's (2016b) Three Body trilogy envisions an alien civilization interfering with human development by altering the results of our scientific experiments in subtle ways we cannot detect.
6. To be sure, there are important cultural variations in attitudes towards autonomy (see Traphagan, 2013).
7. This does not imply that all ethical reasoning is merely opinion or social mores, but a discussion of ethical relativism is far beyond the scope of this chapter (though for an excellent concise discussion of the issue, see Rachels, 2003, chapter 2).

8. I certainly do not mean to suggest that there are *never* good moral grounds for a given decision, even one that goes against a patient's wishes. In principle, certain sorts of patient "reasoning" might give us sufficient grounds to override their wishes. These issues are far too complex to take on in this chapter, where it suffices to note that the public might have perfectly legitimate grounds to worry about METI.

9. This does not mean that those advocating METI are *themselves* immoral, anymore than a physician who takes too much on herself through good intentions is immoral as opposed to misguided. All too often, though, people with the very best of intentions nevertheless act immorally. In the words of Aleksandr Solzhenitsyn (1974, p 168): "If only there were evil people somewhere insidiously committing evil deeds, and it were necessary only to separate them from the rest of us and destroy them. But the line dividing good and evil cuts through the heart of every human being."

10. An interesting example of this attitude can be seen in the creationism documentary, *A Flock of Dodos* (Olson, 2007). In one scene, a group of biologists complain to each other over a game of poker about the stupidity of the public, implying that scientists should be left alone to decide what is best.

11. An initial enthusiasm for paternalism among medical students tends to diminish quickly when they are forced to consider what is actually involved—forcibly sedating a patient for an operation, for example.

12. And also, ironically, because the academic environment allows them greater personal freedom to pursue the research they personally find compelling.

13. This can be a real problem in the creationism debate, with which I am intimately familiar. For example, the denigration of religion that Richard Dawkins and other neo-atheists (Dawkins 2008; Hitchens, 2007) often adopt in defense of science education can be counterproductive, as it causes believers to simply stop listening (Smith, 2011).

14. Especially noteworthy is the famous Asilomar conference on recombinant DNA in 1975, where a group of 140 experts drew up voluntary consensus guidelines to guide genetic research, carefully balancing safety concerns with the need to preserve cutting edge research.

15. Previous attempts to get the International Academy of Astronautics permanent committee on SETI to adopt an explicit policy with regards to METI were unsuccessful, though precisely why is a matter of debate (Brin, 2013, 2014). On the other hand, the United Nations Committee on the Peaceful Uses of Outer Space (n.d.) has a Declarations of Principles that might be at least a promising start. See Newman (2019) for a discussion of soft laws in the context of astrobiology.

16. There seems to be no great obstacle to filling this lacuna, and indeed I am currently collecting data in a small-scale survey of student attitudes concerning METI to begin that process. See also Schwartz (2019) for a summary of existing public survey data on related questions in space science.

17. This is why double-blinded experiments are the gold standard in science—they are designed not to prevent conscious fabrication and falsification but to guard against *unconscious* bias.

18. Stephen Baxter's (2003) *Manifold: Time* imagines a universe filled with super-advanced, galaxy-spanning civilizations and their mega-engineering projects—a prospect many of my students find extremely depressing.

19. During a role-playing simulation of first contact at the CONTACT 2018 conference, one team seemed genuinely surprised that the mere silent approach of their alien vessel could be interpreted as a potential *threat* by the team (on which I served) controlling the human ship. But whatever else it might be, first contact is surely a high-stakes game with unclear rules, and thus we should *expect* even supremely peaceful aliens to be extremely cautious. Wars on Earth often began with just such a mismatch of expectations.

20. By "objective risk" I mean (loosely) a risk that can be adequately assessed by scientific methods alone. Whether there even *is* such a thing as objective risk is a complex question beyond the scope of this chapter, so I will assume without argument that this concept makes sense in order to give the benefit of the doubt to METI proponents. Of course, if there is no such thing as objective risk, their position would be even less tenable.

21. Even this assumes that moral decision-making is simply a subset of social dynamics, which is an extremely controversial claim among professional ethicists. If there are other, uniquely ethical, aspects to the problem (as there likely are), the calculation becomes even more intractable.

22. See Korbitz (2014) for more discussion of risk aversion in the context of METI.

23. Though it could also be argued that an existential risk to all life on Earth is actually *different in kind* from other, more prosaic, risks our decision procedures are designed to consider.

24. An entire subdiscipline of economics (environmental economics) is devoted to investigating the nature of this discount and how it impacts decision-making (Cropper and Oates, 1992).

# References

American Medical Association (2016) "Code of Medical Ethics Consent Opinion 2.1.1: Informed Consent." https://www.ama-assn.org/delivering-care/informed-consent

Barras, Colin (2008) "First Space Ad Targets Hungry Aliens." *Science News.* https://www.newscientist.com/article/dn14130-first-space-ad-targets-hungry-aliens/

Baum, Seth (2016) "The Ethics of Outer Space: A Consequentialist Perspective." In J. Schwartz and T. Milligan (Eds.), *The Ethics of Space Exploration* (Space and Society). Cham, Switzerland: Springer.

Baxter, Stephen (2003) *Manifold: Time.* New York: Del Ray/Penguin Books.

Beauchamp, Tom (2012) *Principles of Medical Ethics.* Oxford: Oxford University Press.

Billingham, John, and James Benford (2014) "Costs and Difficulties of Large-Scale METI, and the Need for International Debate on Potential Risks." *Journal of the British Interplanetary Society,* 67: 22.

Brin, David (2013) "Shouting at the Cosmos." Davidbrin.com. http://www.davidbrin.com/nonfiction/shouldsetitransmit.html

Brin, David (2014) "The Search for Extra-Terrestrial Intelligence: Should We Message ET?" *Journal of the British Interplanetary Society*, 67(1): 8–16. http://www.davidbrin.com/nonfiction/meti.html

Brin, David et al. (2014) "METI: Rebuttals." *Journal of the British Interplanetary Society*, 67: 38–43.

Buck, Pearl (1959) "The Bomb—The End of the World?" *The American Weekly*, March 8.

Cropper, Maureen, and Wallace Oates. (1992) "Environmental Economics: A Survey." *Journal of Economic Literature*, 30(2): 675–740.

Crowe, Micheal (2008) *The Extraterrestrial Life Debate, Antiquity to 1915: A Source Book.* Notre Dame, IN: University of Notre Dame Press.

Dawkins, Richard (2008) *The God Delusion.* New York: Mariner.

Denning, Kathryn (2010) "Unpacking the Great Transmission Debate." *Acta Astronautica*, 67: 1399–1405.

Dick, Steven (1982) *Plurality of Worlds: The Origins of the Extraterrestrial Life Debate from Democritus to Kant.* Cambridge: Cambridge University Press.

Duner, David, G. Holmberg, J. Parthemore, and E. Persson (2013) *The History and Philosophy of Astrobiology: Perspectives on Extraterrestrial Life and the Human Mind.* Cambridge: Cambridge Scholars.

Dyson, Freeman (1964) Letter. *Scientific American*, 210(4): 8–15.

Elliot, Deni, and J. Stern (Eds.) (2011) *Research Ethics.* Dartmouth: Institute for the Study of Applied and Professional Ethics.

Frank, A., and W. T. Sullivan (2016) "A New Empirical Constraint on the Prevalence of Technological Species in the Universe." *Astrobiology*, 16(5). https://doi.org/10.1089/ast2015.1418

Gertz, John (2016) "Reviewing METI: A Critical Analysis of the Arguments." *arXiv*,1605.05663.

Gilster, Paul (2017) "METI: A Longer Term Perspective." Centauri Dreams. https://www.centauri-dreams.org/2017/12/11/meti-a-longer-term-perspective/

Goldschmidt Conference (2018) "Water-Worlds Are Common." Phys.org. https://phys.org/news/2018-08-water-worlds-common-exoplanets-vast-amounts.html

Guzman, Andrew, and Timothy L. Meyer (2010) "International Soft Law." *Journal of Legal Analysis*, 2(1): 171–225.

Hall, Daniel, E., A. V. Prochazka, and A. S. Fink (2012) "Informed Consent for Clinical Treatment." *Canadian Medical Association Journal*, 184(5): 533–540. doi:10.1503/cmaj.112120

Harrison, Albert (2000) "The Relative Stability of Belligerent And Peaceful Societies: Implications for SETI." *Acta Astronautica* 46(10): 707–712.

Hawkes, Alison (2017) "Kepler Has Taught Us That Rocky Planets Are Common." Phys.org. https://phys.org/news/2017-06-kepler-taught-rocky-planets-common.html

Hitchens, Christopher (2007) *God Is Not Great.* New York: Twelve/Hachette Book Group.

Howell, Elizabeth (2017) "How Many Stars Are in the Universe?" Space.com. https://www.space.com/26078-how-many-stars-are-there.html

Institute of Medicine (2009) *US Committee on Conflict of Interest in Medical Research, Education, and Practice* (B. Lo and M. J. Field, Eds.). Washington, DC: National Academies Press https://www.ncbi.nlm.nih.gov/books/NBK22944/

Johnson, Steven (2017) "Greeting, ET (Please Don't Murder Us)." *The New York Times*, June 28. https://www.nytimes.com/2017/06/28/magazine/greetings-et-please-dont-murder-us.html

Korbitz, Adam (2014) "Toward Understanding the Active SETI Debate: Insights from Risk Communication and Perception." *Acta Astronautica*, 105: 517–520.

Lawrence, Peter (2014) *Justice for Future Generations*. Cheltenham, UK: Edward Edgar.

Lemarchand (2000) "Speculations on the First Contact: Encyclopedia Galactica or the Music of the Spheres?" In Allen Tough (Ed.), *When SETI Succeeds: The Impact of High-Information Contact*. Washington, DC: Foundation for the Future.

Liu, Cixin (2016b) *The Dark Forest*. New York: Tor Books.

Liu, Cixon (2016b) *The Three Body Problem*. New York: Tor Books.

Michaud, Michael (2007) *Contact with Alien Civilizations: Our Hopes and Fears About Encountering Extraterrestrials*. New York: Copernicus Books.

Michaud, Michael (2013) "The Ethics of Unintended Consequences." Centauri Dreams. https://www.centauri-dreams.org/2013/10/18/the-ethics-of-unintended-consequences/

Michaud, Michael (2018) Unpublished manuscript.

Nielsen, J. N. (2013) "SETI, METI, and Existential Risk." Centauri Dreams. https://www.centauri-dreams.org/2013/11/08/seti-meti-and-existential-risk/

Newman, Christopher (2019) "Unnatural Selection or the Best of Both Worlds: The Legal and Regulatory Ramifications of the Discovery of Alien Life." In Kelly Smith and Carlos Mariscal (Eds.), *Social and Conceptual Issues in Astrobiology*. Oxford: Oxford University Press.

Nolt, John (2017) "Future Generations in Environmental Ethics." In S. Gardiner and A. Thompson (Eds.), *Oxford Handbook of Environmental Ethics*. Oxford: Oxford University Press. doi:10.1093/oxford/9780199941339.013.28

Olson, Randy (2007) *A Flock of Dodos: The Evolution-Intelligent Design Circus*. Prairie Starfish Productions.

Peters, Ted (2019) "Should We Send Messages to Extraterrestrials?" [Special Issue] *Theology and Science*, 17(1).

Peters, Ted (Ed.) (2017) "Stephen Hawking on METI" [Special Issue] *Theology and Science*, 15(2).

Presidential Commission. (1986) "Report of the Presidential Commission on the Space Shuttle Challenger Accident." https://history.nasa.gov/rogersrep/genindex.htm

Quast, Paul (2018) "A Profile of Humanity: The Cultural Signature of Earth's Inhabitants Beyond the Atmosphere." *International Journal of Astrobiology*. https://doi.org/10.1017/S1473550418000290

Rachels, James (2003) *The Elements of Moral Philosophy*, 4th ed. New York: McGraw-Hill. http://vulms.vu.edu.pk/Courses/ETH202/Downloads/The%20Elements%20of%20Moral%20Philosophy.pdf

"Regarding Messaging to Extraterrestrial Intelligence (METI)/Active Searches for Extraterrestrial Intelligence (Active SETI)."(2015) http://setiathome.berkeley.edu/meti_statement_0.html

Rose, S. L., J. Highland, M. T. Karafa, and S. Joffe (2017) "Patient Advocacy Organizations, Industry Funding, and Conflicts of Interest." *JAMA Intern Medicine*, 177(3): 344–350. doi:10.1001/jamainternmed.2016.8443

Schwartz, James (2019) "Myth-Free Space Advocacy Part IV: The Myth of Public Support for Astrobiology." In Kelly Smith and Carlos Mariscal (Eds.), *Social and Conceptual Issues in Astrobiology*. Oxford: Oxford University Press.

Shostak, Seth (2015) "Should We Keep a Low Profile in Space?" *The New York Times*, May 27.

Shostak, Seth (2017) "Humankind Just Beamed a Signal at Space Aliens. Was That a Bad Idea?" SETI Institute. https://www.seti.org/humankind-just-beamed-signal-space-aliens-was-bad-idea

Siegel, Ethan (2016) "Could the Large Hadron Collider Create and Earth Killing Black Hole?" *Forbes*, March 11. https://www.forbes.com/sites/startswithabang/2016/03/11/could-the-lhc-make-an-earth-killing-black-hole/#789e0f892ed5

Smith, Kelly C. (2011) "Foiling the Black Knight." *Synthese*, 178(2): 219–235.

Smith, Kelly C. (2017) "Hawking and the METI Hawks: Right for the Wrong Reasons." *Theology and Science*. 15(2): 147–149.

Smith, Kelly C. (2018) "A(nother) Cosmic Wager: Pascal, METI and the Barn Door Argument." *Science and Theology*, 17(1): 29–37. https://doi.org/10.1080/14746700.2018.1557393

Solzhenitsyn, Aleksandr (1974) "The Bluecaps." In *The Gulag Archipelago, Part I: The Prison Industry*. New York: Collins.

Soter, Steven (2005). "SETI and the Cosmic Quarantine Hypothesis." *Astrobiology Magazine*. https://www.astrobio.net/alien-life/seti-and-the-cosmic-quarantine-hypothesis/

Suroweicki, James (2005) *The Wisdom of Crowds*. New York: Anchor.

Tipler, Frank (1981) "A Brief History of the Extraterrestrial Intelligence Concept." *Quarterly Journal of the Royal Astronomical Society*, 22: 133.

Traphagan, John (2013) *Rethinking Autonomy: A Critique of Principlism in Biomedical Ethics*. New York: SUNY Press.

Traphagan, John (2017) "Do No Harm? Cultural Imperialism and the Ethics of Active SETI." *Journal of the British Interplanetary Society*, 70: 219–224.

Traphagan, John (2018) "Cargo Cults and the Ethics of Active SETI." *Space Policy*, 46: 18–22. https://doi.org/10.1016/j.spacepol.2018.04.001

United Nations Office for Outer Space Affairs. (n.d.) "Space Law Treaties and Principles." http://www.unoosa.org/oosa/en/ourwork/spacelaw/treaties.html

Vakoch, D. (2016) "In Defense of METI." *Nature Physics*, 12: 890.

Vakoch, Douglas (2014) (Ed.) *Archaeology, Anthropology, and Interstellar Communications* (NASA SP-2013-4413). http://www.nasa.gov/ebooks

Zaitsev, Alexander (2006) "Messaging to Extra-Terrestrial Intelligence." *arXiv:physics*, 0610031.

Zimbardo, Phillip (2007) *The Lucifer Effect: Understanding How Good People Turn Evil*. New York: Random House.

# PART V
# SOCIAL AND LEGAL ISSUES IN ASTROBIOLOGY

# 14

# Earth, Life, Space

## The Social Construction of the Biosphere and the Expansion of the Concept into Outer Space

*Linda Billings*

Philosophers and scientists have been speculating about the existence of other worlds and life on other worlds since the days of ancient Greece. Scientists have been studying the origins and evolution of life on Earth since the beginnings of what we call science. And scientists have been searching in earnest for evidence of extraterrestrial life since the beginning of the space age.

In this chapter I explore whether and how the scientific search for evidence of extraterrestrial life, in the solar system and beyond, has affected our conception of the terrestrial biosphere—and vice versa—and extended the concept of "biosphere" into outer space, expanding the search for habitable environments in the solar system and beyond.[1] My theoretical framework here is social constructivism—the idea that people construct knowledge of the world by means of shared assumptions about reality.[2]

What is the concept of a biosphere? Austrian geologist Eduard Suess originated the term in 1875, describing Earth's biosphere as the area of the planet that supports life. In *Das Antlitz der Erde (The Face of the Earth)*, Suess wrote: "One thing seems to be foreign on this large celestial body consisting of spheres, namely, organic life. But this life is limited to a determined zone at the surface of the lithosphere."[3] Before the beginning of the space age, scientists were building on this concept of the biosphere to establish the field of ecology. In the 1920s, Russian geochemist Vladimir Vernadsky defined ecology as the (interdisciplinary) study of the biosphere.[4] Since then, and especially since the mid-20th century, the concept of the terrestrial biosphere has expanded, as life has been found thriving in the atmosphere,[5] in the deepest parts of the oceans,[6] and in the deep subsurface Earth.[7] It has become the idea of an inhabited planet as a living, changing system. The science of ecology has developed in tandem with the development of an increasingly complex conception of a biosphere.

Thanks to advances in Earth and space science over the past fifty years, the idea that Earth and all of its life have been co-evolving since life began is well established in the science community. In the 21st century, the study of the origins and

Linda Billings, *Earth, Life, Space*. In: *Social and Conceptual Issues in Astrobiology*. Edited by: Kelly C. Smith and Carlos Mariscal, Oxford University Press (2020). © Oxford University Press. DOI: 10.1093/oso/9780190915650.001.0014 evolution of life on Earth and in the universe, of Earth and its sister planets, and of the universe itself are intricately intertwined. With a deeper understanding of the history and nature of the terrestrial biosphere, the community of scientists engaged in space science and exploration recognizes the possibility of other biospheres beyond Earth, in our solar system or in extrasolar planet systems.

From my biased perspective, as a long-time member of the exobiology/astrobiology community,[8] I am convinced that the search for life elsewhere has affected the way we—experts and non-experts alike—think about our home planet and the life on it. Conversely, the way we think about our home planet and the life on it is affecting the way we are going about the search for extraterrestrial life. A thorough review of how this change has occurred—by examining scientific publications, media reports, and other documentation— would be useful. However, such a review is beyond the scope of this chapter. Instead I consider some of the ways in which I believe this change has occurred. While I do not ignore the contributions of space-based Earth observations and human space flight to a change in conceptions of the terrestrial biosphere, I focus on the contributions of astrobiology, addressing research into the origins and evolution of life on Earth, the search for extraterrestrial life, and, to some extent, the search for potentially habitable extrasolar planets (exoplanets).

## Introduction

What follows is a brief and admittedly idiosyncratic review of the parallel and intersecting histories of the study of life on Earth, space exploration and the search for extraterrestrial life, and the deepening understanding of our own biosphere. I briefly address the state of the art in origins of life research, discuss the broadening understanding of the diversity of life on Earth, review the parallel developments of astrobiology and environmentalism, report on the current search for extraterrestrial life, consider conflicting conceptions of biospheres as environments to explore or exploit, and examine social and conceptual issues in astrobiology.

No one knows exactly how terrestrial life began, or precisely where or when, and as the study of the origins and evolution of life expands, theories about origins continue to proliferate.[9] At the same time that astrobiologists are deeply engaged in research into how life began and evolved on Earth, hoping to gain insights into whether and how life might have begun and evolved beyond Earth, they have been considering the possibility that extraterrestrial life could be completely different from terrestrial life—for instance, not carbon-based, using a liquid other than water as a solvent, adapted to environmental conditions more extreme than any found on Earth—so different that it might be unrecognizable to us.[10]

Some forms of life on Earth—that is, carbon-based cellular life—can survive in virtually all known terrestrial environmental extremes—nuclear radiation, permafrost, temperatures above the boiling point of water, the deep subsurface Earth, around deep-sea hydrothermal vents, without sunlight, and so on. Wherever humans or their technological counterparts have gone on Earth, they have found life. Most of these so-called extremophilic life forms are microbes. It is now known that microbial life accounts for a significant portion of the biomass on Earth (though estimates vary), much of it beneath the surface of Earth.[11],[12] This relatively new knowledge of the extent and diversity of microbial life in the terrestrial biosphere has inspired astrobiologists to explore extraterrestrial environments not considered as targets for life detection just a few decades ago as possibly inhabited by microbial life.

At the same time that research into the origin, evolution, and distribution of life as we know it is revealing that life is highly adaptable and resilient, these same lines of research are helping to reveal how life and its environment are deeply interdependent. Some key lines of research in this area—such as understanding the timing and mechanics of the rise of oxygen in the atmosphere of early Earth; the role of the environment in the production of organic molecules; and the co-evolution of climates, atmospheres, interiors, and biospheres—are improving understanding of the evolution of habitability and life on Earth and prospects for the evolution of habitability and life elsewhere, contributing to understanding of global climate history and evolution and at the same time complicating the further study of life, terrestrial or otherwise.

Developed in depth over the past fifty years or so, this modern scientific understanding of the highly interdependent nature of life and its environment—their co-evolution—is changing the way that both experts and non-experts think about the terrestrial biosphere, the place of humanity in it, the possibility of life elsewhere, and even the idea that humans could live on other planets. And over the past twenty years or so, exoplanet searching has introduced new terms to the scientific and public discourse: Earthlike planets, Earth-sized planets, terrestrial or "Terran" planets, rocky planets, super-Earths.[13] It could be argued that the quest to find another "Earth" is a major driver of exoplanet research, whether explicitly or implicitly.

## The Beginnings (of Astrobiology and Environmentalism)

In 1959, NASA funded its first exobiology project, an instrument intended to detect evidence of biological activity on Mars.[14] In 1960 NASA established an exobiology research program to fund studies of the origin and evolution of life on Earth and the search for evidence of life beyond Earth.

A few years later, in 1966, self-described "ecopragmatist" Stewart Brand—founder and editor of the *Whole Earth Catalog* and *CoEvolution Quarterly*—launched a public-awareness campaign aimed at prodding the space community to look at the whole Earth from space. "Why Haven't We Seen a Photograph of the Whole Earth Yet?" Brand was asking, in hopes "that it would stimulate humanity's interest in its mega-habitat" (Brand, 2007).

In their 2018 book, *The Environment: A History of the Idea*,[15] historians Paul Warde, Libby Robin, and Sverker Sorlin explain how the contemporary concept of the environment is only about seventy years old. After World War II, they write, "a new narrative about the effect of people's behavior [on the environment] emerged . . . 'The environment' was at risk, and humans were the cause." This narrative turn

> reversed the usage of the term "environment" that had prevailed for the previous 70 years: the evolutionary idea that a species' or individual's environment could explain its characteristics. . . . The timing of this conceptual shift was not accidental. It linked changes close to home to a new global consciousness.[16]

The first time U.S. astronauts saw and took pictures of "Earthrise" from space, capturing Earth and its Moon in a single frame from lunar orbit, was on the Apollo 8 mission in 1968. In 1972, on NASA's Apollo 17 mission,[17] astronauts took photos of the whole Earth on their way to the Moon. The common wisdom is that the advent of the human ability to see the Earth in its entirety from space helped to inspire Earth Day[18] and the environmental movement.[19]

Environmental historian Neil Maher[20] has argued that the 1968 "Earthrise" photo and the 1972 "Whole Earth" photo played key roles in two opposing cultural narratives. The "Earthrise" photo was one of

> a long line of photographs enlisting nature to support American expansion at home and abroad. . . . By figuratively depicting [President John F.] Kennedy's New Frontier in its sloping lunar surface, the Apollo 8 photograph helped to extend America's Manifest Destiny into the ultimate wilderness—outer space.

NASA's 1972 "Whole Earth" photo "tells a different story," Maher has argued, replacing "the idea of the American frontier with a vision of non-American nature . . . an image of a more global natural environment."[21] The effect of images of Earth from space on public consciousness about the biosphere, though analyzed,[22] has not been precisely measured. Whatever the effect, it arguably is different from the so-called, unmediated, overview effect that some astronauts have reported experiencing, by means of seeing Earth directly from space.[23]

Growing capability to study Earth from space greatly contributed to an understanding of our planet as a global system of systems—one of which is the biosphere. In 1972, NASA launched Landsat I, the first element of a fleet of Earth observation satellites. The Landsat satellites, and other space-based Earth-observing systems after it, have produced a more or less continuous view of the surface of Earth's lithosphere, atmosphere, hydrosphere, and biosphere over more than four decades. Deeper understanding of Earth's biosphere has helped astrobiologists to develop more sophisticated models of what extraterrestrial biospheres might be like. While Earth observations have been providing a picture of how life on Earth is faring (or not), space exploration missions have been looking for evidence of habitability and possibly life, on planetary bodies in our solar system. And now exoplanet researchers are aiming to detect evidence of habitability beyond our solar system.

NASA's Viking landers, launched to Mars in 1976, included three biology experiments designed to look for possible signs of life. Though the scientific consensus is that those experiments did not find any evidence of biological activity on Mars,[24] some scientists are still arguing over the meaning of the results.[25] This dispute persists to some degree because scientists—especially astrobiologists—have been unable to agree on a single, precise definition of what life is and is not.[26]

In 1977, Stewart Brand initiated a public debate, in his magazine *CoEvolution Quarterly*, about expanding human presence into space. With space colonization, he said,

> our perspective is suddenly cosmic, our Earth tiny and precious, and our motives properly suspect.... If we can learn to successfully manage large complex ecosystems in the Space Colonies, that sophistication could help reverse our destructive practices on Earth. And if we fail ... then we will have learned something as basic as Darwin about our biosphere—that we cannot manage it, that it manages us.[27]

(At the time of this writing, more than four decades later, it appears that humankind is not doing well at managing large complex ecosystems on Earth, boding ill for possible space colonies.)

Also in 1977, biophysicist Carl Woese and biochemist George E. Fox announced in the *Proceedings of the National Academy of Sciences* that they had identified a new domain of life on Earth, called archaea—single-celled organisms previously classified as bacteria by the old method of sorting out organisms according to their biochemistry, morphology, and metabolism.[28] Woese and Fox used a new (and now standard) method of classifying organisms by their gene sequences to determine that archaea had evolved on a different pathway from bacteria. Thus scientists redrew the tree of life, depicting the evolution of life by genetic rather than phenotypic relationships. The research that led to this momentous discovery was sponsored by NASA's exobiology program.[29]

In 1978, chemist James Lovelock and biologist Lynn Margulis—both early recipients of research funding from NASA[30]—unveiled their Gaia hypothesis, positing that life is interconnected with its physical environment, making Earth a self-regulating complex system that maintains conditions for life on the planet (if not disturbed by human intervention).[31] As Steven Dick and James Strick[32] have documented in their history of astrobiology at NASA, exobiology program officials at the agency embraced the Gaia hypothesis, at the same time that other disciplinary communities—geology, atmospheric science, climatology, and evolutionary biology—critiqued and attempted to dismiss it. Over time, the Gaia hypothesis was refined, and the science community tamped down its critique,[33] though the hypothesis remains controversial.

In 1979, *Science* magazine published a report of the discovery of life in deep-sea hydrothermal vent systems of the Galapagos Rift.[34] This discovery marked a major advance in scientific understanding of life. Not only was life thriving in environmental conditions considered inhospitable, but also this life depended on chemicals (chemosynthesis), not sunlight (photosynthesis), to survive. This discovery was important to astrobiology, as it showed that some forms of life did not need sunlight to thrive—opening the door to astrobiological conceptions of extraterrestrial biospheres in planetary subsurface environments, even at distances hundreds of millions of miles from the Sun.

During the Viking era, scientists were focused on finding evidence of life on or near the surface of Mars. However, the Viking mission and subsequent missions to Mars revealed that the surface of Mars was inhospitable to life as we know it, primarily due to a lack of liquid water and strong ultraviolet radiation. Subsequent Mars missions have collected evidence of liquid water on the surface billions of years ago, and some scientists speculate that liquid water may exist in the subsurface of the planet today.[35] Now some astrobiologists are exploring the possibility of extant microbial life there—a possible deep-subsurface biosphere on another planet.

In 1996, NASA's exobiology program, which had been funding research on extremophilic life for some years at that point, became an element of a new astrobiology program. This program took a more expansive approach to the study of the origin, evolution, and distribution of life in the universe.[36] The newly configured program was responding not only to thirty-five years of advances in exobiology research but also greatly expanded knowledge of other planetary environments in our solar system as well as the beginning of the era of exoplanet research. The aim of exoplanet research is not only to find planets beyond our solar system but also to find exoplanets that might possibly be habitable (or inhabited).

Barely more than two decades after the official establishment of a NASA astrobiology program, this multidisciplinary, interdisciplinary, transdisciplinary[37] field encompasses the search for and study of potentially habitable environments inside and outside our solar system, the search for evidence of past or present life on Mars, the search for prebiotic chemistry and life on other bodies in our solar system, laboratory and field research into the origins and evolution of life on Earth, and studies of the potential for life to adapt to challenges on Earth and in space. What knits together all of these lines of astrobiology research is the understanding that life and its environment(s) co-evolve. The idea of possible extraterrestrial biospheres is very much alive and well in the space science community.

Thanks to all of these advances, the prospects for finding extraterrestrial life—as we know it, or perhaps as we do not know it—appear to be more promising by the day.

But are they really?

Astrobiologists have spent years on developing a so-called ladder of life detection, "a tool intended to guide the design of investigations to detect microbial life within the practical constraints of robotic space missions."[38] This tool shows how challenging it will be to test and prove, definitively, a claim of evidence of life:

> The direct detection of extant life has not been attempted by NASA since the Viking Missions in the late 1970s. NASA's Ladder of Life Detection was generated to stimulate and support discussions among scientists and engineers about how one would detect extant life beyond Earth but within our Solar System (particularly on Europa and the other "Ocean Worlds"). In creating the Ladder, we started with [this] definition of life, "Life is a self sustaining chemical system capable of Darwinian evolution" and considered the specific features of [Earth] life. . . . The rungs of the Ladder were assembled from features that can be used to access (1) potential habitability, (2) suspicious biomaterials that could be biogenic or abiogenic, and (3) active processes of life. The lowest rungs are the least directly related to extant life and in some cases are the easiest to measure. For each rung (feature), the target and potential flight instruments for measurement were identified. Our ability to detect and properly interpret a measurement was evaluated in terms of how specific the feature was for Terran-type life, how likely the feature could be produced abiotically (called ambiguity), how likely the measurement would be a false positive due to contamination or measurement interference, how likely the measurement would be a false negative (missing life when it is present), and easy the measurement is to make (detectability).[39]

All of this said, extraterrestrial life detection is, nonetheless, a driving goal of astrobiology today.

## Exploration or Exploitation?

Questions about life—the origin and evolution of life on Earth and in the universe, the fate of life on Earth, the possibility of life elsewhere—have driven space exploration from its beginnings. The search for evidence of extraterrestrial life is a primary focus of NASA's planetary exploration program. This search proceeds on the assumption that extraterrestrial environments are and will remain pristine, untainted by terrestrial biology. Now that astrobiologists are zeroing in on several solar system environments that might be habitable (or inhabited), the search for extraterrestrial life is coming into conflict with the drive for human exploration.

Spacecraft have flown by, orbited around, or landed on Mercury, Venus, Mars, Jupiter and several of its moons, Saturn and several of its moons, the dwarf planet Pluto and its moons, and the dwarf planet Ceres. A myriad of spacecraft have orbited Earth to study the home planet. Comparative planetology is a thriving field. Ocean worlds in our solar system—in particular, Jupiter's moon Europa and Saturn's moons Enceladus and Titan—are top targets for astrobiological investigations of prebiotic chemistry, habitability, and possible life. Many astrobiologists are interested in exploring the possibility of extant life in the deep subsurface of Mars.

The conception—that is, the social construction—of the solar system and beyond as an environment to exploit, for living space and resources, preceded the human ability to travel into space but has recently been picking up steam as the pace of space exploration accelerates. The Russian scientist and mystic Konstantin Tsiolkovsky (1857–1938) is known for promoting the idea that humanity is destined to spread itself into outer space.[40] In the United States, the social construction of the solar system as an environment to exploit (as humanity has done with its own home planetary environment) was energized in the Reagan era, reinvigorated during the George W. Bush years, and furthered by the Obama administration, which promoted the goals of asteroid mining and human settlements on Mars. This conception of the solar system also has been embraced by the Trump administration, which has called for more "commercial" development of space and human missions to the Moon and Mars.[41] Recently, the governments of Luxembourg and Belgium signed an agreement to collaborate on the development of an international agreement for the exploration, exploitation, and utilization of space resources.[42]

This conception of outer space as an environment to exploit depends on an embrace of a belief system that I would describe as "dominionist" or "manifest destiny," an ideology establishing that our home planet, and now our home solar system, are resources here for humanity to use as it likes. This idea of dominion or manifest destiny is much older than 19th century U.S. government policy or the space age. It is a deeply Christian belief, brought to North America by the Puritans in the 17th century and refreshed by advocates of American expansionism in the 19th century. "The world as God's 'manifestation' and history as predetermined 'destiny' had been ideological staples of the strongly providentialist period in England between 1620 and 1660," writes historian Anders Stephanson,[43] the period when English Puritans migrated to North America. The related belief in "right"—the right of white Europeans, given by God, to possess North America—is at least as old. These beliefs came to underlie a U.S. national narrative of exploration, expansion, and exploitation. In the 21st century, when further U.S. expansionism on Earth is politically unacceptable, expansion into space appears attractive to those who still believe that the United States is destined to expand.

At the same time, space exploration and the environmental movement have spawned the idea of "astroenvironmentalism"—a call to preserve pristine extraterrestrial environments for their own sake. The principles of astroenvironmentalism, as detailed by one environmental writer,[44] include "considering space and the celestial bodies pristine wildernesses that need to be protected," requiring environmental impact statements for space missions, treating planetary bodies "as wildernesses that need to be protected," and creating ethical guidelines for protecting life elsewhere. Astrobiologists Charles Cockell and Gerda Horneck have proposed the creation of "planetary parks" in extraterrestrial environments that might be habitable, preserving these spaces as pristine sites for the scientific search for evidence of extraterrestrial life.[45]

The idea of preserving pristine extraterrestrial environments for their own sake is controversial in some circles of the space community. It conflicts with the dominant narrative of space exploration, which posits that humans should be—even must be, are destined to be—colonizing outer space. This Western-centric narrative of conquest and exploitation has been propagated by individuals and organizations advocating for the human settlement of space. The National Space Society, for example, promotes a "vision" of "people living and working in thriving communities beyond the Earth, and the use of the vast resources of space for the dramatic betterment of humanity." Its mission is "to promote social, economic, technological, and political change in order to expand civilization beyond Earth, to settle space and to use the resulting resources to build a hopeful and prosperous future for humanity."

The society's rationale for its mission includes

> survival of the human species and Earth's biosphere. . . . The human species is encountering increased natural, man-made, and extraterrestrial threats, including disease, resource depletion, pollution, urban violence, terrorism, nuclear war, asteroids, and comets. . . . Many forms of animal and plant life on Earth are suffering increased loss of population and quality habitat because of the growing presence of humans on planet Earth, via expansion, pollution, deforestation, fishing, farming, mining, and promotion of certain species of animals and plants. Space technology provides both means to monitor threats to life on Earth and ways to help curtail them. Space industrialization and settlement provide safety valves to relieve the pressures that cause Earth-bound threats.[46]

And then there is the Space Frontier Foundation, whose goals include "protecting the Earth's fragile biosphere *and* [emphasis added] creating a freer and more prosperous life for each generation by using the unlimited energy and material resources of space. Our purpose is to unleash the power of free enterprise and lead a united humanity permanently into the Solar System."[47]

This conflict between the goal of protecting pristine environments and the goal of exploiting space resources has heated up in recent years, fueled by proponents of the human settlement of Mars. During Obama's term and into Trump's, advocates of human exploration and exploitation have been waging a campaign to press NASA to relax planetary protection requirements for human missions into space, requirements intended to preserve pristine extraterrestrial environments for scientific exploration—requirements based on a conception of possible extraterrestrial biospheres as environments to be preserved and protected, not exploited. Planetary protection requirements—compliance with which is mandated by NASA policy—are intended to "support the scientific study of chemical evolution and the origins of life in the solar system."[48] Planetary protection policy already imposes stringent cleanliness requirements on robotic missions that will be exploring potentially habitable environments (such as certain regions on Mars). Presumably policy would establish even more stringent requirements for human missions to potentially habitable environments. The conflict between planetary protection proponents human exploration advocates recently led NASA to ask the U.S. National Academies' Space Studies Board to conduct a review and assessment of planetary protection policy development processes.[49]

## Social and Conceptual Issues in Astrobiology

The divergent goals of astrobiological exploration and human exploration and settlement raise legal, ethical, and philosophical questions about the status of potential extraterrestrial biospheres. As their subject is life in the universe, and as this subject has been of interest not only to philosophers and scientists but also to public audiences over centuries and even millennia,[50] astrobiologists have long considered the broader impacts of their work. However, projects addressing the so-called societal implications of astrobiology have been, until recently, sparse, sporadic, and disconnected.[51] In addition, they have tended to focus on possible responses to the discovery of extraterrestrial intelligence and involved primarily a small community of researchers largely engaged in the search for extraterrestrial intelligence (SETI).[52] Those efforts proceeded on the assumption that the discovery and verification of an extraterrestrial intelligence (ETI) signal would be world-changing, an assumption that may not be widely held outside the SETI community. In 2018, the Breakthrough Listen initiative, a privately funded SETI project,[53] conducted a "Making Contact" workshop that engaged scholars of anthropology, feminist epistemology, future studies, indigenous studies, and other social scientific and humanistic fields in a dialogue with SETI scientists that explored more broadly and deeply questions about the role and effects of SETI in and on human cultures.[54]

Today, the field of astrobiology is focused on the search for evidence of past or present microbial life in the solar system and possible habitable environments beyond the solar system, not for evidence of ETI. Thus, in recent years, discussion of ethical, philosophical, theological, and legal issues relating to astrobiology has been broadening and focusing accordingly.

Following the 1996 publication of claims of fossil evidence of past microbial life in a martian meteorite fragment, the NASA astrobiology program, in the face of growing scientific, political, and public interest in the possible existence of such life, focused some of its attention on social, ethical, and philosophical questions relating to the discovery of extraterrestrial microbial life, funding efforts to introduce astrobiology to the broader scientific community and to public audiences as well. For example, the program cosponsored a series of workshops organized by the American Association for the Advancement of Science's newly established Dialogue on Science, Ethics, and Religion[55] on the philosophical, ethical, and theological implications of astrobiology, held in 2003–2004.[56]

The NASA astrobiology community published its first science "roadmap" in 1998, followed by updated roadmaps in 2003 and 2008. All of these roadmaps articulated four basic principles to guide implementation of NASA's astrobiology program, including the principle that the astrobiology community recognizes a broad interest in its work, especially in areas such as the search for extraterrestrial life, achieving a deeper understanding of life, the potential to engineer new life forms adapted to live on other worlds, the broader implications of discovering life beyond Earth, and envisioning the future of human life on Earth and in space.

The NASA astrobiology roadmap has since evolved into a lengthy science strategy document. The 2015 NASA astrobiology strategy does not specifically identify goals, objectives, and questions relating to social, cultural, ethical, and theological issues arising in the study of the origins of life and the search for evidence of extraterrestrial life, because the community has embraced this endeavor as part of its ongoing work. Among five broad goals identified in this strategy is to

> enhance societal interest and relevance. Astrobiology recognizes a broad societal interest in its endeavors, especially in areas such as achieving a deeper understanding of life, searching for extraterrestrial biospheres, assessing the social implications of discovering other examples of life, and envisioning the future of life on Earth and in space.[57]

Coincident with the development of the 2015 strategy, the NASA astrobiology program initiated a number of short-term activities intended to broaden and diversify the community of scholars participating in the ongoing dialogue about astrobiology in culture and to focus this dialogue more sharply on the possible cultural impacts of the discovery of extraterrestrial microbial life—which most members of the astrobiology community believe is more likely, and more imminent, than contact with ETI life.[58] These activities included the Baruch S. Blumberg NASA/Library of Congress Chair in Astrobiology, the Center of Theological Inquiry 2015-017 study-in-residence project, "Inquiry on the Societal Implications of Astrobiology," and a 2015–2016 NASA Astrobiology Debates project.

The Blumberg Chair (an ongoing activity) was created in 2012 to support scholars interested in the intersection of the sciences and humanities in the field of astrobiology. According to the Kluge Center of the Library of Congress, which houses the Blumberg Chair,[59] "the program makes it possible for a senior researcher to be in residence at the Kluge Center, to make use of the Library of Congress collections, and to convene programs that ensure the subject of astrobiology's role in culture and society receives considered treatment each year in Washington, D.C."[60] The first Blumberg Chair, astrobiologist David Grinspoon, devoted his year at the Kluge Center to a study of comparative planetology—comparing the history of Earth with the history of other planets, focusing on the era of Earth's history called the Anthropocene—the period when human activity began to reconfigure the face, and the climate, of our planet, altering the biosphere in unnatural ways. Exploration of other planets, especially Venus and Mars, those planets closest to Earth, has led astrobiologists to speculate about whether those planets have ever been, or perhaps are now, habitable or inhabited, and if not, why. Grinspoon writes of the need to "listen to the planets," learning about how other planets in our solar system are so different from Earth today, and in the process understanding how to preserve and protect Earth's biosphere while searching for other biospheres.[61]

The inquiry into the societal implications of astrobiology conducted by the Center of Theological Inquiry (CTI) in Princeton, New Jersey, for the 2015–2016 academic years, focused on this question: "The discovery of another form of life, whether microbial or complex, would change how we see ourselves and our world. How would theology, the humanities, and the social sciences relate life as we know it to this background of other possibilities?" Astrobiologists participated in this inquiry as "visitors," spending time at CTI to meet with scholars selected for the inquiry.

The 2015–2016 NASA Astrobiology Debates was a year-long academic project for university and high-school students involving in-person and online debate tournaments, speech competitions, public exhibition debates, topic-expert panels for student-debater audiences, and student-debater interviews with a cross-disciplinary group of subject matter experts. The aim of the debates project was to stimulate student, teacher, and school research and dialogue on astrobiology in preparing for these events and at the events themselves. The project's managers, at George Washington University in Washington, D.C., estimated that they involved over 2,000 students during the 2015–2016 academic year.

Most recently, a nascent organization of social scientific and humanistic scholars called the Society for Social and Conceptual Issues in Astrobiology[62] is coming together. An initial workshop held in 2016 drew around 30 participants. A second workshop held in 2018 involved more than 100 scholars.[63] These workshops involved anthropologists, bioethicists, philosophers and others in wide-ranging discussions about topics such as the moral status of extraterrestrial life and the place of human life in the solar system. A third workshop will be held in 2020.

## Closing Thoughts

It would be difficult—though interesting and potentially fruitful—to document exactly how the search for extraterrestrial life has changed the way experts and non-experts think about the biosphere (or biospheres). Such a project would require far more in-depth research than this chapter has entailed. The evidence I have presented here points to the conclusion—at least for me—that the quest to find evidence of extraterrestrial life has affected our conception of the biosphere, the way we think about our home planet and our place on (or in) it, and our perspective on the possibility of extraterrestrial biospheres nearby and far away. Expert understanding of life and environment, at the macro and the micro level, has evolved over the past fifty years. Astrobiology, planetary exploration, and exoplanet science have made significant contributions to this changing understanding.

Considering the biosphere at the microscopic scale, we now know that life teems kilometers beneath the surface of Earth, in deep-sea environments, inside Antarctic rocks and at the bottom of perpetually ice-covered Antarctic lakes,[64] in highly acidic hot springs and radioactive water, by chemosynthesis as well as photosynthesis. Scientists have come to understand that microbial life, extremophilic or otherwise, dominates our biosphere. As some astrobiologists have put it, "For more than 3.5 billion years, microbes of untold diversity have dominated every corner of [it]."[65] Microbes play a key role "in geochemical cycling, biodegradation, and the protection of entire ecosystems from environmental insult . . they control global utilization of nitrogen through nitrogen fixation, nitrification, and nitrate reduction; and they drive the bulk of carbon, sulfur, iron, and manganese biogeochemical cycles." Most importantly, "the continued survival of later evolving multicellular plants and animals is completely dependent upon interactions with microorganisms."[66] In the search for evidence of extraterrestrial life in our solar system, astrobiologists agree that at the same time that complex life is highly unlikely, microbial life is appearing to be more and more likely. The idea of life in other planetary environments barely requires a leap of imagination these days.

Considering the biosphere at the planetary scale, we now know that the more we explore extraterrestrial environments, the more we learn about how different they are from Earth. And at the same time we find ways in which they are similar too. We now routinely look at pictures of Earth and other planets taken from space and remark upon their resemblances and their differences. We know that planets and planetary environments evolve over time. We know that, on Earth, life and environment have coevolved, and we assume that in other planetary environments that might be habitable, such coevolution would occur as well. And given what they know about the diversity of life and environments on Earth, astrobiologists are considering the possibility of a diversity of biospheres beyond Earth, from the ice-covered global ocean of Europa to exoplanets that are not Earthlike but nonetheless considered potentially habitable. Thus, concepts— social constructions—of environment, nature, and the biosphere in what sociologist Steven Yearley[67] has called "advanced modernity" have been, and are being, extended from Earth into space. We have yet to identify another biosphere beyond Earth. But the search for potential extraterrestrial biospheres is underway.

The more we look, the more we find that life is everywhere on, in, and above[68] Earth. The more we look, the more we find that the basic ingredients for life as we know it are everywhere, not only throughout our solar system but also throughout interstellar space. We have also found that much of life on Earth is "weird" extremophilic microbial life, adapted to niche environments that humans and their multicellular relatives could not tolerate. And astrobiologists are thinking about how they might be able to detect, definitively, extraterrestrial life that is like Earth life, or not. While astrobiologists are identifying biosignatures for detecting extraterrestrial life as we know it[69]—carbon-based, cellular life—they are also working on developing a set of biosignatures to use in searching for signs of life as we do not know it—for example, life that does not need water as a solvent or life that is not made of carbon-based compounds.[70]

Can—does—the search for extraterrestrial life help humans to figure out how to live together on their home planet? Astrobiologist David Grinspoon has observed, "We can conceive of a truly intelligent, sustainable communicating society. But we don't know if we can become one. So we search the skies for confirmation of a hopeful image of ourselves."[71] Astrobiologist Ian Crawford has argued that the field of astrobiology "cannot help but engender a worldview infused by cosmic and evolutionary perspectives. Both these attributes of the study of astrobiology are, and will increasingly prove to be, beneficial to society regardless of whether extraterrestrial life is discovered or not."[72]

Stewart Brand has continued to think about the terrestrial biosphere from the perspective of space. In the journal *Nature*,[73] he asserted that to better understand our home planet, we must integrate data gathered by Earth-observing satellites with genetic data gathered by microbiologists—a merger of the macro with the micro. "Metagenomics is giving us detailed access to the genes and gene communities of bacteria and archaea," he wrote. "A unifying body of data, ideas, models and images of the whole-Earth system could inspire the public and may shift scientific thinking. In studying the energy dynamics of the Earth–Sun system while learning how our microbial partners manage to keep this planet comfortably terraformed for life, we would begin to step up to the full meaning of Earth stewardship."

From a different perspective, political scientist Walter McDougall has considered how space exploration and the search for life elsewhere has affected our views of life on Earth and has concluded that we are not getting the message. "The greatest icon bequeathed by space technology" to humankind thus far, McDougall claims, is the Apollo 8 "Earthrise" photo. "It has become a cliche to observe how the photo *Earthrise* inspired environmentalists since it vividly depicted our biosphere as finite and fragile."[74] Just as important as this new perspective on our biosphere, he says, "was the urgent if obvious revelation that the natural, holistic earth seen from space is free of political, racial, and religious boundaries. The sum of those two perceptions must be a Spaceship Earth mentality transcending mundane considerations of geopolitics and geoeconomics."

But "no such transcendence has begun to occur," McDougall concluded. "Even after the end of the Cold War, so often blamed for perverting the dream, astronautics has worked no metamorphosis, no paradigm shift, in human behavior. Conceptions of extraterrestrial worlds as our property to exploit and as pristine environments to protect are in competition for a central role in U.S. space policy." Regrettably, more than a decade after McDougall offered those views, space exploration and development continues to be driven by a strongly libertarian ideology of conquest and exploitation.[75]

As David Grinspoon has observed,

A gradual change to a planetary world view is aided by the proliferation of views of Earth from orbit, and the experiences that some human beings have had of actually physically going into space.... Stimulated by the sight of Earth looking alive, fragile, and achingly beautiful..., they report a powerful sense of identity with the entire human race, the entire biosphere, and the entire planet.... When humans go into space, the biosphere is extending a fragile eye and looking down at itself.[76]

In the 1980s, rhetorical critic Janice Hocker Rushing speculated that the post-Apollo focus of space exploration on the search for evidence of extraterrestrial life is a product of a widespread understanding that humankind exists in a universe, not only on planet Earth. A contemporary narrative of space exploration might better reflect this understanding by telling a story of "a spiritual humbling of self" rather than "an imperialistic grabbing of territory."[77] Space is too big to be conquered, she pointed out.

Opposing perspectives on the place of humans in space are reflective of two parallel and conflicting cultural narratives of space exploration: outer space as a sort of supermarket of resources, open to exploitation by whoever gets there first, and outer space as a pristine wilderness to be studied and appreciated but left unaltered.

I am inspired by the recent work of scholars who are engaged with or interested in the search for evidence of life on Mars, not necessarily opposed to the idea of human settlements on other planets but concerned about the ways in which such activities might be conducted—safely, ethically, with appropriate concern for the protection of extraterrestrial environments—in other potential biospheres.

Serving as Blumberg Chair in Astrobiology in 2017–2018, astrophysicist Lucianne Walkowicz organized two events designed to address ethical, philosophical, and other issues relating to human settlement of Mars. "Decolonizing Mars: An Unconference on Inclusion and Equity in Space Exploration" brought together a diverse group of scholars working "at the intersection of astrobiology, anthropology, social justice, and space exploration" for a wide-ranging discussion. "The term decolonization," according to Walkowicz, "refers to undoing the legacy of colonialism." "Decolonizing Mars" was designed to "examine how using a colonialist framework in space reproduces past harm from humanity's history on Earth [and to envision] fresh pathways for thinking about space exploration by stepping away from the ways we usually talk about space, which by definition is 'decolonizing' the topic."[78] Her "Becoming Interplanetary" conference was a public event designed to address "what living on Earth can teach us about living on Mars."

Another multidisciplinary group of scholars has produced a "manifesto for governing life on Mars."[79] The manifesto is a response to "the recent expansion of human-led activities in space, high-profile declarations of ambitions to develop settlements on Mars, and growing media interest in these developments. Its purpose is to provoke a richer set of debates in mainstream thinking about the social and political dimensions of establishing a permanent human presence on Mars."[80]

Which vision of the human future in space will win out? Will it be the aim of conquering and exploiting extraterrestrial environments, perpetuating the destructive practices that human cultures have inflicted on our home planet and extending them into extraterrestrial biospheres? Or will it be the quest to find life elsewhere, the careful robotic exploration of extraterrestrial environments to look for life, the preservation of extraterrestrial biospheres? Will it be possible for people on Earth to embrace a vision of humanity's peaceful coexistence on Spaceship Earth and the need to work together to preserve life here and look for life out there? More to the point, will it be possible for governments and corporations to embrace such a vision?

My hope is that astrobiology ultimately will win out, prodding space-faring cultures to focus on a careful, mindful search for life elsewhere rather than a mindless quest for "gold."

## Notes

1. An early version of this chapter was presented at the 23d annual conference of the International Association for Science, Technology, and Society, February 1, 2008, Baltimore, MD.
2. See Berger, P.L. and Luckmann, T. (1966). *The Social Construction of Reality: A Treatise in the Sociology of Knowledge*. New York: Doubleday.
3. https://en.wikipedia.org/wiki/Eduard_Suess.

4. The Vernadsky Institute of Geochemistry and Analytical Chemistry, established by the Soviet Union in 1947, has developed experiments and instrumentation for numerous Soviet/Russian planetary exploration missions. http://www.geokhi.ru/en/default.aspx.

5. Morrison, J. (2016). Living bacteria are riding Earth's air currents. Smithsonian.com, January 11. https://www.smithsonianmag.com/science-nature/living-bacteria-are-riding-earths-air-currents-180957734/.

6. Zierenberg, R.A.; Adams, M.W.W.; and Arp, A.J. (2000). Life in extreme environments: hydrothermal vents. *Proceedings of the National Academy of Sciences* 97(24): 12961–12962. https://doi.org/10.1073/pnas.210395997.

7. Parnell, J. and McMahon, S. (2016). Physical and chemical controls on habitats for life in the deep subsurface beneath continents and ice. *Philosophical Transactions of the Royal Society: Mathematical, Physical and Engineering Sciences.* https://doi.org/10.1098/rsta.2014.0293.

8. The distinction between exobiology and astrobiology is slight. NASA established an exobiology program in 1960. In the 1990s, NASA expanded the purview of the program, relabeling it astrobiology and including in it an element called exobiology. In this chapter, I use the term "astrobiology" to refer to the study of the origin, evolution, and distribution of life in the universe.

9. For an excellent review of theories of the origins of life, see Gollihar, J.; Levy, M.; and Ellington, A.D. (2014). Many paths to the origin of life. *Science* 343: 259. Also see Scharf, C. et al. (2015). A strategy for origins of life research. *Astrobiology* 15(12): 1031.

10. See the National Academies report, *The Limits of Organic life in Planetary Systems,* 2007, also known as the "weird life" report. https://www.nap.edu/catalog/11919/the-limits-of-organic-life-in-planetary-systems.

11. See, for example, Fredrickson, J.K. and Balkwill, D.L. (2006). Geomicrobial processes and biodiversity in the deep terrestrial subsurface. *Geomicrobiology Journal* 23(6). https://doi.org/10.1080/01490450600875571.

12. See Kallmeyer, J.; Pockalny, R.; Adhikari, R.R. et al. (2012). Global distribution of microbial abundance and biomass in subseafloor sediment. *Proceedings of the National Academy of Sciences* 109(40): 16213–16216, www.pnas.org/cgi/doi/10.1073/pnas.1203849109.

13. See, for example, Summers, M. and James Trefil, J. (2017). *Exoplanets: Diamond Worlds, Super Earths, Pulsar Planets, and the New Search for Life Beyond Our Solar System.* Washington, DC: Smithsonian Books. None of these terms have precise definitions. See Siegel, E. (2018). Sorry, super-Earth fans, there are only three classes of planet. *Forbes,* March 2, https://www.forbes.com/sites/startswithabang/2018/03/02/sorry-super-earth-fans-there-are-only-three-classes-of-planet/#5bc17cdc78c4; Oppenheimer, R. (2016). Making sense of the exoplanet zoo. *Science* 353(6300): 644–645.

14. The instrument, known as the "Wolf Trap," after its principal investigator Wolf Vishniac, was intended to fly on NASA's Viking mission to Mars. Ultimately, it was dropped from the mission.

15. Baltimore, MD: Johns Hopkins University Press.

16. Warde, P.; Robin, L.; and Sorlin, Sverker (2019). "The environment" is only 70 years old. *The Scientist,* February, p. 61.

17. Apollo 17 was NASA's last human mission to the Moon.

18. The first Earth Day was celebrated on April 22, 1970. The unofficial flag of Earth Day featured an image of Earth taken from space. In 1974, the United Nations declared June 5 World Environment Day. An "Earth anthem" composed for World Environment Day features these lyrics: "Our cosmic oasis, cosmic blue pearl, the most beautiful planet in the universe, all the continents and the oceans of the world united we stand, as flora and fauna united we stand as species of one earth. black, brown, white, different colors, we are humans, the earth is our home."

19. See, for example, the Report of the World Commission on Environment and Development: Our Common Future, 1987, in which commission chair Gro Harlem Brundtland wrote: "In the middle of the 20th century, we saw our planet from space for the first time. Historians may eventually find that this vision had a greater impact on thought than did the Copernican revolution of the 16th century, which upset the human self-image by revealing that the Earth is not the centre of the universe. From space, we see a small and fragile ball dominated not by human activity and edifice but by a pattern of clouds, oceans, greenery, and soils. Humanity's inability to fit its activities into that pattern is changing planetary systems, fundamentally."

20. Maher, N. (2004). Neil Maher on shooting the Moon. *Environmental History,* July 2004. Http://www.historycooperative.org/journals/eh/9.3/maher.html.

21. Ibid.

22. See, for example, Potter, C. (2018). *The Earth Gazers: On Seeing Ourselves.* New York: Pegasus Books; Marina Benjamin, M. (2003). *Rocket Dreams: How the Space Age Shaped Our Vision of a World Beyond.* New York: Free Press.

23. Seehttps://en.wikipedia.org/wiki/Overview_effect;Calderon,J.(2017).Whathappens when astronauts see Earth from space for the first time. *Tech Insider*, August 15. http://www.businessinsider.com/overview-effect-nasa-apollo8-perspective-awareness-space-2015-8.

24. Committee on Planetary Biology and Chemical Evolution, Space Science Board, National Research Council (1977). Post-Viking biological investigations of Mars. Washington, DC: National Academies Press.

25. See, for example, Levin, G.V. (1997). The Viking labeled release experiment and life on Mars, Proceedings of SPIE–The International Society for Optical Engineering, "Instruments, Methods, and Missions for the Investigation of Extraterrestrial Microorganisms, July 29– August 1, San Diego, CA, http://www.gillevin.com/Mars/Reprint107-spie-files/Reprint107-spie.htm; Miller, J.D.; Straat, P.A.; and Levin, G.V. (2002). Periodic analysis of the Viking lander labeled release experiment. *Proceedings Volume 4495, Instruments, Methods, and Missions for Astrobiology IV*, International Symposium on Optical Science and Technology, 2001, San Diego, CA, doi:10.1117/12.454748, https://www.spiedigitallibrary.org/conference-proceedings-of-spie/4495/0000/Periodic-analysis-of-the-Viking-Lander-Labeled-Release-experiment/10.1117/454748.short?SSO=1.

26. See, for example, Scharf, C.; Virgo, N.; Cleaves II, H.J.; et al. (2015). A strategy for origins of life research. *Astrobiology* 15(12): 1031–1042. doi:10.1089/ast.2015.1113

27. Brand, S. (Ed.) (1977). *Space Colonies: A Coevolution Book*. New York: Penguin Books, p. 72.

28. Woese, C.R. and Fox, G.E. (1977). Phylogenetic structure of the prokaryotic domain: The primary kingdoms. *Proceedings of the National Academy of Sciences* 74(11): 5088–5090. https://doi.org/10.1073/pnas.74.11.5088

29. The study of archaea is still a thriving field. See, for example, Dance, A. (2018). Archaea family tree blossoms, thanks to genomics. *The Scientist*, June 1, https://www.the-scientist.com/?articles.view/articleNo/54649/title/Archaea-Family-Tree-Blossoms--Thanks-to-Genomics/

30. Both Lovelock and Margulis were involved in NASA's early efforts to study the origins and evolution of life and to search for extraterrestrial life. In the 1970s, Lovelock played a role in planning for the Viking life-detection mission to Mars, and Margulis received funding from NASA's exobiology program for research into endosymbiosis, the means by which an independent single-celled organism comes to live in another single-celled organism and evolves into an organelle.

31. Margulis, L. and Lovelock, J.E. (1978). The biota as ancient and modern modulator of the Earth's atmosphere. *Pure and Applied Geophysics* 116(2–3): 239–243.

32. Dick, S.J. and Strick, J.E. (2004). *The Living Universe: NASA and the Development of Astrobiology*. Rutgers, NJ: Rutgers University Press.

33. Ibid., 116–118.

34. Corliss, J.B. et al. (1979). Submarine thermal springs on the Galapagos Rift. *Science* 203: 1073–1083.

35. See, for example, Greshko, M. (2018). Huge water reserves found all over Mars. *National Geographic*, January 11, https://news.nationalgeographic.com/2018/01/mars-buried-water-ice-subsurface-geology-astronauts-science/?beta=true; https://en.wikipedia.org/wiki/Water_on_Mars..

36. Dick and Strick, 2004.

37. Astrobiology has been characterized as multidisciplinary in its content and interdisciplinary in its execution. It has also been described as transdisciplinary. It has been noted that "the terms multidisciplinary, interdisciplinary and transdisciplinary are increasingly used in the literature, but are ambiguously defined and interchangeably used" (Choi, B.C. and Pak, A.W. [2006]. Multidisciplinarity, interdisciplinarity and transdisciplinarity in health research, services, education and policy: 1. Definitions, objectives, and evidence of effectiveness. *Clinical and Investigative Medicine*, 29(6): 351–364). They are not interchangeable, though some in the astrobiology community may use them as such. For the purposes of this chapter, it will suffice to say that the three terms "refer to the involvement of multiple disciplines to varying degrees on the same continuum" (ibid.) Recognizing a need to think beyond disciplines, in recent years the NASA astrobiology program has adopted a systems-science approach to solving complex problems in the field, organizing research collaboration networks to address specific questions.

38. Neveu, M. et al. (2018). The ladder of life detection. *Astrobiology*, June 4, https://doi.org/10.1089/ast.2017.1773.

39. See https://astrobiology.nasa.gov/research/life-detection/ladder/

40. http://www.russianspaceweb.com/tsiolkovsky.html

41. The Reagan, George H.W. Bush, and George W. Bush administrations also advocated for sending people back to the Moon and on to Mars.

42. The Government of the Grand Duchy of Luxembourg and the Kingdom of Belgium (2019). The Grand Duchy of Luxembourg and Belgium join forces to develop the exploration and utilization of space resources. January 23. https://space-agency.public.lu/dam-assets/press-release/2019/2019-01-23-ENG-joint-press-release-BE-LU.

43. Stephanson, A. (1995). *Manifest Destiny: American Expansion and the Empire of Right*. New York: Hill and Wang.

44. Miller, R.W. (2001). Astroenvironmentalism: The case for space exploration as an environmental issue, Electronic *Green Journal*, December 15, 2001. Also see Hargrove, E.C. (1986). *Beyond Spaceship Earth: Environmental Ethics and the Solar System*. San Francisco, CA: Sierra Club Books.

45. Cockell, C.S. and Horneck, G. (2006). Planetary parks—formulating a wilderness policy for planetary bodies. *Space Policy* 22(4): 256–262.

46. http://www.nss.org/about/philosophy.html#vision

47. https://spacefrontier.org

48. https://sma.nasa.gov/sma-disciplines/planetary-protection

49. Committee on the Review of Planetary Protection Policy Development Processes, Space Studies Board (2017). *Review and Assessment of Planetary Protection Policy Development Processes*. Washington, DC: National Academies Press.

50. See, for example, Dick, S.G. (1982). *The Extraterrestrial Life Debate from Democritus to Kant*. Cambridge. U.K.: Cambridge University Press; Dick, S.J. (1998). *Life on Other Worlds: The 20th Century Extraterrestrial Life Debate*. Cambridge, U.K.: Cambridge University Press.

51. Since the 1970s, the Royal Society of London and the Vatican Observatory have organized periodic discussions and published reports about how the discovery of extraterrestrial life might affect human civilization. In 1974, the Royal Society of London sponsored a discussion meeting on "the recognition of alien life." In 2010, the Society held a discussion meeting on "the detection of extraterrestrial life and the consequences for science and society." In 2009, the Vatican Observatory and the Pontifical Academy of Sciences held a "study week on astrobiology" in Rome that drew many scientists and global media coverage. Also in 2010, the Royal Society held a discussion meeting on the detection of extraterrestrial life and the consequences for science and society, which also drew global media coverage. In 2014, the Observatory cosponsored a conference with the University of Arizona's Seward Observatory on "the search for life beyond the solar system."

52. The earliest record of a NASA-sponsored activity that pulled together a multidisciplinary group of scholars to address "the social, philosophic, and humanistic impact" of the discovery of extraterrestrial life was a symposium held at Boston University on November 20, 1972—before the Viking mission to Mars. The symposium was convened to address the question: "How might human beings react to the discovery of life beyond Earth?" Discussion focused on the possible discovery of evidence of ETI life. Symposium participants were anthropologist Ashley Montagu; physicist and "SETI pioneer" Philip Morrison; planetary scientist and SETI advocate Carl Sagan;

theologian Krister Stendahl, then dean of the Harvard School of Divinity; and biologist George Wald (Nobel laureate, 1967). See Berendzen, R. (Ed.). *Life Beyond Earth and the Mind of Man.* National Aeronautics and Space Administration, Scientific and Technical Information Office, Washington, DC, 1973; https://ntrs.nasa.gov/archive/nasa/casi.ntrs.nasa.gov/19730022075.pdf

53. https://breakthroughinitiatives.org/initiative/1

54. I participated in this workshop.

55. DOSER was established in 1995: https://www.aaas.org/page/about-doser

56. Workshop Report: Philosophical, Ethical and Theological Questions of Astrobiology, American Association for the Advancement of Science, Washington, DC, 2007. A workshop on the "societal implications of astrobiology," held at NASA Ames Research Center in California in 1999, addressed "implications of astrobiology for human psychology, society, and culture, and the contributions that the social sciences can make to the field of astrobiology." Questions to guide future research that were identified at this workshop included: "What are the biological, psychological, and cultural factors that compel humankind to envision life beyond our planet's surface? Why do we seek evidence of extraterrestrial life and intelligence, and why do we strive to establish a continuing human presence off of our home planet?" This workshop was not sponsored by the NASA astrobiology program. In 2009, the SETI Institute in California held a workshop to develop a roadmap of societal issues relating to astrobiology. This workshop was funded by the NASA Astrobiology Institute (NAI), not by the NASA astrobiology program. In 2012, the NAI approved the formation of a focus group on the societal implications of astrobiology. The focus group was dissolved in 2015.

57. https://astrobiology.nasa.gov/research/astrobiology-at-nasa/astrobiology-strategy/

58. The scope of NASA's activities is specified by law (the 1958 National Aeronautics and Space Act, as amended), and thus these activities were not intended to lead to an ongoing NASA program of research into ethical, theological, and social issues relating to astrobiology, which would be beyond the scope of the 1958 law. However, one of NASA's statutory objectives is to conduct "studies of the potential benefits to be gained from, the opportunities for, and the problems involved in the utilization of aeronautical and space activities for peaceful and scientific purposes." (Section 102(c)(4), https://history.nasa.gov/spaceact.html)

59. https://www.loc.gov/today/pr/2016/16-046.html

60. The Chair creates an opportunity to study the range and complexity of issues related to how life begins and evolves and to examine the philosophical, religious, ethical, legal, cultural, and other concerns arising from scientific research on the origin, evolution, and nature of life. Blumberg Chairs work at the Kluge Center of the Library of Congress and are expected to meet with members and staff in Congress to discuss the work they are doing. One result of such meetings was a series of three hearings on astrobiology held by the House Science, Technology, and Space Committee. Other products of this initiative include three books: Dick, S.J. (Ed.) (2015) (Blumberg Chair, 2014–2015). *The Impact of Discovering Life Beyond Earth.* Cambridge, U.K.: Cambridge University Press; Dick, S.J. (2018). *Astrobiology, Discovery, and*

*Societal Impact.* Cambridge, U.K.: Cambridge University Press; and Grinspoon, D. (2016). (Chair, 2012–2013). *Earth in Human Hands: The Rise of Terra Sapiens and Hope for Our Planet.* New York, NY: Grand Central.

61. Grinspoon, D. (2016). *Earth in Human Hands: Shaping Our Planet's Future.* New York, NY: Grand Central.

62. https://socia.space

63. I participated in both the 2016 and 2018 workshops.

64. See, for example, Friedmann, E.I.; Hua, M.; and Ocampo-Friedmann, R. (1988). Cryptoendolithic lichen and cyanobacterial communities of the Ross desert, Antarctica, *Polarforschung* 58(2–3): 251–259; Squyres, S.W.; Andersen, D.W.; Nedell, S.S.; and Wharton, R.A. (1991). Lake Hoare, Antarctica: Sedimentation through a thick perennial ice cover. *Sedimentology* 38(2): 363–379.

65. Sogin, M.L. and Jennings, D.E. (2003). Introduction. *Biological Bulletin* 204: 159.

66. The U.S. National Institutes of Health's Human Microbiome Project has shed a lot of light on this codependence.

67. Yearley, S. (2008). Nature and the environment in science and technology studies, pp. 921–947 in Hackett, E.: Amsterdamska, O.; Lynch, M.; and Wajcman, J. (eds.), *The Handbook of Science and Technology Studies* (3rd ed.). Cambridge, MA: MIT Press.

68. See, for example, DeLeon-Rodriguez, N.; Latham, T.L.; Rodriguez, L.M. et al. (2013). Microbiome of the upper troposphere: Species composition and prevalence, effects of tropical storms, and atmospheric implications. *Proceedings of the National Academy of Sciences* 110(7): 2575–2580, https://doi.org/10.1073/pnas.1212089110

69. See, for example, Krissansen-Totton, J.; Olson, S. and Catling, D.C. (2018). Disequilibrium biosignatures over Earth history and implications for detecting exoplanet life. *Science Advances* 4; Life detection ladder, https://astrobiology.nasa.gov/research/life-detection/ladder/, 2017.

70. Space Studies Board, National Research Council (2007). *The Limits of Organic Life in Planetary Systems.* Washington, DC: National Academies Press.

71. Grinspoon, D. (2003–2004). *Lonely Planets: the natural philosophy of Alien Life.* New York: HarperCollins, p. 414.

72. Crawford, I.A. (2018). Widening perspectives: The intellectual and social benefits of astrobiology (regardless of whether extraterrestrial life is discovered or not. *International Journal of Astrobiology* 17(1): 57–60.

73. Brand, S. (2007). Whole earth comes into focus. *Nature* 450(797).

74. McDougall, Walter A. (2007). *The Space Age That Never Arrived: A Meditation on the 50th Anniversary of Sputnik 1.* Foreign Policy Research Institute, Philadelphia, PA, November. Http://www.fpri.org/enotes/200711.mcdougall.sputnikanniversary.html.

75. Billings, L. (2017). "What's the National Space Council for?" https://doctorlinda.wordpress.com/2017/10/23/whats-the-national-space-council-for/, October 23; Witze, A. (2017). Trump tells NASA to return to the Moon. *Nature*, December 11, https://www.nature.com/articles/d41586-017-08473-1.

76. Grinspoon, D. (2016). *Earth in Human Hands: Shaping our Planet's Future.* New York: Grand Central, p. 439.

77. Rushing, J.H. (1986). Mythic evolution of "The New Frontier" in mediated rhetoric. *Critical Studies in Mass Communication* 3(3): 265–296, 284.

78. https://www.decolonizemars.org/motivation/. I participated in "Decolonizing Mars."

79. Cowley, R. (Ed.) (2019). *A Manifesto for Governing Life on Mars*. London: King's College London.

80. Ibid., 1.

# 15

# Myth-Free Space Advocacy Part IV

## The Myth of Public Support for Astrobiology

*James S.J. Schwartz*

## Introduction

A common refrain among astrobiology advocates is that astrobiology seeks to answer questions of great importance to all of humanity: How did life originate? Are we alone in the universe? If we are not alone, what other kinds of life are out there? However, such claims are seldom based on anything resembling sociological research and instead tend to be purely anecdotal. If we are rationally to believe that there is broad and deep interest in answering the major questions of astrobiology, then anecdotes will not suffice—genuine evidence is required. This chapter is intended to provide an overview of what is known (or perhaps to give away the ending, an overview of how *little* is known) about the public's interest in and support for astrobiology.

To provide a few examples of the kinds of claims I have in mind, consider Mark Brake et al., who claim that "[n]o subject captures the public's scientific imagination more than astrobiology" (Brake et al. 2006, p. 321). According to Ian Crawford (2018),

> familiarity with the cosmic and evolutionary perspectives provided by astrobiology, powerfully reinforced by actual views of the Earth from space, can surely also act to broaden minds in such a way as to make the world less fragmented and dangerous. Astrobiology as a discipline can play a major part in achieving this, not least because . . . much astrobiology research is of wide public interest and often in the public eye. (p. 58)

Similarly, Bruce Jakosky (2006) claims that "the public has always been interested in knowing whether life exists elsewhere in the universe" (p. 5), later explaining in more detail that

> Astrobiology and astrophysics address questions that are close to universal, both to scientists and to the public. We look for answers to questions about

Jim S.J. Schwartz, *Myth-Free Space Advocacy Part IV*. In: *Social and Conceptual Issues in Astrobiology*. Edited by: Kelly C. Smith and Carlos Mariscal, Oxford University Press (2020). © Oxford University Press.
DOI: 10.1093/oso/9780190915650.001.0015

how the universe formed and evolved; how galaxies and stars form, evolve, and die; how planets form, behave, and evolve, and whether they are widespread; whether Earth-like planets exist elsewhere; how life originates and whether microbial life exists elsewhere; and whether intelligent life is unique, rare, or common in the universe. These questions touch us deeply as humans. They get at the basic issue of how we, both collectively and as individuals, relate to our surroundings. (p. 99)

Mark Sephton (2014) reiterates the issue of curiosity:

astrobiology provides a particular stimulus for human curiosity being directed at the subject of alien life. . . . The detection of alien life will have consequences far exceeding those of the "Earthrise" photograph taken during the Apollo eight mission that captured all of humanity and its environments with a single image. (p. 147)

And finally, Richard Randolph, Margaret Race, and Christopher McKay state that "[t]he Pathfinder mission is just one of several recent events—both scientific and cultural—that reveal this deep and almost unquenchable curiosity about space—and the possibility that life is 'out there' " (1997, p. 1).

Claims such as these fit a pattern common to much of space advocacy—the promulgation of assertions about the benefits of space exploration, the public's support for space exploration, and so on, without the provision of adequate evidence—and in many cases, without the provision of any evidence whatsoever. There is, to be sure, a considerable amount of information available about the public's views on space exploration and on extraterrestrial life (ETL). Most surveys on ETL have been conducted in order to determine the educational effects of including astrobiology in school curricula, or to gauge how the public might react to the discovery of ETL. Seldom has anyone gathered data on the public's interest in and support for specifically *astrobiological* attempts to answer questions about ETL. This is important for assessing the level of support for astrobiology, because one could be interested in life's extent and origin without being at all interested in what scientists have to say about these issues. Thus, even if it is true that humans are broadly interested in the origin and extent of life, it does not follow that the public is greatly supportive of the *scientific* search for ETL, that is, as it is conducted by astrobiologists.[1]

In order to make some headway on this issue, I compare what is known about public opinion on ETL with what is known about public opinion on evolution and on space exploration, since both evolution and space exploration are implicated in astrobiology and the search for ETL. Following this, I relay some findings about the influence of religion on one's views about evolution. Unsurprisingly, at least in the United States, religiosity has a clear influence on views about evolution. I anticipate that religiosity similarly influences views about ETL, insofar as the existence of life elsewhere might be perceived as implicating evolutionary theory. Next, I relay some findings about the history of U.S. public opinion on space exploration. Though the U.S. public largely approves of NASA, few think that it deserves funding increases. I anticipate that individuals will feel similarly about the search for ETL, insofar as it implicates space exploration.

I then discuss the results of a handful of surveys that have asked questions specifically about beliefs about ETL and about the scientific search for ETL. Though all results appear consistent with the suspicions of the previous paragraph, they only serve to muddy the waters further. With one exception, none of these surveys attempted to control for beliefs about nonintelligent ETL versus beliefs about intelligent ETL (ETI). It is consequently unclear how to interpret most of the available information on the public's interest in, and willingness to fund, the search for ETL. It could be that the public is primarily interested in searching for ETI and not much interested in the search for simpler forms of ETL. If so, then astrobiology does not receive the warm and universal support often claimed of it. I close the chapter by highlighting a series of concerns that should be addressed as part of future data collection.

It is worth expressing up front that I ardently support astrobiology (as well as space science more generally).[2] So my quibble in this chapter is not with astrobiology itself but instead with certain aspects of astrobiology *advocacy*. In the spirit of the "myth-free" series of papers of which this is the fourth installment,[3] I would implore astrobiologists, as scientists, to model good scientific practice when speaking of the importance or value of astrobiology. This means refraining from promulgating rationales that have little to no backing in evidence. If my suspicions are correct, we cannot support any robust conclusions about the level of public support for astrobiology, which means that we are not in a position to confirm (or deny) the universal appeal of astrobiology. Even if, at the end of the day, it is true that there is wide, enthusiastic support for astrobiology, the problem is that at the moment we do not have sufficient *evidence* for believing that this is the case.

## Views on Religion and Evolution

One hypothesis worth exploring is that views on evolution are connected to views on ETL. The former, notably, tend to be highly influenced by religion. Thus, it is plausible that one's religious views influence one's views on the possibility of ETL and the importance of the scientific search for ETL. According to Pew Research Center's 2014 Religious Landscape Study of U.S. adults (sample size = 35,071), 33% believe that humans evolved due to natural processes, 25% believe that humans evolved "due to God's design," and 34% believe that humans always existed in their present form.[4] This latter figure is down from the 42% who denied evolution in 2005, and the combined figure for believers in evolution (58%) is up from 48% over the same time period.[5]

As Table 15.1 shows, views on evolution vary considerably by religious tradition. Among those who believe in intelligent design, 90% say their religion is either somewhat or very important in their lives. Of those who believe that humans always existed in their present form, 93% say religion is either somewhat or very important in their lives. Meanwhile, 52% of those believing in evolution due to natural processes say that religion is either somewhat or very important in their lives.[6] There is rather little variation for how often one "feels a sense of wonder about the universe" based on one's views about evolution. Sixty two percent of those believing in evolution due to natural processes felt this way at least one a month, while 37% felt this way several times a year or less. For those believing in intelligent design, the percentages are 64% and 36%, respectively. Meanwhile, for those believing that humans always existed in their present form, the percentages are 60% and 38%, respectively.[7]

The 2014 Pew survey also noted that the more frequently one attends religious services, the more likely one is to believe that humans always existed in their present form, or that humans evolved due to God's design.[8] A similar pattern holds for individuals based on the strength of their belief that scripture is the

Table 15.1. Views on Evolution Based On Religion: 2014 Pew Religious Landscape Study.

| Religion | Evolved (natural selection) | Evolved (intelligent design) | Always existed in present form |
|---|---|---|---|
| Unaffiliated | 63% | 14% | 15% |
| Catholic | 31% | 31% | 29% |
| Evangelical Protestant | 11% | 25% | 57% |
| Black Protestant | 16% | 31% | 45% |
| Mainline Protestant | 28% | 31% | 30% |
| Muslim | 25% | 25% | 41% |
| Orthodox Christian | 29% | 25% | 36% |
| Jewish | 58% | 18% | 16% |

literal word of God.[9] This corroborates an hypothesis of (Freeman and Houston, 2009) that it is neither religion per se nor political affiliation that influences one's views on evolution but rather the *nature* of one's religious beliefs:

> those [conservatives] who accept the inerrancy of the Bible, are more likely to reject the validity of evolution. But political conservatives who do not accept this face of orthodox Christian doctrine do not differ from political moderates. This suggests that a rejection of evolution is no part of a conservative ideology, but is instead a part of ones orthodox beliefs. . . political liberals who accept the Bible as the inerrant word of God are less likely than others to accept the validity of evolution. (p. 68)

It should be noted, however, that a 2013 Pew study found that partisan differences remain even when controlling for religion.[10] Though not included in the 2014 study, respondents to the 2005 Pew survey were asked "How certain are you about how life developed on Earth?" Among those believing that humans always existed in their present form, 63% were "very certain about how life developed" (69% of biblical literalists), meanwhile, 39% of believers in intelligent design were very certain, and only 28% of believers in evolution due to natural processes were very certain.

Now what might this have to do with support for astrobiology? It must be admitted up front that one's views on evolution may not be predictive about one's views on astrobiology or on ETL. After all, people are known to possess inconsistent belief sets. Nevertheless, it is initially plausible that that skepticism about evolution correlates with a lack of interest in *scientific* answers about, for example, the origin of life or the existence of ETL. That is, those with extreme doubts about evolution via natural processes might not only lack interest in astrobiology; they might also prefer that astrobiologists not engage in research on the origin and extent of life. This could be evidenced by the fact that the majority of creationists are "very certain" in their creationism and so presumably are not at all interested in what biologists have to say on the issue. The cogency of this hypothesis depends on the degree to which evolution, astrobiology, and the search for ETL are linked together in the eyes of the public—an interesting avenue for further research, to be sure.

## Views on Space Exploration

A second hypothesis worth exploring is that the public's views on astrobiology are influenced by their views on space exploration more generally. As Roger Launius has on several occasions noted, there has been and there continues to

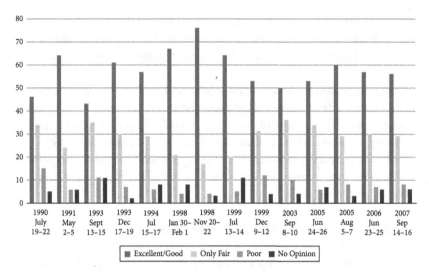

**Figure 15.1** Answers by percentage to the question "How would you rate the job being done by NASA—the U.S. space agency? Would you say it is doing an excellent, good, only fair, or poor job?" from a series of Gallup polls from 1990 to 2007.

be a mismatch between, on the one hand, the public's approval of NASA and the space program and, on the other hand, the public's willingness to support funding increases for space exploration.[11] The public's approval of NASA has always been relatively high. For instance, a series of Gallup polls between 1990 and 2007 asked respondents to "rate the job being done by NASA—the U.S. space agency" as doing either an excellent, good, only fair, or poor job. During this time (see Figure 15.1) an average of 57.6% said that NASA was doing an excellent or good job; 28.5% said NASA was doing only a fair job; 7.8% said NASA was doing a poor job; and 6% had no opinion.[12]

More recently, a 2015 Pew Research Survey found that 68% of Americans have a favorable opinion on NASA, with this position positively correlated with educational attainment (see Table 15.2).[13] NASA was not the most highly rated federal agency—that title belongs to the Center for Disease Control, with a 70% favorable rating.[14] And as the General Social Surveys (GSS) from between 2008 and 2014 show (see Figure 15.2), an average of 67.2% of Americans were either very or moderately interested in space exploration, while an average of 32.1% were not at all interested.[15]

Meanwhile, a different picture emerges when the public is asked specifically about funding for NASA. Figure 15.3 captures GSS responses to the question: "Are we spending too much, too little, or about the right amount on" space exploration? As can be seen, the numbers from 2000 onward are more friendly

**Table 15.2.** 2015 Pew Research Survey Responses to the Question "Is Your Overall Opinion of NASA Very Favorable, Mostly Favorable, Mostly Unfavorable, or Very Unfavorable?" Sorted by Educational Attainment

|                    | Very / Mostly Favorable | Very / Mostly Unfavorable | No Opinion |
| ------------------ | ----------------------- | ------------------------- | ---------- |
| Total              | 68%                     | 17%                       | 16%        |
| Postgraduate       | 78%                     | 11%                       | 11%        |
| College Graduate   | 71%                     | 16%                       | 13%        |
| Some College       | 71%                     | 13%                       | 16%        |
| High School or Less | 61%                    | 21%                       | 18%        |

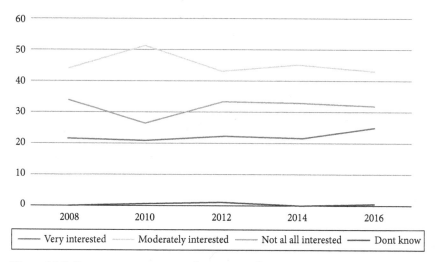

**Figure 15.2** Percentage responses to the question: "Are you very interested, moderately interested, or not at all interested in issues about space exploration?" from the GSS intspac data set (in use 2008 to 2016).

than the numbers since 1973 when the GSS began asking about space. Moreover, the numbers from the most recent GSS are even more friendly than the average of the period between 2000 and 2016. As Bainbridge (2015) argues, this is most likely due to holdover animosity to the relatively large expenditures on space during the Apollo era. Figures 15.4 and 15.5 show how opinions on space funding have fluctuated over time, both during the entire period of the GSS and since 2000.[16] These figures also provide a comparison with NASA's share of the federal budget over the same period.

**Table 15.3.** A Comparison of Responses to the NATSPAC and NATSPACY GSS Questionnaires on Funding for "Space Exploration" (NATSPACY) versus "the Space Exploration Program" (NATSPAC), 1984–2016

|  | NATSPACY | NATSPAC | Difference |
|---|---|---|---|
| Too little | 13.8% | 13.8% | 0.01% |
| About right | 41.2% | 41.9% | 0.66% |
| Too much | 38% | 37.3% | 0.67% |
| Don't know | 6.1% | 6.7% | 0.60% |

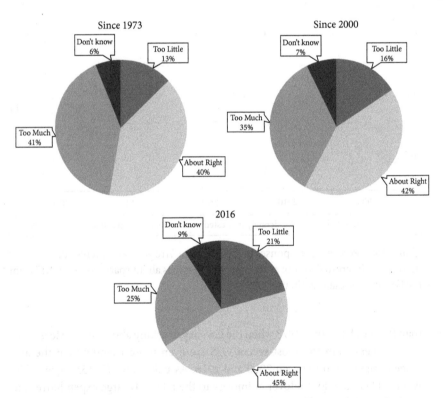

**Figure 15.3** Total responses to the question "are we spending too much, too little, or about the right amount" on space exploration? Percentages are based on combining totals from both the natspac (1973 onward) and natspacy (1984 onward) GSS data sets.

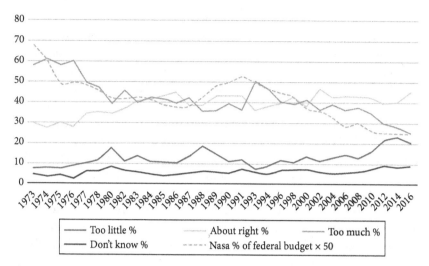

**Figure 15.4** Perceptions on space exploration spending from 1973 to 2016, based on combining totals from both the natspac (1973 onward) and natspacy (1984 onward) GSS data sets. NASA's share of the federal budget, which varied from 1.35% to 0.5% during this period, has been multiplied by 50 to better visualize a comparison of it with the results of the GSS questionnaire.

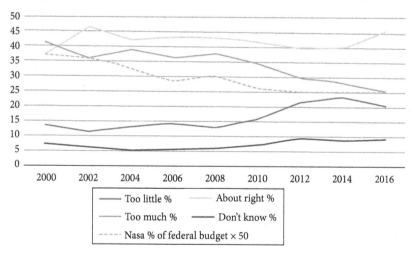

**Figure 15.5** Perceptions on space exploration spending from 2000 to 2016, based on both the natspac and natspacy GSS data sets. As in Figure 15.4, NASA's percentage of the federal budget, which varied from 0.75% to 0.5% during this period, has been scaled up by a factor of 50 to help visualize a comparison of it with the results of the GSS questionnaire.

Focusing on the period between 2006 and 2010, William Bainbridge (2015) finds evidence of the influence of scientific literacy and religion on beliefs about space funding. Among the 52.7% of GSS respondents who correctly identified that "human beings, as we know them today, developed from earlier species of animals," 18.2% said there was too little space funding; meanwhile, only 10.7% of those denying this claim thought there was too little funding for space (p. 122). Similarly, among the 50.4% who said that "the universe began with a huge explosion," 20.2% said there was too little space funding; of those denying this claim, only 9% say there is too little space funding (p. 122). Scientific literacy may be the primary culprit here, since other questions seemingly unrelated to any potential conflict between science and religion revealed similar differences in attitudes toward space. For instance, of those correctly denying that lasers work by focusing sound waves, 20.1% say there is too little space funding; meanwhile of those incorrectly agreeing that lasers work by focusing sound waves, only 9.8% say there is too little funding for space (p. 122).

Nevertheless Bainbridge does find data that bear directly on the influence of religion and, in particular, on the strength of one's religious convictions. Of those who "know God really exists" and who have no doubts about it, only 11.7% say there is too little funding (pp. 124–125). Compare this with 22.4% of atheists and 24.9% of agnostics (p. 125). A similar trend emerges when examining frequency of attendance of religious services (see Table 15.4): of those who say they attend religious services more than once per week, only 7% say there is too little space funding. Compare this with 16.9% of those who never attend religious services.

According to Joshua Ambrosius' (2015) analysis of data from the GSS and from several other surveys, we must also be mindful of religious tradition. Just as religious tradition appears to influence one's views on evolution, so too does religious tradition seem "to affect space knowledge, policy support, and the general

Table 15.4. Answers to the Question "Are We Spending Too Much, Too Little, or About the Right Amount" on Space Exploration? Based on Frequency of Attending Religious Services

|  | More than once a week | Once a week | Once month | a Never |
|---|---|---|---|---|
| Too little | 7% | 9.3% | 13.4% | 16.9% |
| Too much | 57.5% | 48.4% | 45.7% | 43.6% |

Source. Data from Bainbridge (2015, pp. 124–125).

benefits of space exploration" (p. 22). As with evolution, Evangelical Christians stand out:

> Evangelicals are indeed less knowledgeable (even if unwilling to admit their knowledge), interested, and supportive of space/space policy than the population as a whole and/or other religious traditions. This is a problem for the future of space exploration because Evangelicals make up more than one-quarter of the U.S. population . . . and thus a significant share of potential space-minded constituents. (p. 25)

Meanwhile, those identifying as Jewish, Hindu, or Buddhist had greater than average interests in space (p. 23). Thus, part of the explanation for why religiosity correlates with pessimism about space must reference Evangelicals, who are more likely than other religious groups to attend services once a week or more and to believe that scripture is the literal word of God.

To summarize, space exploration is widely popular, but this popularity does not extend to increasing funding for space exploration. What does this mean for astrobiology? A natural hypothesis is that, as with space exploration, most Americans are interested in astrobiology and the scientific search for ETL but that few Americans are willing to increase the amount of funding for astrobiology and the search for ETL. Nevertheless, extra care must be taken when speculating about the public's interest in or approval of astrobiology. To reiterate the primary concern of the previous section, insofar as astrobiology is part and parcel with evolution by natural selection, it might be that those skeptical of evolution by natural selection are similarly skeptical of, and unwilling to support, astrobiology. Indeed, educational attainment and scientific literacy positively correlate both with belief in evolution by natural selection and with willingness to increase space funding, and religiosity negatively correlates with belief in evolution by natural selection and with willingness to increase space funding. Thus, astrobiology might be subject to acute levels of disapprobation, at least among those with less education or those with higher religiosity, and especially among Evangelicals. So it is not safe to assume that astrobiology inherits the same degree of popularity as space exploration more generally. Astrobiology could be an exception in any number of ways. It would be worth determining whether the public distinguishes astrobiology from other forms of space exploration, as well as where the public's priorities in space lie. It would also be worth learning how the public prioritizes both space exploration generally and astrobiology specifically in the context of their views on science and in the context of their views on social issues. What, then, is known about astrobiology specifically?

## Astrobiology Surveys

Next I review and discuss the small number of surveys that have collected data specifically relating to the public's interest in, and support for, astrobiology. A complicating issue is that most papers in this area aim primarily to report on the outcomes of various educational experiments (e.g., including astrobiology in school curricula; hosting public educational outreach events). Such studies typically involve samples consisting either of students learning about astrobiology or of self-selected outreach participants, and these individuals may not be representative of the general public when it comes to their views on astrobiology.[17] And, while it is certainly important to learn what might influence individuals, especially schoolchildren, to become more interested in astrobiology, that is not the goal of this chapter, which is rather to ascertain what is known about the public's interest in astrobiology—and not what the public *would think* if its members were exposed to various educational stimuli. It is granted that many of the educational outcomes are probably relevant for support for astrobiology. Indeed, as I argued in Part III of this series, there is evidence of a causal link *from* scientific literacy *to* support for spaceflight.[18] If including astrobiology in the school curricula succeeds in increasing students' scientific literacy, that may well increase students' support for astrobiology.

## Swami et al. (2009)

Swami et al. (2009) conducted a survey of 577 Austrian and British individuals from various backgrounds to ascertain their beliefs about ETL, broadly speaking. That is, their survey aimed to gather data not only on beliefs about, for example, microbial life in the Solar System, but other forms of ETL as well, including ETI. Respondents were asked to rate their level of agreement on 37 statements using a 7-point Likert scale, where a value of 1 signifies disagreement and a value of 7 signifies agreement. Results were reported on the basis of nationality and gender. The results for questions that pertain directly to views about the scientific search for life follow.

With respect to the statement, "the search for extraterrestrial life is a serious and important scientific endeavor," the average scores were as follows:

| Austrian Women | Austrian Men | British Women | British Men |
|---|---|---|---|
| 4.38 | 3.97 | 3.67 | 4.00 |

Thus, among the sample there is only a very slight agreement with the claim. This statement was reverse-coded as "the search for extraterrestrial life is a waste of time and money" with the following average scores:

| Austrian Women | Austrian Men | British Women | British Men |
|---|---|---|---|
| 3.61 | 4.04 | 4.44 | 4.17 |

Thus, among the sample there is somewhat higher agreement with this negative claim about the search for ETL. It would seem that at least some individuals responded inconsistently to these two statements. Meanwhile, respondents were somewhat inclined to agree that "the search for extraterrestrial life is a pseudoscience":

| Austrian Women | Austrian Men | British Women | British Men |
|---|---|---|---|
| 3.36 | 3.93 | 3.70 | 3.81 |

Swami et al. do not indicate whether any of the inconsistencies in these responses are significant. Even so, it may be that the responses were subject to framing effects. The description of the search for life as a "scientific endeavor" may have contributed to a higher agreement rate to the first statement, whereas the absence of this description may have contributed to higher rate of agreement with the reverse-coded second statement as well as the third statement. After all, it is possible that when respondents viewed the search for ETL as a *scientific* endeavor they were thinking primarily of, for example, attempts to discover signs of microbial life on Mars; and when respondents viewed the search for ETL as a waste of time and money or as a pseudoscience that they were thinking primarily of the more controversial search for ET*I* (or even of alien visitation). Nevertheless, consistent with polls of American attitudes about space exploration funding, the respondents generally disagreed with the claim that "governments should direct more funding to the scientific search for extraterrestrial life":

| Austrian Women | Austrian Men | British Women | British Men |
|---|---|---|---|
| 3.08 | 2.40 | 2.32 | 2.44 |

Swami et al. provide a possible explanation for the moderate responses concerning the search for life—that "although participants believed in the possibility of extraterrestrial life, the lack of conclusive findings to date has resulted in a somewhat negative perception of astrobiology and related sciences" (p. 40). They also venture a further hypothesis, related to the above discussion of religion:

> Our results also suggest that extraterrestrial beliefs were related to participants' religiosity and political orientation. In general, higher religiosity and more right-wing political orientation were associated with decreased belief in extra- terrestrial life. The former finding is consistent with the suggestion that there is an inverse relation between extraterrestrial beliefs (or other paranormal beliefs) and religious beliefs. (p. 40)

This suggestion highlights the importance of distinguishing clearly between var- ious categories of ETL. This is because it is possible that religious individuals feel differently about the possibility of nonintelligent ETL (e.g., Martian microbes) than they do about the possibility of ETI.

Given that the bulk of this survey asked specifically about ET*I*, as well as the paranormal (e.g., UFOs, visitation, abduction), it is possible that these results are only descriptive of beliefs about ETI. This, however, does not automatically in- dicate that results concerning, for example, the search for microbial life on Mars or Europa, would be more positive. The discovery of ETI could be much more exciting to the general public than the discovery of microbial life. So even if the public feels that we are more likely to discover microbial life, they may be less en- thusiastic about searching for it.

## Pettinico (2011)

Pettinico (2011) relays a variety of results from a 2005 telephone survey of a random sample of 1,000 U.S. adults, originally intended for use in the 2005 tel- evision documentary *Extraterrestrial*. The survey was a joint project between the National Geographic Channel, the SETI Institute, and the Center for Survey Research and Analysis at the University of Connecticut. Several of the results discussed by Pettinico are worth mentioning here.

In response to the question "Do you believe that there is life on other planets in the universe besides Earth?," 60% of the sample said yes; 32% said no; and 8% were not sure (p. 104). Belief in ETL correlated negatively with frequency of reli- gious service attendance: Only 45% of those attending weekly services believed in life on other planets, whereas 70% of those rarely or never attending services believed there was life on other planets (p. 104). Pettinico also reports a positive correlation with belief in life on other planets and household income (p. 106).

The 68% who did not reject belief in ETL were asked to identify the likelihood of the existence of a number of categories of ETL, which is summarized in Table

15.5, recreated from Pettinico (2011, p. 111).[19] Of the 32% not open to ETL, 56% cited religion as a major reason (p. 113). Among this group (about 18% of the total sample), frequency of attending religious services correlated positively with the identification of religion as a major reason for rejecting the possibility of ETL, with 72% of those attending services weekly giving this reason compared to only 31% of those attending rarely or never (p. 114).

This information might lend credence the analogy between evolution and astrobiology, since religiosity is negatively correlated with both belief in evolution by natural selection and belief in the possibility of ETL. Nevertheless, these results do not provide definitive insight into the public's interest in and support for astrobiology, if only because of the curious spread of beliefs about the likely nature of ETL shown in Table 15.5, where those open to ETL view human-like and superior ETI as more likely to exist than complex, nonintelligent ETL. Pettinico ventures an explanation:

> It makes sense that the public feels very simple life forms such as microbes are the most probable extraterrestrial life forms, since most space scientists would generally agree—at least that extraterrestrial microbes would probably be more common than more advanced life forms. However, the public is more likely to believe in the probability of advanced life forms . . . than they are to believe in the probability of plant-like or animal-like life forms. This may be, in part, due to the impact of the media, which tends to emphasize human-like or advanced extraterrestrial life forms. When average Americans think about aliens, they

**Table 15.5.** Response Rates to the Question "In Your Opinion, How Likely Is it That Some of These Life Forms on Other Planets Would Be _____? Would You Say It Is Very Likely, Somewhat Likely, or Not At All Likely?"

|  | Very Likely | Somewhat Likely | Not Likely At All |
|---|---|---|---|
| Similar to single cell or few cell organisms like microbes or bacteria | 45% | 42% | 8% |
| Similar to plants like trees and flowers | 25% | 54% | 17% |
| Similar to animals, like birds, lizards or mammals | 21% | 48% | 25% |
| Similar to humans | 30% | 46% | 21% |
| Intelligent life forms more advanced than humans | 39% | 41% | 16% |

*Source.* From Pettinico (2011, p. 111).

may more easily envision *Star Trek*'s Klingons than they do some sort of lower-level animal. (p. 111)

Of course, it is important to ask why it is that, for example, media portrayals of ETL tend to be ETI and very often human-like ETI. Clearly, human-like ETI are easier to portray in film. But it also could be that ETI, and human-like ETI, are simply more interesting to humans than other forms of ETL. Thus, it could be that what drive the responses in the cases of human-like and superior ETI are not scientifically grounded opinions but instead *preferences* based on what the respondents hope is the case or what they would find most exciting. Again, it is important to attempt to control for this potential difference in enthusiasm about ETL. It is possible that those who are enthusiastic about the search for life are primarily enthusiastic about the potential discovery of human-like or superior ETI, and less so concerning "simpler" forms of ETL.

## Oreiro and Solbes (2017)

Oreiro and Solbes (2017) conducted a survey of 89 students (averaging 15 years in age) attending urban schools in Spain with the goal of ascertaining both students' knowledge of astrobiology, and the quality of students' reasoning on astrobiological issues. The authors asked three questions that are relevant in the present setting.

The first of these questions (question 6 from their survey) is: "Do you think there exist living forms in other places of the universe, outside our solar system? Why?" Answers were sorted based on both agreement and rationale. A total of 68.6% of the students thought that ETL probably exists, giving what the authors took to be a correct argument (p. 95).[20] Meanwhile, 10.1% thought ETL probably does not exist, giving what the authors took to be a correct argument;[21] and 21.3% responded (either positively or negatively) without explaining, with an incorrect explanation, with doubt, or with no answer (p. 95).

The second question of interest (question 7 from the survey) asked: "Do you think it is important to figure out whether intelligent civilizations exist anywhere? Why?" A large majority, 86.5%, said it was important to determine whether ETI's exist; 7.9% said it was not important; 3.4% expressed doubt; and 2.2% did not answer (p. 95). Among the 86.5% agreeing that it is important to figure out whether ETI exist, 53.9% said this was because of the potential for advances in science and technology; 11.2% expressed "intrinsic interest toward other species"; and 4.5% wanted "to learn whether or not we are alone" (p. 95).

The final question of interest (question 8 from the survey) asked: "Do you believe UFOs exist? That is, intelligent beings visiting our planet? Justify the answer." Most students, 53.9%, gave a negative answer; 21.3% expressed doubt; and 20.2% said they believed that ETI have visited Earth (p. 95).

Assuming consistency on the part of the students' responses (i.e., assuming no student said they believed in ETI but said they did not believe in ETL), it would seem that belief in ETL is at least somewhat independent from belief in alien visitation. Again, assuming consistency among the responses, it would appear that 48.4% of the sample (70.6% of those believing in ETL) believes that ETL exists but does not believe that any ETI has ever visited Earth (and it would be interesting to learn how many of those answering "yes" to question 8 also gave "correct arguments" for question 6!).

Most noteworthy here is the large majority of respondents to question 7 (86.5%) who expressed an interest in figuring out whether ETI exists. It is unclear what this means for astrobiology. Since the question was asked about ETI specifically, and not ETL more generally, it is not clear what the same individuals would say were they to be asked about the importance of, for example, searching for microbial life on Mars, Europa, or Enceladus. Even so, only a very small percentage (4.5%) provided a rationale—to learn whether we are alone in the universe—that is often heralded by astrobiology advocates as being of paramount importance. Moreover, question 8, which asks specifically about alien visitation, does not serve as a perfect indicator of beliefs about ETI. After all, one might be open to the possibility of ETI but doubt that any ETI has visited Earth or otherwise contacted humans in any way. So it is not possible to tell, based on the results of the survey, how many students open to the possibility of ETL were also open to the possibility of ETI. This again highlights the need to carefully distinguish between these possibilities when surveying the public.

## Persson, Capova, and Li (2019)

Persson, Capova, and Li (2019) conducted a written survey of 512 high school and university students in Sweden. Among the respondents, 90% answered "yes" when asked "Do you think there is life outside our own planet?" (p. 281). Table 15.6 shows the overall responses to the question, "How important do you think it is to search for extraterrestrial life?" A total of 52% felt it was at least somewhat important; 24% were indifferent; and 23% did not think the search was important.[22] Respondents were also asked "What is the primary reason for your answer to [the previous question]?," with the sorted responses given in Table 15.7.

Persson, Capova, and Li explain that the "interesting" and "uninteresting" responses indicate intrinsic interest or excitement (or lack thereof); the "useful" and "useless" responses indicate perceived scientific, social, or personal utility (or disutility); and the "economic reasons" responses indicate a feeling that the search for ETL would be wasteful (p. 284). Presumably, the 37% who thought

**Table 15.6.** Answers to the Question: "How Important Do You Think It Is to Search For Extraterrestrial Life?"

| | |
|---|---|
| Very important | 15% |
| Quite important | 37% |
| Does not matter | 24% |
| Something we should not prioritize | 20% |
| Something we should avoid | 3% |

**Table 15.7.** Respondents' Reasons for the Answers Reported in Table 15.6

| What is the main reason for your answer [about the importance of the search for life]? | |
|---|---|
| Interesting | 37% |
| Uninteresting | 8% |
| Useful | 20% |
| Useless | 24% |
| Economic Reasons | 3% |

the search for ETL was interesting and the 20% who felt the search was useful overlap significantly with the individuals responding that the search for ETL was either quite important (37%) or very important (15%).

Persson, Capova, and Li recognize that their survey did not attempt to control for students' views on ETL versus ETI, or for their views on ETL and ETI versus alien visitation:

> Even though we explicitly asked about life outside our planet and the question was asked in a questionnaire dealing with attitudes towards the scientific search for extraterrestrial life, we can of course not know whether some of the respondents were actually thinking of extraterrestrial life visiting the Earth rather than about the life we are looking for off our planet. (p. 285)

They later on state that they "found a need to look more deeply into what kind of life students think of when asked about extraterrestrial life" (p. 287).

Consequently it is unclear what activities are supported by those who are in favor of the search for ETL. It is quite possible that varying degrees of importance would be attached to, e.g., the search for microbial life on Mars, listening for ETI signals, detecting exoplanets, uncovering the truth about Roswell, and so on.

## General Comments

It appears there is evidence corroborating the idea that one's religious views influence one's views on the search for ETL. As Pettinico (2011) suggests, religiosity is negatively correlated with being open to the possibility of ETL. And, as the surveys discussed in this section suggest, more individuals claim to be interested in the search for ETL than individuals claiming otherwise. It is possibly important that the surveys described in Swami et al. (2009), Oreiro and Solbes (2017), and Persson et al. (2019) were conducted in Europe, where religion appears less likely to influence public opinion on science and evolution. Thus, support for the search for ETL may be lower in the United States, where religion exerts a stronger pull on public opinion on science and evolution. At the same time, given that the United States spends considerably more on space than Europe, it may be that the average European is less aware of, for example, the European Space Agency and its activities than is the average American with respect to NASA and its activities.[23] If so, then it would not be surprising if the U.S. public is more supportive than Europe when it comes to the search for ETL.

## Lessons

As should be clear, not much is known with confidence about the public's views on astrobiology and the search for ETL. This is perfectly understandable given the limitations on funding not only for astrobiology but also for sociological study related to astrobiology. Whether there is need for a substantive study on these topics hinges on for what purposes such information might be used. If indeed the public is exceptionally interested in astrobiology and the search for ETL, this would be relevant for space policy and for decisions about space exploration budget allocations. Nevertheless, my aim here has not been to advocate for astrobiology but instead to determine whether there is evidence supporting a common talking point of astrobiology advocacy—that astrobiology's search for ETL seeks to answer questions of great interest to all of humanity. While it is consistent with this information that a majority of the U.S. public is interested in the search for ETL, many factors have yet to be disentangled in a satisfying way:

- **Interests in ETL are diverse.**
  Interest in ETL could come from interest in the possibility of microbial life in the Solar System, or from interest in the possibility of intelligent life elsewhere in the universe, or from interest in the paranormal (e.g., UFOs and alien visitation). These interests are independent. A person could be interested in the paranormal and not at all interested in, for example, microbial ETL. Similarly, a person could be very interested in the possibility of life on Mars but not at all interested in possible biosignatures from exoplanets. With the exception of Pettinico (2011), I am not aware of any attempt to gauge public opinion at this level of detail.

- **Beliefs about ETL are not identical to beliefs about the importance or value of searching for ETL.**
  Just because someone believes that ETL exists does not mean one thinks it is important to search for evidence of ETL. Moreover, just because someone believes that ETL does not exist does not mean one thinks it is not important to search for evidence of ETL. The public's beliefs about the extent of ETL are not known to be reliable indicators the public's views about attempts to uncover evidence of the existence of ETL.

- **Interest in ETL is not tantamount to interest in what *science* has to say about ETL.**
  It is possible that someone could be interested in ETL but indifferent or opposed to the *scientific* search for ETL. Some religious individuals are interested in life's origin but equally uninterested in anything science has to say about the topic. It is plausible that the same holds for some people when it comes to ETL. That is, that many individuals interested in ETL may have little interest in what, for example, astrobiology uncovers. Two obvious examples would be conspiracy types who believe in alien visitation despite the privation of good evidence for such beliefs, as well certain religious individuals who believe (and have little doubt) either that God only created life on Earth or that God created life wherever it exists.

- **Interest in ETL, and even interest in the science surrounding ETL, is not tantamount to willingness to increase funding for the scientific search for ETL.**
  The search for ETL, like space exploration more generally, no doubt attracts many fans. But if the analogy with space exploration is apt, then we should

expect few individuals to be supportive of funding increases for the search for ETL. It is important to ask both sorts of questions when soliciting the public's views on the search for ETL.

- **Absolute interest in ETL is not tantamount to relative interest in ETL or to viewing the search for ETL as a priority.**
  It is possible that those who are very interested in ETL, and even those who think it deserves funding increases, nevertheless do not view the search for ETL as a priority compared to other interests. As much holds for views on space exploration more generally. Thus, it is not enough to merely ask, in isolation, whether one thinks the search for ETL is important. Rather, it should be determined how important the search for ETL is compared both to other space exploration objectives as well as to other projects, scientific or otherwise.

It also bears remarking that human curiosity is highly idiosyncratic. Some people are more prone to engage in sensation-seeking (e.g., social or physical thrill-seeking); others are more prone to engage in information-seeking (e.g., to acquire new knowledge). There is no single issue that every person is curious about—and certainly not every person is curious about space or about the possible existence of ETL.[24] Admittedly, there is more to astrobiology than the search for ETL, and thus, interest in astrobiology may not imply interest in the search for ETL. Nevertheless the search for ETL is a core item on the astrobiology agenda, and one that astrobiologists are not shy about promoting. Presumably, the individuals working on astrobiology projects are fascinated by questions about the origin and extent of life. Advocates who presume that the public shares their deep interest in the origin and extent of life may be exhibiting the false-consensus effect.

We know a fair bit about what the public thinks about space exploration. And we know some things about what the public believes about ETL. But we do not know very much at all about how interested the public is in supporting science's search for answers to questions about ETL. This should give astrobiology advocates pause when proclaiming the universal importance of astrobiology to answering "life's big questions." I, of course, have no interest in discounting the importance of astrobiology as a scientific discipline. I simply would prefer that we all do a better job of modeling good philosophical and scientific practice on the occasions when we articulate the value of astrobiology (and space exploration more generally). Far too many arguments are promulgated not because they have merit but rather because they have become part of the standard advocacy package for astrobiology. If evidence matters to the practice of astrobiology, it ought also to matter in discussions about the importance and value of astrobiology.

# Acknowledgments

A presentation based on this chapter was delivered to the 2018 Social and Conceptual Issues in Astrobiology Meeting on April 14, 2018, at the University of Nevada-Reno. Thanks to three anonymous referees for helpful comments.

# Notes

1. There is some debate as to whether, for example, the search for extraterrestrial intelligence and messaging extraterrestrial intelligence fall under the remit of astrobiology. While I intend to remain agnostic on these matters, my focus here is on the demarcation of public opinion concerning the (biologically oriented) search for *nonintelligent* ETL through, for example, Solar System exploration, exoplanet observation, and so on.
2. A thorough defense of the value (intrinsic as well as instrumental) of space science is the subject of Schwartz (2020).
3. The first three installments are Schwartz, (2017a, 2017b, 2018)–but see also Chapter 1 of Schwartz (2020), which dramatically improves upon the data analysis from Schwartz (2018).
4. "Religious Landscape Study," http://www.pewforum.org/religious-landscape-study/. All respondents were asked "Which comes closer to your view? [That] humans and other living things have evolved over time [or that] humans and other living things have existed in their present form since the beginning of time?" Those selecting the first choice were given a follow up: "And do you think that [either] humans and other things have evolved due to natural processes such as natural selection [or] a supreme being guided the evolution of living things for the purpose of creating humans and other life in the form it exists today?"
5. Those believing that humans evolved due to natural processes increased from 26% to 33%; those believing in intelligent design increased from 18% to 25%. See "Religion: A Strength and Weakness for Both Parties," http://www.people-press.org/2005/08/30/religion-a-strength-and-weakness-for-both-parties/. The 2005 and 2014 questionnaires used nearly identical wording, although the 2005 questionnaire prefaced the evolution question with "Some people think that humans and other living things [have evolved over time]. Others think that humans and other living things [have existed in their present form since the beginning of time]."
6. These numbers come from comparing responses to the 2014 Pew survey's evolution question with another question on the survey, which asked: "How important is religion in your life—very important, somewhat important, not too important, or not at all important?"
7. These numbers come from comparing responses to the 2014 Pew survey's evolution question with another question on the survey, which asked: "Now, thinking about some different kinds of experiences, how often do you feel a deep sense of wonder

about the universe—would you say at least once a week, once or twice a month, several times a year, seldom, or never?"

8. This comes from comparing answers to the evolution question to another question from the survey, which asked: "Aside from weddings and funerals, how often do you attend religious services: more than once a week, once a week, once or twice a month, a few times a year, seldom, or never?"

9. This is based on comparing answers to the evolution question with two further questions, the wording of which varied based on the religion of the respondent: "Which comes closest to your view? The [holy book] is the word of God [or] the [holy book] is a book written by men and is not the word of god." Those picking the first option were given a follow up: "And would you say that the [holy book] is to be taken literally, word for word, [or] not everything in the [holy book] should be taken literally, word for word."

10. See "Public's Views on Human Evolution," http://www.pewforum.org/2013/12/30/publics-views-on-human-evolution/.

11. See, e.g., Launius (2003).

12. See "Americans Continue to Rate NASA Positively," http://news.gallup.com/poll/102466/americans-continue-rate-nasa-positively.aspx.

13. See "NASA Popularity Still Sky-High," http://www.pewresearch.org/fact-tank/2015/02/03/nasa-popularity-still-sky-high/.

14. The Department of Defense was a close third, with a 65% approval rating. Unsurprisingly, the Internal Revenue Service was the least favorably rated, at only 45%.

15. See "GSS Data Explorer: Interested in Space Exploration," https://gssdataexplorer.norc.org/variables/3459/vshow.

16. Figures 15.3, 15.4, and 15.5 use data aggregated from both the NATSPAC and NATSPACY GSS data sets. The sole difference between these questionnaires was that the NATSPAC questionnaire (in use since 1973) asked about funding for "the space exploration program" whereas the NATSPACY questionnaire (in use since 1984) asked about funding for "space exploration." As Table 15.3 shows, from 1984 onwards there is very little difference between the NATSPAC and NATSPACY wordings. This means it is likely that the responses during the period from 1973 to 1983 regarding "the space exploration program" are representative of how individuals would have responded during this time had they been asked about "space exploration." Thus, aggregation of the NATSPAC and NATSPACY responses is appropriate.

17. Studies and reports in this category include Brake et al. (2006); Oliver and Fergusson (2007); Foster and Drew (2009); Fergusson, Oliver, and Walter (2012); Hansson and Redfors (2013); and Cockell et al. (2018).

18. See Schwartz (2018) and Chapter 1 of Schwartz (2020).

19. Without being able to access the data it is not possible to identify any correlations between these categories and, for example, education level or religiosity.

20. Rationales counted as correct arguments included: "the infinite Universe . . . the adaptability of life to a wide variety of ambient conditions . . . the notion that if life

emerged on Earth, it can emerge elsewhere . . . and the notion that 'we can't be the only living beings in the universe" (p. 95).

21. Rationales counted as correct arguments included the lack of clear evidence for ETL, and doubts about the likelihood of the conditions necessary for life occurring elsewhere (p. 95).

22. For native English speakers, the options listed ("very important," "quite important," "does not matter," "something we not should prioritize," "something we should avoid") may not appear to correspond to a linear, 5-point Likert scale—especially the fourth option, which could indicate either mild support or mild opposition to the search for ETL. It is worth mentioning that the survey was conducted in Swedish. Erik Persson has indicated in personal communication (February 26, 2018) that this issue is merely an artifact of translating the survey results into English, and that they used standard Swedish wording for strong positive through strong negative response options.

23. See Ottavianelli and Good (2002) for a survey of European students' views of and knowledge about space.

24. See Schwartz (2017a) for further discussion and references regarding human curiosity and space exploration.

# References

Ambrosius, Joshua. 2015. Separation of Church and Space: Religious Influences on Public Support for U.S. Space Exploration Policy. *Space Policy* 32: 17–31.

Bainbridge, William S. 2015. *The Meaning and Value of Spaceflight: Public Perceptions*. New York, NY: Springer.

Brake, Mark, et al. 2006. Alien Worlds: Astrobiology and Public Out-Reach. *International Journal of Astrobiology* 5: 319–324.

Cockell, Charles, et al. 2018. The UK Centre for Astrobiology: A Virtual Astrobiology Centre. Accomplishments and Lessons Learned, 2011–2016. *Astrobiology* 18: 224–243.

Crawford, Ian. 2018. Widening Perspectives: The Intellectual and Social Benefits of Astrobiology (Regardless of Whether Extraterrestrial Life is Discovered or Not). *International Journal of Astrobiology* 17: 57–60.

Fergusson, Jennifer, Oliver, Carol, and Walter, Malcolm. 2012. Astrobiology Outreach and the Nature of Science: The Role of Creativity. *Astrobiology* 12: 1143–1153.

Foster, Jamie, and Drew, Jennifer. 2009. Astrobiology Undergraduate Education: Students' Knowledge and Perceptions of the Field. *Astrobiology* 9: 325–333.

Freeman, Patricia, and Houston, David. 2009. The Biology Battle: Public Opinion and the Origins of Life. *Politics and Religion* 2: 54–75.

Hansson, Lena, and Redfors, Andreas. 2013. Lower Secondary Students' Views in Astrobiology. *Research in Science Education* 43: 1957–1978.

Jakosky, Bruce. 2006. *Science, Society, and the Search for Life in the Universe*. Tucson: University of Arizona Press.

Launius, Roger. 2003. Public Opinion Polls and Perceptions of US Human Spaceflight. *Space Policy* 19: 163–175.

Oliver, Carol, and Fergusson, Jennifer. 2007. Astrobiology: A Pathway to Adult Science Literacy? *Acta Astronautica* 61: 716–723.

Oreiro, Raquel, and Solbes, Jordi. 2017. Secondary School Students' Knowledge and Opinions on Astrobiology Topics and Related Social Issues. *Astrobiology* 17: 91–99.

Ottavianelli, G., and Good, M. 2002. Space Education: A Step Forward. *Space Policy* 18: 117–127.

Persson, Erik, Capova, Klara, and Li, Yuan. 2019. Attitudes Towards the Scientific Search for Extraterrestrial Life Among Swedish High School and University Students. *International Journal of Astrobiology* 18: 280–288.

Pettinico, George. 2011. American Attitudes About Life Beyond Earth: Beliefs, Concerns, and the Role of Education and Religion in Shaping Public Perceptions. In *Civilizations Beyond Earth: Extraterrestrial Life and Society*, edited by Douglas Vakoch and Albert Harrison, 102–117. New York, NY: Berghahn Books.

Randolph, Richard, Race, Margaret, and McKay, Christopher. 1997. Reconsidering the Theological and Ethical Implications of Extraterrestrial Life. *Center for Theology and the Natural Sciences Bulletin* 17: 1–8.

Schwartz, James. 2017a. Myth-Free Space Advocacy Part I: The Myth of Innate Exploratory and Migratory Urges. *Acta Astronautica* 137: 450–460.

Schwartz, James. 2017b. Myth-Free Space Advocacy Part II: The Myth of the Space Frontier. *Astropolitics* 15: 167–184.

Schwartz, James. 2018. Myth-Free Space Advocacy Part III: The Myth of Educational Inspiration. *Space Policy* 43: 24–32.

Schwartz, James. 2020. *The Value of Science in Space Exploration*. New York: Oxford University Press.

Sephton, Mark. Astrobiology Can Help Space Science, Education and the Economy. *Space Policy* 30: 146–148.

Swami, Viren, et al. 2009. The Truth Is Out There: The Structure of Beliefs About Extraterrestrial Life Among Austrian and British Respondents. *The Journal of Social Psychology* 149: 29–43.

# 16

# Unnatural Selection or the Best of Both Worlds?

## The Legal and Regulatory Ramifications of the Discovery of Alien Life

*Christopher J. Newman*

## Introduction

The discovery of alien life will be a seminal event in the history of the human race. It will produce a myriad of scientific, philosophical, and theological questions that will have a fundamental and far-reaching impact upon society. This discussion examines some of the legal mechanisms that can be deployed to deal with challenges of the discovery of extraterrestrial life. The law in such discovery events will operate in a number of ways, defining the roles of different terrestrial agencies, protecting the scientific integrity of any discovered life, and providing valuable protection for the newly discovered life form. This discussion looks at the way in which law can be developed to assist our society as it begins to grapple with the first tentative contact between humans and nonterrestrial life forms.

Such a discussion might once have sat within the realms of science fiction. Now, missions of unprecedented scope and technical ambition,[1] coupled with the discovery of liquid water on Mars,[2] means that serious consideration must be given to the prospect of discovering alien life. In exploring these legal issues, the chapter starts by examining the existing body of space law to establish whether there is any immediate guidance to be found on how the discovery of life should be managed. The difficulty in establishing a binding legal framework to govern the relationship between humans and nonterrestrials is highlighted with reference to the problems in regulating the environmental issue of space debris.

It is clear, even from the outset, that law and legal precedent on human/alien relations will be sparse. In order to see whether any consensus on how to manage the discovery of alien life exists currently, the discussion looks beyond "hard" treaty law. An examination of the nonbinding planetary protection guidelines will establish whether the building blocks of a governance framework for the

Christopher J. Newman, *Unnatural Selection or the Best of Both Worlds?* In: *Social and Conceptual Issues in Astrobiology.* Edited by: Kelly C. Smith and Carlos Mariscal, Oxford University Press (2020). © Oxford University Press.
DOI: 10.1093/oso/9780190915650.003.0016

discovery of alien life already exist. A crucial element of a discovery event is that it is likely to generate significant media interest, and there may be pressure on the scientific community to act quickly. It is unlikely that there will be time for the negotiation and ratification of an appropriate treaty. The discussion therefore makes an assessment on the efficacy of seeking to build consensus on the regulation of humans relations with extraterrestrial life through such things as voluntary agreements and codes of practice in the hopes of establishing a normative position.

Finally, an attempt is made to suggest some ways forward to ensure that the seminal discovery event does not lead to conflict on Earth, the loss of irreplaceable scientific information, or even the destruction of the alien life. At the heart of the discussion is that, at present, there is little consensus as to how to start formulating laws to govern relations with alien lifeforms, no matter what their composition. While a binding treaty may currently be beyond the international community there will need to be some form of legal response to protect both humanity and the alien lifeform.

## Establishing the Contours of the Legal Response

The purpose of this discussion is to identify the legal mechanisms that could be put in place to deal with the discovery of alien life. Yet a significant definitional issue becomes immediately apparent. The existence of alien life is, as yet, unproven. A central assumption of the discussion is that if humans are to discover "life," then Mars is the likeliest place where it will be found. As a species we therefore know where we are looking. But the question remains: What is it we are looking for—what constitutes life? The notion of "life" covers a huge spectrum of organisms, from single-celled creatures such as amoebae through to human beings and who knows what beyond that. In general, laws on Earth afford the greatest protection to humans while offering minimal protection to other "things" that might be considered alive.

This discussion does not entangle itself in deciding *what* the laws governing extraterrestrial contact *should* look like. The result of discovery on microbial life on Mars may mean that special sanctity may need to be given to Martian microbes where it would not even be considered in relation to terrestrial microbes. While there has been considerable academic consideration of this,[3] it would be overstating the case to say that there exists any consensus on the value (intrinsic or instrumental) that the law should place on martian or indeed any other form of extraterrestrial life. There is, however, a pressing need to identify the legal mechanisms that could be introduced to ensure orderly management of the discovery of extraterrestrial life.

Such a discovery event could, however, take a number of different forms.[4] By far the most disruptive and traumatic would be the arrival of a species on Earth who are able to announce themselves as being extraterrestrial in origin and with whom discussion and negotiations are possible. Similarly, there exists the remote possibility of interplanetary microbial travel, where extremophile organisms may travel through space and arrive on Earth.

In both cases, the law will need to work with unprecedented speed and agility to adapt to changing circumstances. It will also be very difficult to coordinate an international response should the discovery event occur within the boundaries of a nation-state. In such circumstances the response to a discovery event will be in the hands of a nation-state (and that response will be hostage to a whole host of geopolitical, cultural, political, and possibly even religious factors[5]). For that reason, this discussion instead focuses on the discovery of nonterrestrial life away from planet Earth by an exploratory, non-human mission. As stated, this is the type of discovery event that the scientific community believes is the most likely,[6] and it is the one where a coordinated, international response will be most efficacious.

## Whose Law Is It Anyway?

There has been considerable literature[7] dedicated to identifying values that should inform and underpin space activity focusing both on the altruistic drive to explore and the corporate imperative to make profit, both of which will be at play when considering the discovery of alien life. A report written for the UNESCO World Commission on the Ethics of Scientific Knowledge and Technology[8] identified that a crucial issue that needs to be addressed is one of leadership: who is to take the lead in determining the priorities and choices of science and space technologies and on the basis of which objectives? This question applies to space exploration generally but is particularly relevant for the purpose of this inquiry: who then should formulate the rules regarding managing contact with alien life?

Unsurprisingly, the UNESCO report advocates that the establishment of ethical norms in space should be undertaken collectively by the international community. This means it should be nation-states, under the umbrella of the United Nations (UN), who will be expected to draft a legally binding framework for managing the discovery of alien life based on consultations and negotiations. Without the necessary international consensus, each spacefaring nation will pursue its own agenda, which could have serious implications for continued peaceful cooperation in space.[9]

According to von der Dunk, "in view of the major political and social overtones of any game-changing discovery of extra-terrestrial life, such a [coordination] role would seem to fit . . . a body representative of all states of the world."[10] While it is an imperfect intergovernmental organization,[11] undoubtedly the UN would be involved in any discovery event. Specifically, The UN Committee for the Peaceful Uses of Outer Space (UNCOPUOS) may well provide the starting point for discussions should alien life be discovered. UNCOPUOS was formed in 1958 when the General Assembly of the UN created a bespoke committee tasked with fostering international consensus on matters of space law and policy.

Even at the outset of the space age, it was recognized by the international community that a specialist committee would be needed to deal with the unique issues posed by human space activity[12]. UNCOPUOS, therefore, was the forum in which the existing space law treaties emerged at the start of the space race and operates by consensus.[13] As a logical first stage in the inquiry, consideration is now given to the existing corpus of space law. Exploring this will illustrate what, if any, law or precedent exists to assist in establishing the legal issues arising from the discovery of alien life.

## Extant Space Law: Consultation in the Alien Void

When discussing any aspect of the legal framework governing human space activity, the Outer Space Treaty 1967 (OST)[14] is the single most important provision of international law to consider. It provides the basic principles governing the behavior of humans in space. It is a binding treaty and a source of international law providing the codified framework by which current activities in space are regulated.[15] Recognized as the key development in the creation of a set of binding principles underpinning space governance and the cornerstone of international space law conventions,[16] the OST draws on a number of previously nonbinding UN Resolutions in respect of space exploration.[17] As such, it is widely accepted by the international community as providing the legal basis for space activity, having been ratified by over 100 nations.[18]

At the time of negotiating the OST, there was no evidence to suggest that the search for alien life would be successful, much less any consideration of the ramifications that such a discovery would have upon society.[19] It is not surprising, therefore, that there is nothing in the treaty that describes what should be done in the event of discovery of alien life. Instead, the OST codifies the consensus that exists around the prohibition of national appropriation of outer space[20] and the need for outer space to remain peaceful if not free from the military.[21] The OST also places responsibility upon nation-states for their own space activities in respect of authorization, supervision, and liability for any objects launched (Articles VI to VIII). Accordingly, it would seem that the OST has little to offer in respect of seeking guidance on how to even start to regulate human–alien relations.

Despite being limited in its coverage, there is a nod toward the fragility of the space environment to be found in Article IX of the OST. Although substantial, it is worth considering in its entirety as it might provide some legal basis for an international approach to regulating the actions of state and non-state actors upon the discovery of life beyond Earth. Article IX states;

> In the exploration and use of outer space, including the moon and other celestial bodies, States Parties to the Treaty shall be guided by the principle of co-operation and mutual assistance and shall conduct all their activities in outer space, including the moon and other celestial bodies, with due regard to the corresponding interests of all other States Parties to the Treaty. States Parties to the Treaty shall pursue studies of outer space, including the moon and other celestial bodies, and conduct exploration of them so as to avoid their harmful contamination and also adverse changes in the environment of the Earth resulting from the introduction of extra-terrestrial matter and, where necessary, shall adopt appropriate measures for this purpose. If a State Party to the Treaty has reason to believe that an activity or experiment planned by it or its nationals in outer space, including the moon and other celestial bodies, would cause potentially harmful interference with activities of other States Parties in the peaceful exploration and use of outer space, including the moon and other celestial bodies, it shall undertake appropriate international consultations before proceeding with any such activity or experiment. A State Party to the Treaty which has reason to believe that an activity or experiment planned by another State Party in outer space, including the moon and other celestial bodies, would cause potentially harmful interference with activities in the peaceful exploration and use of outer space, including the moon and other celestial bodies, may request consultation concerning the activity or experiment.[22]

At first glance, the protection offered by Art. IX is extremely limited. Instead of forbidding any contamination, the only prohibition is on "harmful contamination." There is no definition within the treaty as to what amounts to contamination and when it will be harmful.[23] Butler states that "Article IX is too general to compel any action on the part of a ratifying nation unless that nation passes specific, concrete domestic laws to implement its provisions."[24] Bohlmann further emphasizes the impotence of Article IX, stating it "heavily lacks specificity" and argues that the effect of it is "minimal."[25] Certainly, the provisions of Article IX have done little to protect low Earth orbit from an accumulation of human-made debris, and there is little evidence that it is considered in respect of planetary protection generally.[26]

Yet, when trying to manage the potential issues that could arise upon the discovery of the existence of extraterrestrial life, Article IX may be of some utility. It may not, of itself, provide any legal guidance on what states must or must not do when faced with a discovery event. Nor does it provide any sanctions against states for noncompliance. It does, however, stipulate that any activity that may cause harmful interference should be subject to consultation with other states. While it does not define what harmful interference is,[27] it still places consultation and cooperation at the heart of any activity. How that consultation manifests itself in respect of a discovery event is yet to be determined.

As with so much, the OST is more about broad principles than specific granular detail. Any arbitrary experimentation, near or at where alien life has been detected or is suspected, has the potential to cause harmful interference. States would be well within their right to request consultation over any such activity. Similarly, if the experimentation was carried out by a private company, under Article VI, it would require authorization from a state and could be required, via the authorizing state, to engage in consultation.

Irrespective of this, it is unlikely that Article IX of the OST would be interpreted in a way that gives sufficient clarity and responsibility. Rather than reinterpret existing law, the ideal solution is to create a new legal instrument which spells out obligations, procedures, and limitations of activity when faced with alien life. This would be binding on signatory states who, upon ratification, would pass appropriate laws to fulfill their supervisory duties over state and nonstate actors within their jurisdiction.

The construction of a new treaty must employ a number of crucial assumptions. First, it is assumed that the nature of the law that is being discussed is a formal, social construct and that it can be understood and explained by reference to the social practices of a community.[28] This means, when discussing the laws relating to relations with alien life, the international community provides those agreed rules that have legal force and that have been agreed upon by an appropriate authority. This is clearly within the realm of international law as it would comprise a part of "the rules or usages which civilized States have agreed shall be binding upon them in their dealings with one another."[29] Ideally, therefore, the new legal order would be codified in a multilateral international treaty. Defined as "an international agreement concluded between states in written form and governed by international law,"[30] treaties are the product of negotiations between states, and, in all likelihood, such a treaty would emanate from UNCOPUOS.

The second assumption is ideological in nature: that the foundations of lawful authority come from the legitimacy of stable democratic states and not through dictatorship, theory of divine right, or any other form of absolute rule. In such a democracy, law reflects the consensus that exists about the procedures by which political decisions are taken.[31] Namely, laws governing human activity must in some way reflect a commonly held set of values. Human weaknesses and self-interest mean that there needs to be laws in place to define and in some cases limit the scope of human activity. As Bradley and Ewing have pointed out "the rules of football are often broken. But if we shoot the referee and tear up the rules, football as an organised activity ceases to exist."[32]

While football may seem something of a trivial analogy, the same underlying principle is also true of human activity in space, and especially in respect of a discovery event. There is much that will be of value to consider in respect of alien life; the worth of the scientific data alone could be incalculable, and that is before considering any commercial implications in the field of biotechnology. Without establishing "a guide to what we should and should not do in space,"[33] then space exploration will quickly become untenable and there could be a scientific "gold rush" to try and harvest anything of value from the newly discovered life-forms. National, commercial, and in some cases individual short-term priorities may well end up trumping all other considerations.

## Orbital Debris and the Consensus Conundrum

While an overarching, binding international treaty would appear to provide an optimal way to manage the discovery of extraterrestrial life, such an approach is not without its problems. Indeed, the chances of such a treaty being realized in the near future are remote. There is no established consensus on what such a treaty should even look like. Yet that is not the most serious obstacle, as establishing consensus is by no means a guarantee to securing a binding international agreement. The case of space debris provides a clear illustration of the way in which geopolitics and institutional inertia can inhibit the creation of a treaty even where there is a clear need for coordinated international action.

At first sight, orbital debris may seem unconnected to the legal issues relating to extraterrestrial life, although others have made this comparison[34] and the governance of both issues bears examination.[35] The problem of space debris was outlined in 1978 when Donald Kessler identified the dangers posed by the detritus of human activity in space. As the amount of debris in any individual orbit increases, so will the probability of collisions. Accordingly, if a large piece of debris hits a satellite, the collision will shatter it and produce thousands of new pieces of debris, which could in turn impact other satellites in a chain reaction. Eventually, human space activity will become unsustainable, as increasingly congested orbits may lead to the creation of debris belts beyond which rockets and spacecraft cannot penetrate.[36]

There is little doubt that the spacefaring community have shared values on the need to limit and eventually decrease the amount of orbital debris. It must be said that the consensus on the need to manage the debris created by space activity around the orbit of the Earth is heartening. This emerging set of values, however, is based on simple pragmatism—the need to keep the orbit of the Earth useful. Even with such an instrumentalist approach to the need to protect the orbit of the Earth (and with support from the commercial space sector), there is no indication that these values will become enshrined into law. Instead, it was not until three decades later that UNCOPUOS promoted the UN Debris Mitigation Guidelines.[37]

The reason behind the lack of a treaty is due largely to the reluctance of states to commit to binding international treaties. This presents a significant stumbling block in any attempt to codify binding international law on the discovery of extraterrestrial life. The aforementioned Moon Agreement has not succeeded because states did not wish to subscribe to binding commitments on the distribution of resources.[38] It is instructive to consider that the Moon Agreement opened for signature over four decades ago, and this represents the last attempt at producing a binding international treaty.

In concluding the assessment of existing space law, the OST was written at a time when preservation of life on Earth was the priority, not the discovery of life on other planets. Article IX of the OST may provide a basis for consultation and opening negotiations, but it is unlikely it will be enough to resist the clamor from all sides for lucrative data. In any event, there is no flesh on the bone to describe how a discovery event should be managed. Ideally, therefore, a new legal regime regulating relations with alien life will be the product of consensus and this consensus will be based on a commonly held set of values.

As can be seen, however, the threat posed by orbital debris shows the difficulties inherent in creating binding treaty law, even where overwhelming scientific evidence points to the need for action. In respect of alien life, there is little concrete evidence of a shared set of values that nation-states could fix upon to inform any new legal framework. The next phase of this inquiry examines whether there are any existing practices that can point to an emerging consensus. In this respect, therefore, the focus shifts to the search for extraterrestrial life and measures to ensure planetary protection from contamination.

## Planetary Protection and the Safeguarding of Science

The current position on the "protection of life" is woven within the earliest iterations of space law. From the start of human space activity, the scientific community has been aware of the potential dangers of contamination. In its various iterations over the last six decades, the Committee on Space Research (COSPAR) has developed a well-established planetary protection policy that has gained widespread acceptance throughout the spacefaring world.[39] The policy has been the subject of much academic discussion,[40] and it is not the intention of this chapter to provide a detailed examination of the policy. A brief examination of the operative provisions illustrates "a consistent and highly developed system of recommendations"[41] that form the international and scientific consensus around the steps that should be taken to protect both Earth and other planets from cross-contamination.

The approach and underlying philosophy behind the policy is undeniably anthropocentric, and, given the lack of evidence for the existence of life beyond Earth, this is understandable. The policy is predicated on the search for life rather than providing a robust and enduring template for what should happen upon the discovery of life. COSPAR planetary protection policy may not provide any real guidance as to the shape of future law because of the lack of regard paid to the value of alien life. C. S. Cockell provides a useful overview of the two competing approaches when assessing the value of alien life.[42] One approach is to see things as having intrinsic value, that is, a value inherent in their existence; this is considered later on. The second approach is seeing alien life as having instrumental value, that is, an assessment of the value of the discovered alien life to humanity. It might be, for example, that discovery of any type of alien life is of immense scientific utility and may well have commercial value in terms of the creation of new drugs and so on.[43] As will be seen from the following discussion, COSPAR planetary protection policy and the underlying rationale for the policy is almost completely framed from an instrumental value perspective.

The notion that human space exploration could potentially interfere with attempts to engage in scientific study of other planets predates the launch of the first satellite.[44] From as early as 1963, NASA had adopted policies to prevent contamination of the Moon, Mars, and Venus. COSPAR followed this in 1964, and by 1967 there was a comprehensive, quantitative framework providing suggested "limits on the probabilities of carrying viable organisms aboard spacecraft to planetary bodies or producing accidental impacts." By 1983, the policy was reframed to break down missions into five discrete categories and "provide detailed anticontamination measures calibrated to the nature and destination of every space mission."[45] The overarching intent of the policy is that

> The conduct of scientific investigations of possible extra-terrestrial life forms, precursors and remnants must not be jeopardized. In addition, the Earth must be protected from the potential hazard posed by extra-terrestrial matter carried by a spacecraft returning from an interplanetary mission.

This statement codifies the concerns of the scientific community that have been present since the start of the Space Age.[46] It places planetary protection into two distinct areas: the need to keep alien environments pristine so scientific investigation is not compromised and the need to protect Earth from the risk of contamination by alien infections. It does this by promoting certain combinations of planets and missions as having great propensity to contaminate areas of special scientific interest and in need of specific measures (such as adopting certain trajectories, requirements for clean-room assembly, bioload reduction).[47]

Dealing first with situations where samples are brought back to Earth, the COSPAR guidelines for Earth-return missions can be found under Category V of the planetary protection policy. The aim of these guidelines is the protection of Earth from the threat of alien biomatter being introduced into the Earth ecosystem (so-called back contamination). The arrival of a "space plague" is the staple of many science fiction stories, and, as Butler highlights, human history is replete with warnings as to potential devastation that the arrival of extraterrestrial microbes could cause to our ecosystem, which has not evolved a suitable biological defense.[48] The oft-cited case of the devasting effect of Variola Major[49] on the indigenous population of North America coupled with a lack of substantial data on what space microbes might actually look like means that COSPAR has put in place robust protocols for space missions that aim to bring material back from other planets.

Such sample return missions are relatively few in number.[50] They are expensive and fraught with technical difficulties[51] and as such have been restricted in their scope. The most notable of these are, of course, the Apollo missions, which brought back significant amounts of lunar material. Sample return missions have also brought back particles from low Earth orbit, cometary dust, and particles from the solar wind.[52] The protective mandate of Category V means that measures that need to be put in place for the returning sample are heavily contingent on where the samples have been taken from. All of the sample return missions to date have been to celestial bodies that science deems to be devoid of life and not likely to contain threatening pathogens.

The second limb of the planetary protection policy, and of most interest for those considering the existence of life beyond Earth, concerns itself with so-called forward contamination (i.e., infection of alien environments with biological material from Earth) and was a crucial driver behind the creation of COSPAR. The protection against forward contamination examines the type of mission and the target of the mission and adjusts the requirements based upon this mission/target body combination. So, missions that are fly-by or orbital in nature will have less stringent standards of decontamination than lander missions. Missions to planets that are not of interest for understanding the process of chemical evolution or the origin of life have no requirements for protection of such a body from terrestrial microbial contamination.[53] Missions to planets such as Mars, where it is felt the highest chance of finding life exists, are subject to the most stringent protection requirements.

The fear of scientists from the very beginning of space exploration was encapsulated by former NASA Planetary Protection Officer Dr. J. Rummel, who stated, "the best way to find life on Mars . . . is to bring it from Florida."[54] The seeding of a planet with human microbes runs the risk of creating false positives. Williamson goes on to explore this, recognizing that most space professionals would encourage planetary protection measures as contamination could invalidate science data collected by spacecraft and damage any indigenous life-forms, thereby invalidating any experiments.[55]

There is recognition among those involved in the formulation of planetary protection policy that current policies for dealing with sample returns are out of date and need revisiting.[56] As plans for human missions to Mars start to gain momentum, there will need to be a fundamental revision of the planetary protection policy, which should "include plans to engage with other nations on the policy and legal implications of missions to Mars."[57] Most significantly, the report suggests that there may come a time when, as capabilities increase and knowledge of solar system environments grow, "it is conceivable that there may be a lesser need for strict policies." As human activity in space continues to develop, the pressure to remove the regulatory (and financial) burden is only likely to increase.

Already, however, the pressure to amend planetary protection policies is starting to emerge. The recent inquiry into planetary protection policy by the National Academy of Sciences in the United States found that the current "planetary protection policy development process is inadequate to respond to progressively more complex solar system exploration missions, especially in an environment of significant programmatic constraints."[58] The paradigm shift in space exploration has seen commercial actors becoming involved to an extent never previously imagined.[59] This involvement is not just limited to the apparent commercial exploitation of low Earth orbit. Instead, commercial companies (most notably Elon Musk's Space X) has promulgated plans to colonize Mars.[60]

It has also been highlighted that dispensing with COSPAR's contamination and protection procedures could reduce the cost of a high-risk mission by 10%.[61] For both nations and commercial operators alike, planetary protection measures could be one of the casualties in an effort to make the budget go further. As private-sector firms motivated by profit look to explore space, even the most persuasive guideline may be ignored with little by way of consequences.

When examining the COSPAR planetary protection measures from the standpoint of the governance of the space environment, therefore, one issue is immediately apparent: the policy serves only as a guideline. It has no legal force, and there are certainly no "protection police" to ensure compliance. The policy exists by virtue of acceptance by the scientific community. It also exists because, otherwise, there is no other mechanism, voluntary or otherwise, that requires space explorers to consider issues of cross-contamination. In essence, the guidelines represent more of an "industry standard" that scientific and state actors choose to adhere to, rather than a regulatory regime with any attendant sanctions.

It is here that the "regulatory gap" between governmental and nongovernmental actors could lead to significant issues. It is easy to envisage that the multisectored space community will pay less heed to the stringent COSPAR requirements, relying on co-opted, for-hire scientific opinion to provide a pretext for short-circuiting the planetary protection guidelines for budgetary purposes. It has been suggested that planetary protection needs to be given the force of international law.[62] The argument then follows that, given the consensus that already exists, amending the OST should be a relatively straightforward exercise. Yet as can be seen from the discussion on the Moon Agreement earlier, this is by no means an easy task. Although circumstances may change, the current geopolitical circumstances mean that the chances of successfully bringing together a binding international treaty or even amending the existing ones are remote.

## First Principles: Shaping First Contact

For the vast majority of this discussion, the model of governance advocated for dealing with the discovery of alien life has been a very traditional one: the binding, international treaty. There is also considerable pessimism about the likelihood of such an agreement ever coming to fruition. Yet the situation is not completely bleak, and there are other alternatives. As can be seen from the previous discussion, both planetary protection and the problems of orbital debris have been addressed with a degree of success by using the nonbinding guidelines. Despite concern surrounding the lack of enforcement, the planetary protection guidelines have been praised as being "a very consistent and highly developed system of recommendations by an independent and international body of scientists"[63]

Nonbinding guidelines, voluntary codes, and self-regulation lack the compulsion of hard international law. Instead, they are "soft law" and part of the matrix of nontreaty relationships that are crucial in the management and governance of contemporary space activities.[64] There are a myriad of soft instruments that regulate space activity, such as the UN Debris Mitigation Guidelines[65] and the International Telecommunication Union regulatory framework governing the regulation of communications satellites.[66] That the COSPAR guidelines are nonbinding is an essential element of such agreements.

Such agreements establish a broad consensus without requiring the commitment of a more binding treaty. Space exploration is a collaborative venture, and the discovery of extraterrestrial life will have profound implications for all of humanity. A soft, nonbinding code could be a viable method for establishing the normative rules when substantial disagreement occurs among states and when roles and responsibilities are unclear.[67] The lack of appetite on behalf of states to engage in binding treaties means that soft law solutions could be used to plug the gap and provide "all of the functional aspects of substantive discussion without the legal import of a binding treaty framework, [and] these non-treaty agreements are generally recorded in declarations that evince a workable compromise."[68]

Whether by means of a legally binding treaty or a series of guidelines that provide best practice, there are two requirements that can be deduced from Article IX of the OST. The first of these is the protection of the new life and ensuring that any scientific value of the discovery is not diminished or destroyed by premature and unregulated interference. The precautionary nature of the COSPAR planetary protection policy may not protect alien life for its intrinsic value. The value of alien life is clearly predicated on its instrumental utility. Nonetheless, it is clear that any law or guideline regulating first contact with any form of new life must have this precautionary principle as a central thread.

While there may be clamor on Earth for action, a discovery event on another celestial body means access to that body can be regulated. A coordinated, measured response to the discovery event can be promoted, and the response can be part of a collaborative mission of exploration. This is the second of the broad principles that can be drawn from the OST and COSPAR planetary protection policy: that of cooperation and collaboration once a discovery event has occurred.

Additionally, as well as promoting the safety of the newly discovered life forms from human interference, the new governance mechanism must consider the effect of a discovery event upon society. There will doubtless be considerable speculation among governments and, indeed, the population at large. While it may be tempting for the discoverer to hoard the data surrounding new life, the governance protocols must discourage this as far as possible. The sharing of data and information will be crucial to the creation of transparent and inclusive scientific cooperation and should be at the heart of any governance framework.

## Conclusion

Having examined the treaty provisions of international space law and the nonbinding but persuasive guidelines of COSPAR's planetary protection policy, there is little to suggest any recognition of alien life beyond its instrumental, scientific value. The OST is completely silent on extraterrestrial life, and COSPAR provides little more than a series of procedures designed to minimize the risk of cross con-tamination. The discussions about reforming COSPAR planetary protection means that even the current precautions may be diminished[69] in the near future.

Yet, the existing governance framework for planetary protection does provide some instruction for the way in which to manage a discovery event. The view of extraterrestrial life from the perspective of their scientific utility may not be an appropriate basis upon which to start defining the relationship between hu-manity and a new, alien species, but it could be a starting point for building up consensus on managing the first encounter.

More prosaically, a soft-law solution may be the only way to generate *any* sort of agreement. Treaties are almost always only signed after lengthy nego-tiations and, as such "ambient developments in technology, science and engi-neering capability can shift the fragile ground upon which such negotiations are based."[70] It may be decades, or even centuries, until humans discover life on other planets. Nation states, through COPUOS, should work collaboratively to-ward establishing protocols to be followed upon the discovery of life. Achieving widespread support for these first critical procedures on the discovery of life is a crucial first step leading to the establishment of norms of behaviour on this matter and reducing the inevitable uncertainty such an event would precipitate. A widely supported framework will ensure that first contact for humans will herald the start of a new chapter in the history of our species and not the last chapter in the history of our new celestial neighbors.

## Notes

1. See, for example, the statement of intent signed in April 2018 by the European Space Agency and NASA to investigate the possibility of a sample-return mission: https://www.esa.int/Our_Activities/Human_Spaceflight/Exploration/ESA_and_NASA_to_investigate_bringing_martian_soil_to_Earth

2. Anja Diez, "Liquid Water on Mars," *Science*, 361 (2018), 448–449

3. Kelly C. Smith, "The Curious Case of the Martian Microbes: Mariomania, Intrinsic Value and the Prime Directive," in James Schwartz and Tony Milligan (eds.), *The Ethics of Space Exploration* (Berlin: Springer International, 2016), 195–208 and also an earlier study Charles S. Cockell, "The Rights of Microbes," *Interdisciplinary Science Review*, 29 (2004), 141–150.

4. For a full discussion see Paul Davies, *Are We Alone? Philosophical Implications of the Discovery of Extra-Terrestrial Life* (New York: Penguin, 1995).

5. Davies, *Are We Alone?*, 131–139.

6. Charles S. Cockell, "The Ethical Status of Microbial Life on Earth and Elsewhere: In Defence of Intrinsic Value," in Schwartz and Milligan, *The Ethics of Space Exploration*, 167–178, 167.

7. There is a rich body of work on the legal, ethical, and societal extent of human activity in space. See Mark Williamson, *Space: The Fragile Frontier* (AAIA, 2006), which approaches space activity from a sustainability perspective, George S. Robinson, "Forward Contamination of Interstitial Space and Celestial Bodies: Risk Reduction, Cultural Objectives and the Law," *Zur Kontamination des Weltraums*, 55 (2006), 380–399 provides a legal and philosophical approach to planetary protection. Perhaps the most concentrated discussion is to be found in Schwartz and Milligan, which explores a number of themes germane to this discussion.

8. Alain Pompidou, *The Ethics of Space Policy*, UNESCO World Commission on the Ethics of Scientific Knowledge and Technology (Paris: COMEST, 2000), http://unesdoc.unesco.org/images/0012/001206/120681e.pdf.

9. Christopher J. Newman, "The Undiscovered Country: Establishing an Ethical Paradigm for Space Activities in the 21st Century," in Alan Lawton, Leo Huberts, and Zeger van der Wal (eds.), *Ethics in Public Policy and Management* (New York: Routledge, 2015), 301.

10. Frans G. von der Dunk, "Shaking the Foundations of the Law: Some Legal Issues Posed by a Detection of Extra-Terrestrial Life," in James Schwartz and Tony Milligan (eds.), The *Ethics of Space Exploration* (Berlin: Springer International, 2016), 251–263, 257.

11. This is not the appropriate place to engage in a full-scale dissection of the difficulties faced by the UN in the modern world. For a comprehensive discussion on the role of the UN in the current geopolitical climate see Thomas G. Weiss, *What's Wrong with the United Nations and How to Fix It*, 3rd ed. (New York: Polity Press, 2016).

12. Francis Lyall and Paul B. Larsen, *Space Law: A Treatise*, 2nd ed. (Boca Raton, FL: CRC Press, 2017).

13. Frans von der Dunk, "International Space Law," in Frans von der Dunk and Fabio Tronchetti, *Handbook of Space Law* (Cheltenham, UK: Edward Elgar, 2015), 43.

14. Treaty on Principles Governing the Activities of States in the Exploration Use of Outer Space including the Moon and Other Celestial Bodies (known as the Outer Space Treaty [OST]) 18 UST 2410, 610 UNTS 205 was adopted by the General Assembly of the UN on December 19, 1966, by virtue of Resolution 2222 (XXI). It opened for signature on January 27, 1967, and entered into force on October 10, 1967.

15. Lyall and Larsen, *Space Law.*

16. Bin Cheng, *Studies in International Space Law* (New York: Oxford University Press, 1997).

17. See, for example, Resolution 1962 (XVIII) of December 13, 1963, A/RES/1962 Declaration of Legal Principles Governing the Activities of the States in the Exploration and Use of Outer Space, which prefaced many of the principles that would find their way into the OST.

18. As of July 2017, 107 states have ratified the OST with a further 23 nonratified signatories, according to the UN Office for Outer Space Affairs http://www.oosa.unvienna.org/oosa/en/SpaceLaw/treatystatus/index.html.

19. The definitive description of the background to the negotiations of the early space law treaties can be found in Cheng, *Studies in International Space Law*, Part III, 215–382.

20. Article II of the OST 1967 states, "Outer space, including the Moon and other celestial bodies, is not subject to national appropriation by claim of sovereignty, by means of use or occupation, or by any other means."

21. Article IV of the OST prohibits outright the positioning and use of nuclear weapons and weapons of mass destruction in orbit. As with much of the OST, this provision has been criticized for tacitly permitting a substantial range of other military activity. See Blair Stephenson Kuplic, "The Weaponization of Outer Space: Preventing an Extra-terrestrial Arms Race," *North Carolina Journal of International Law and Commercial Regulation*, 39 (2014), 1123.

22. OST, Article IX.

23. Williamson, *Space*, 160.

24. Jeb Butler, "Unearthly Microbes and the Laws Designed to Resist Them," *Georgia Law Review*, 41 (2007), 1355–1394, 1376.

25. Ulrike M. Bohlmann, "Planetary Protection in Public International Law" *Proceedings of the 46 Colloquium on the Law of Outer Space*, 18 (2003), 19.

26. There was an attempt in 1979, with the Moon Agreement to expand on the limited environmental provisions of Article IX; however, only 17 countries have currently ratified the agreement and no states with human spaceflight capability (specifically the United States, Russia, or China) are party to the treaty. The Moon Agreement lacks consensus and traction in international space law and the provisions are actively repudiated by the major space powers (see Christopher Newman and Mark Williamson, "Space Sustainability: Reframing the Debate," *Space Policy* (2018).

27. Lotta Viikari, *The Environmental Element in Space Law* (Leiden, The Netherlands: Martinus Nijhoff, 2008), 60.

28. Raymond Wacks, *Understanding Jurisprudence*, 5th ed. (New York: Oxford University Press, 2017), Ch 1.

29. M. R. Lord Denning, in *Trendtex Trading Corp v Central Bank of Nigeria, Lloyd's Report*, 581 (1977), 600.

30. Vienna Convention on the Law of Treaties, 1155 U.N.T.S. 331, 8 I.L.M. 679, entered into force January 27, 1980, Art. 2(1).

31. Anthony Bradley, Keith Ewing, and Christopher Knight, *Constitutional and Administrative Law*, 17th ed. (New York: Pearson, 2018), 3.

32. Ibid., 3.

33. Williamson, *Space*, 182

34. Butler, "Unearthly Microbes," 1377–1378.

35. For discussion on the legal issues facing the removal of debris see Stephan Hobe, "Environmental Protection in Outer Space: Where We Stand and What Is Needed to Make Progress with Regard to the Problem of Space Debris," *Indiana Journal of Law and Technology*, 8 (2012), 1–10.

36. Donald J. Kessler and Burton G. Cour-Palais, "Collision Frequency of Artificial Satellites: The Creation of a Debris Belt," *Journal of Geophysical Research*, 83.A6 (1978), 637–646.

37. Debris Mitigation Guidelines of the Committee on the Peaceful Uses of Outer Space, as annexed to UN doc. A/62/20, Report of the COPUOS (2007). For discussion on

these guidelines see Stephan Hobe and Jan Helge Mey, "UN Space Debris Mitigation Guidelines," *Zur Kontamination des Weltraums*, 58 (2009), 388–403.

38. See Fabio Tronchetti, "The Moon Agreement in the 21st Century: Addressing Its Potential Role in the Era of Commercial Exploitation of the Natural Resources of the Moon and Other Celestial Bodies," *Journal of Space Law*, 36 (2010), 489–524.

39. The latest iteration of the COSPAR planetary protection policy was promulgated in 2017: https://cosparhq.cnes.fr/sites/default/files/pppolicydecember_2017.pdf

40. See, for example; Butler, "Unearthly Microbes," 1362, and for an authoritative discussion on the issues surrounding the forward contamination in particular see George S. Robinson, "Forward Contamination of Interstitial Space and Celestial Bodies: Risk Reduction, Cultural Objectives and the Law," *Zur Kontamination des Weltraums,* 55 (2006), 380–399

41. Bohlmann, "Planetary Protection," 19.

42. Cockell, "The Ethical Status," 168.

43. Ibid.

44. "In 1956, the International Astronautical Federation attempted to coordinate efforts to prevent interplanetary contamination" quoted in National Academies of Sciences, Engineering and Medicine, *The Goals, Rationales, and Definition of Planetary Protection: Interim Report* (Washington, DC: National Academies Press, 2018).

45. Butler, "Unearthly Microbes," 1359.

46. National Academies of Sciences, Engineering and Medicine, *The Goals*, 2.

47. National Academies of Sciences, Engineering and Medicine, *The Goals*, 3.

48. Butler, "Unearthly Microbes," 1362.

49. See Butler, "Unearthly Microbes," at fn 58 for a fascinating discussion on whether Variola Major may be classed as an invasive species.

50. Mike Wall, "Pieces of Heaven: A Brief History of Sample Return Missions," Space. com, September 8, 2016, https://www.space.com/34002-sample-return-space-missions-history.html

51. See, for example, the difficulties encountered by the NASA Genesis mission in Michael Ryschkewitsch "Genesis Mishap Investigation Board Report, Volume I," June 13, 2006: http://www.nasa.gov/pdf/149414main_Genesis_MIB.pdf

52. Wall, "Pieces of Heaven."

53. COSPAR Planetary Protection Policy, Category I.

54. Butler, "Unearthly Microbes," 1366.

55. Williamson, *Space*, 160.

56. National Academies of Sciences, Engineering, and Medicine, *Review and Assessment of Planetary Protection Policy Development Processes* (Washington, DC: National Academies Press, 2018), 2.

57. Ibid., Recommendation 5.1 in Chapter 5.

58. Ibid., 2.

59. https://spacenews.com/report-recommends-nasa-revise-its-planetary-protection-policies/

60. Mike Wall, "SpaceX's Mars Colony Plan: How Elon Musk Plans to Build a Million-Person Martian City," Space.com, June 14, 2017, https://www.space.com/37200-read-elon-musk-spacex-mars-colony-plan.html

61. Butler, "Unearthly Microbes," 1389.

62. Butler, "Unearthly Microbes," 1389–1390.

63. Bohlmann, "Planetary Protection," 24.

64. Newman, "Undiscovered Country," 309.

65. Debris Mitigation Guidelines of the Committee on the Peaceful Uses of Outer Space, as annexed to UN doc. A/62/20, Report of the COPUOS (2007).

66. https://www.itu.int/en/about/Pages/default.aspx

67. Newman, "Undiscovered Country," 309.

68. G.M. Goh, "Softly, Softly, Catchee Monkey: Informalism and the Quiet Development of International Space Law," *Nebraska Law Review,* 87.3 (2008), 725–746, 728.

69. National Academies of Sciences, Engineering, and Medicine, Recommendation X.

70. Newman, "Undiscovered Country," 309.

# List of Contributors

Linda Billings
Director of Science Communication with the Center for Integrative STEM Education at the National Institute of Aerospace, and Communications Consultant for NASA's Astrobiology Program and Planetary Defense Coordination Office
billingslinda1@gmail.com

T.D.P. Brunet
D. Phil. Candidate
Department of History and Philosophy of Science
Cambridge University

Luis Campos
Regents' Lecturer and Associate Professor, Department of History
University of New Mexico and former Baruch S. Blumberg NASA/Library of Congress Chair in Astrobiology
luiscampos@unm.edu

Steven J. Dick
Former NASA Chief Historian and Former Baruch S. Blumberg NASA/Library of Congress Chair in Astrobiology
stevedick1@comcast.net

Derek Malone-France
Associate Professor, Departments of Religion and Philosophy
George Washington University
dmf@gwu.edu

Brian Patrick Green
Director of Technology Ethics at the Markkula Center for Applied Ethics, Santa Clara University
bpgreen@scu.edu

Jason J. Howard
Professor of Philosophy, Dept. of Ethics, Culture and Society at Viterbo
University
jjhoward@viterbo.edu

Carlos Mariscal
Ph.D. in philosophy from Duke University
Assistant Professor
Department of Philosophy; Ecology, Evolution, and Conservation Biology
Program; Integrative Neuroscience Program
University of Nevada, Reno
carlos@unr.edu

Cole Mathis
Post Doctoral Researcher, School of Chemistry, University of Glasgow
Cole.Mathis@glasgow.ac.uk
Sean McMahon
UK Centre for Astrobiology, School of Physics and Astronomy and School of
Geosciences, University of Edinburgh
sean.mcmahon@ed.ac.uk

Lucas Mix
Episcopal Priest and Research Scholar at the Ronin Institute
lucas@flirble.org

Christopher J. Newman
Professor, Space Law, Northumbria University
christopher.newman@northumbria.ac.uk

Emily C. Parke
Senior Lecturer, Philosophy, School of Humanities
Te Ao Mārama—Centre for Fundamental Inquiry, University of Auckland
e.parke@auckland.ac.nz

Adam Potthast
Director of Assessment and Faculty Development, Saint Mary's University of
Minnesota
apotthas@smumn.edu

Kelly C. Smith
Professor and Chair, Department of Philosophy & Religion, Clemson University
kcs@clemson.edu

James S.J. Schwartz
Department of Philosophy, Wichita State University
james.schwartz@wichita.edu

# Index

Tables and figures are indicated by *t* and *f* following the page number

*For the benefit of digital users, indexed terms that span two pages (e.g., 52–53) may, on occasion, appear on only one of those pages.*